OPTICAL TURBULENCE
Astronomy Meets Meteorology

OPTICAL TURBULENCE
Astronomy Meets Meteorology

Proceedings of the Optical Turbulence Characterization for Astronomical Applications

Sardinia, Italy 15 – 18 September 2008

Editors

Elena Masciadri
Instituto Nazionale di Astrofisica, Italy

Marc Sarazin
European Southern Observatory, Germany

ICP Imperial College Press

Published by

Imperial College Press
57 Shelton Street
Covent Garden
London WC2H 9HE

Distributed by

World Scientific Publishing Co. Pte. Ltd.
5 Toh Tuck Link, Singapore 596224
USA office: 27 Warren Street, Suite 401-402, Hackensack, NJ 07601
UK office: 57 Shelton Street, Covent Garden, London WC2H 9HE

British Library Cataloguing-in-Publication Data
A catalogue record for this book is available from the British Library.

OPTICAL TURBULENCE
Astronomy Meets Meteorology
Proceedings of the Optical Turbulence Characterization for Astronomical Applications

ISBN-13 978-1-84816-485-7
ISBN-10 1-84816-485-8

Printed in Singapore by World Scientific Printers

CONTENTS

Preface xi

Conference Group Picture xvii

Organizers xix

In Memory of A. Erasmus xxi

Scientific Program xxiii

Optical Turbulence in High Angular Resolution Techniques in Astronomy 1
J. M. Beckers

An Astronomer's View of Optical Turbulence 13
R. Racine

Using the Scintillation of Extended Objects to Probe the Lower Atmosphere 23
J. M. Beckers

Optical Turbulence Profiles at CTIO from a 12-element Lunar Scintillometer 26
P. Hickson, T. Pfrommer and A. P. Crotts

HVR-GS at Mt. Graham: Optical Turbulence Vertical Distribution with Standard and High Resolution 34
E. Masciadri, J. Stoesz, F. Lascaux and S. Hagelin

Profilometry for the Lower Terrestrial Atmosphere 42
J. Borgnino, A. Berdja, A. Ziad and J. Maire

SLODAR Turbulence Monitors 50
R. W. Wilson, T. Butterley and J. Osborn

High Resolution SLODAR Measurements on Mauna Kea 58
T. Butterley, R. W. Wilson, M. R. Chun, R. Avila and J.-L. Aviles

Vertical Turbulence Profiles at Canary Islands Astronomical Sites ... 66
B. Garcia-Lorenzo, J. J. Fuensalida, J. Castro-Almazan and M. A. C. Rodriguez-Hernandez

C_n^2 Profile from Shack-Hartmann Data: First Steps for Co-slidar Data Processing 74
C. Robert, N. Védrenne, V. Michau and J.-M. Conan

How We Can Understand the Antarctic Atmosphere? 82
J. W. V. Storey, M. C. B. Ashley and J. S. Lawrence

TMT Site Testing Survey: Instruments, Methods and Operations 90
R. L. Riddle, W. Skidmore, T. Travouillon, M. Schöck and S. Els

TMT Site Testing Survey: Calibration and Results 98
M. Schöck, S. Els, R. Riddle, W. Skidmore and T. Travouillon

Wavefront Characterization Campaign at Paranal Using GSM, MOSP, DIMM-MASS, LUSCI and SCIDAR 108
W. Dali Ali, A. Ziad, A. Berdja, J. Maire, J. Borgnino, M. Sarazin, G. Lombardi, J. Navarrete, H. V. Ramio, M. Reyes, J. M. Delgado, J. J. Fuensalida, A. Tokovinin and E. Bustos

Turbulence Outer Scale for High Angular Resolution Techniques 120
A. Ziad, J. Borgnino, J. Maire, A. Berdja and F. Martin

Retrieving High Layer Atmospheric Turbulence Statistics on E-ELT Scales 128
C. Arcidiacono, R. Ragazzoni, J. Farinato, G. Gentile, A. Baruffolo, M. Dima, C. Metti, V. Viotto and E. Diolaiti

The Paranal Surface Layer 136
J. Melnick, M. Sarazin, G.-L. Lombardi and J. Navarrete

The Mesoscale Meteorological Models Meso-NH and AROME 144
C. Lac, T. Maric, J. P. Pinty, J. Cuxart, V. Masson and P. Tulet

Introduction to Data Assimilation in Meteorology 156
P. Brousseau and L. Auger

Adaptation of Force Restore ISBA Model to Polar Conditions 165
P. Le Moigne, J. Noilhan, E. Masciadri, F. Lascaux and I. Pietroni

Mesoscale NWP Over Antarctica: AMPS and Support for Ground-Based Astronomy 173
J. G. Powers

ForOT: A New Approach for the Optical Turbulence Studies Applied to the Ground-based Astronomy 184
E. Masciadri, F. Lascaux, S. Hagelin and J. Stoesz

The Mauna Kea Weather Center: A Case for Custom Seeing Forecasts 196
T. Cherubini, S. Businger and R. Lyman

Numerical Simulations of the Wintertime Optical Turbulence in Antarctica with the Mesoscale Model Meso-NH 208
F. Lascaux, E. Masciadri, S. Hagelin and J. Stoesz

Meteorological Study of an Event of Occurrence of Optical Turbulence at the TMT Candidate Sites 216
O. Cuevas, M. Cure, A. Chacón, S. Els, M. Schöck, R. Riddle, W. Skidmore and T. Travouillon

The Ground Layer Turbulence Observed by MASS-DIMM During the TMT Site Testing and Its Relation to Meteorological Parameters 224
S. G. Els, M. Schöck, R. Riddle, W. Skidmore and T. Travouillon

Astroclimatological Analysis of Ground Based Observatories 232
G. Lombardi, V. Zitelli and S. Ortolani

Comparison of the Atmosphere Above the South Pole, Dome C and Dome A: First Attempt 240
S. Hagelin, E. Masciadri, F. Lascaux and J. Stoesz

Correlating SODAR Turbulence Measurements with Ground Heat Flux Data 248
T. Travouillon, R. Riddle, W. Skidmore, M. Schöck and S. Els

Climatology at the Roque de los Muchachos Observatory 256
A. M. Varela and C. Muñoz-Tuñón

SNODAR II: Probing the Atmospheric Boundary Layer on the Antarctic Plateau 264
C. S. Bonner, M. C. B. Ashley, J. S. Lawrence, D. M. Luong-Van and J. W. V. Storey

Adaptive Optics and Interferometry: Present and Future Systems and Impact of Turbulence 271
P. Wizinowich

A New Sensor for Laser Tomography on ELTs 283
E. Gendron, M. Brangier, G. Chenegros, F. Vidal, Z. Hubert, G. Rousset and F. Pouplard

IMAKA: Imaging from Mauna Kea 291
O. Lai, J.-C. Cuillandre, M. R. Chun, R. Carlberg and H. B. Richer

Dealing with Turbulence: MCAO Experience and Beyond 299
R. Ragazzoni, Y. Momany, C. Arcidiacono, R. Falomo, J. Farinato, M. Gullieuszik, A. Moretti, E. Diolaiti, M. Lombini, G. Piotto, R. Turolla, E. Marchetti and R. Donaldson

Pupil Slicing Adaptive Optics: Making Extremely Large Telescopes Diffraction Limited at Short Wavelenghts 307
J. M. Beckers

Why is the VLT Very Efficient? 308
F. Comerón

Future-Look Science Operations for the LBT 316
R. F. Green

Preparing for Operations with the European Extremely Large Telescope (E-ELT) 324
F. Comerón

Evaluation of the GFS and MM5 Meteorological Models for Paranal, Macon and Pachon Astronomical Zones 334
A. Chacón, O. Cuevas and M. Curé

The Impact of Seismicity on High Angular Resolution Astronomy: The Case of the Canary Islands 342
A. Eff-Darwich, B. García-Lorenzo, L. Bonatto, L. Hernández-Gutiérrez, R. Viñas, J. A. Rodriguez-Losada, M. J. Blanco and C. Muñoz-Tuñón

Turbulence Structure and Meteorological Conditions at Teide and Roque de los Muchachos Observatories (Canary Islands) 350
J. A. Castro-Almazán, B. García-Lorenzo and J. J. Fuensalida

Studying the Relationship Between the Average Velocity of the Turbulence and High Altitude Winds at the Teide Observatory 358
B. García-Lorenzo, C. del Giorgio-Castiglione, A. Eff-Darwich and J. J. Fuensalida

The Antarctica Polar Vortex: Study of Winter 2005 366
F. Lascaux, E. Masciadri, S. Hagelin and J. Stoesz

Surface Layer SLODAR 371
J. Osborn, R. Wilson and T. Butterley

Giant Magellan Telescope Site Testing and Characterization at Las Campanas Observatory 379
J. E. Thomas-Osip and G. E. Prieto

Author Index 387

PREFACE

The International Conference *Optical Turbulence - Astronomy meets Meteorology* was held at the Nymphes Bay, Alghero, Sardinia, Italy on the 15-18 September 2008. The present volume contains the texts related to most of the invited talks, standard talks and poster presented at the Conference. An on-line version of this Conference Proceedings, including figures in color, can be found at the sites: *http : //forot.arcetri.astro.it/otam08* and *http : //www.eso.org/otam08*. A Conference report can be found on *The Messanger* - ESO Newsletter - December 2008, N. 134, pag. 53.

Motivations

Optical turbulence (OT) is one of the main causes limiting the spatial resolution attainable in ground-based visible and infrared astronomical observatories. It is certainly one of the principal obstacles to be overcome in achieving the performances potentially accessible by the overwelming facilities planned for the next generation ground-based astronomy, the Extremely Large Telescopes (ELTs). The success of these facilities strongly depend on our ability to:

1. Characterise optical turbulence at astronomical sites from a qualitative as well as quantitative point of view,

2. Improve our knowledge of the mechanisms that produce and develop optical turbulence,

3. Predict 3D maps of optical turbulence to optimise flexible-scheduling of scientific programs and instruments,

4. Correct wavefront perturbations produced by the atmospheric turbulence.

Many of the most challenging scientific programmes to be carried out with ground-based telescopes and aiming to enhance our understanding of Universe requires excellent turbulent conditions to be successful. The competitiveness of ground-based astronomy with respect to the space-based astronomy is strictly related to our ability to identify and predict temporal windows of favourable atmosphere conditions in the most accurate way. New and sophisticated Adaptive Optics (AO) techniques, assisted by either natural or laser guide stars (such as Multi-Conjugate AO (MCAO), Ground-Layer AO (GLAO) and Laser Tomographic AO (LTAO)), are conceived to optimise perturbed wavefronts corrections on different fields of view but, to optimise their efficiency, they will also require a detailed knowledge of the vertical distribution of the OT (and not simply integrated values). The new generation of AO requires a detailed understanding of the connections between the turbulence spectrum and the shape of the point-spread function (PSF) over the entire field of view. Some specific topics such as the precise nature and role played by the spatial coherence outer scale in the high angular resolution (HAR) techniques and the turbulence spectrum features in non-Kolmogorov regimes, are still active research topics at the frontiers of the theory in this field.

From the meteorological side, Operational Numerical Weather Prediction (NWP) systems at medium and mesoscale range might play an important role for the ground-based astronomy over the next few decades. 4D-Var Assimilation Data* employing satellites measurements have greatly improved the quality of the medium range weather forecasts recently. A new challenge for the meteorology appeared on the horizon: the Mesoscale Data Assimilation. This consists of a network of surface stations and an assimilation system with a resolution of a few kilometers. Such a system is mandatory to improve the ability of mesoscale models in reconstructing the unresolved physical parameters (such as the OT) evolving at spatial and temporal scales smaller than the resolution of the General Circulation Model† and to improve the accuracy of meteorological weather forecast models that extend over limited surface areas. How this can be set up in remote regions of the Earth, such as those that are typically of interest to astronomers, is an important question. This International conference was aimed to all these topics. The original intention of the Conference was to attempt to link the two communities of astronomers and meteorologists. This step is fundamental to guaranteeing the success of dedicated systems conceived for the prediction of the optical turbulence (seeing and related integrated astroclimatic parameters such as isoplanatic angle, wavefront

*In meteorology, Assimilation Data is the procedure that provides the distribution in space and time of the status of a set of variables supposed to describe the atmosphere in a given volume. The accuracy of this description depends on the nature and density of observations (radio-soundings, satellites, etc.) and it is fundamental for better descriptions of the initialization of a model.

†General Circulation Model (GCM) are models that extend over the whole Earth and are used for weather forecasting.

coherence time, etc.) above astronomical sites to be done with mesoscale atmospheric models. The reason is simple. We need to apply an investigation tool developed in meteorology (atmospheric models) to do science (the characterisation of optical turbulence) in the astronomical field.

Conference Results

The Conference started with a general introduction by two Emeritus Professors. Jacques Beckers introduced the topic of optical turbulence in high angular resolution techniques, and Rene Racine provided a personal and provocative vision of the problem of the turbulence characterisation in astronomy. Subsequent sessions were dedicated to the characterisation of turbulence from measurements. The number and type of optical instruments to measure the vertical distribution (vertical profilers) has suddenly increased in recent years and many of them are conceived to monitor dedicated regions of the troposphere. Among the most interesting results it is worth to highlight the proliferation of vertical profilers dedicated to measuring and characterising the OT with high vertical resolution near the surface. From a typical resolution of the order of 1 km (typical of Generalized Scidar (GS)) we have moved on to resolving thinner vertical slabs of up to a few tens of meters. Concepts on which the instruments are based and/or results obtained in first site-testing campaigns have been presented for the solar and lunar scintillometers called SHABAR (SHAdow Band and Ranger), respectively from Beckers and Paul Hickson, the SLODAR (SLOpe Detection and Ranging) by Richard Wilson and Tim Butterley; HVR-GS (High Vertical resolution Generalised Scidar) and LOLAS (Low LAyer SCIDAR), were presented by Elena Masciadri and Remy Avila, representing two different ways of improving the resolution of a GS near the surface. A version of the SODAR (SOnic Detection and Ranging) at high vertical resolution called SNODAR (Surface layer Non-Doppler Acoustic Radar) by Colin Bonner and even a vertical profiler based on the measurements of the wavefront angle-of-arrival statistic (Julien Borgnino) were also discussed. Most of these instruments have been employed in these last years to characterise the first kilometer above several astronomical sites (Mt. Graham, Mauna Kea, Paranal, Cerro Tololo) providing fundamental results for the optimisation of many of the GLAO systems that are under feasibility study for the existing facilities. On the topic of the surveys we highlight the conclusion of the analysis of the extended site-testing campaigns made by the TMT group on a set of pre-selected sites in the world (Matthias Schoeck) and the presentation of the preliminary results of a cross-correlation analysis between many different optical instruments performed at Paranal in December 2007 as part of the FP6 ELT Design Study (Sarazin). Once more, the spatial coherence outer scale was confirmed to be a key astroclimatic parameter for the astronomical sites (Aziz Ziad). On the meteorological side, a very comprehensive description of a couple of European mesoscale atmospheric models, such as Meso-Nh (Non-hydrostatic Meso-scale atmospheric model) and AROME (Applications of Research to Operations at Mesoscale) (Christine Lac) was presented as well as the American Weather Research and Forecasting model (WRF) (Jordan Powers). Beside, it is worth to highlight the description (Pierre Brousseau) of the status of art of the Data Assimilation systems employed for mesoscale models. As explained previously, the performances of such models strongly depend on our ability to set the initial conditions in the most possible detailed way (that is on the Assimilation Data). Concerning the dynamic and optical turbulence simulations, considerable progress has been made in the last ten years. A key role in this section has been played by the activities of ForOT (Masciadri). ForOT aims to continue the path undertaken by several years ago by Masciadri that led to many relevant results in this discipline and, in particular, to prove that a mesoscale model can reconstruct the optical turbulence above an astronomical site with an accuracy that is not worse than that achievable with measurements. Among the activities of this group, the interesting first simulations of the C_N^2 ever done above Antarctica with good reliability of the model done in statistical terms (Franck Lascaux) are highlighted. Main goal for this research group is to be a reference and a support for observatories in developing turbulence prediction systems above astronomical sites. It is worth noting the creation of the Mauna Kea Weather Center where astronomers hired meteorologists to make an operational forecasting system of the atmosphere above the Mauna Kea summit (Steven Businger). The general impression has been that this research field is gaining interest among the astronomers and this, once more, supports the thesis that it is time to boost actions to support benchmark site-testing campaigns expressly conceived to validate the atmospheric model above astronomical sites as proposed by the ForOT group. There were several contributions aimed at the study the correlation between OT and the meteorological parameters that frequently provide valuable inputs on the OT characteristics. On the front of the Adaptive Optics and Interferometry, we report a few among the most significative results concerning the implications for the turbulence

constraints. In the field of MCAO, a detailed investigation on the limits of the Taylor hypothesis validity would provide useful insights on ways to improve the sensitivity of the MCAO with natural guide stars (Roberto Ragazzoni). For GLAO systems, if the vertical structure of the turbulence decays sufficiently sharply above an astronomical sites, GLAO systems in the visible can be applied over an extremely large field of view (Olivier Lai). We also discovered that new wavefront sensors concepts, such as the Differentiation Wavefront Sensor, reported on by Eric Gendron, might be used to characterize the turbulence in more efficient way than a Shack Hartmann. An exhustive overview (Peter Wizinovich) depicted the main turbulence constraints dependings on the typology of astronomy (Galactic, extragalactic and solar ones) and the typology of observational techniques. The final session was dedicated to science operations. The studies related to OT do have a direct impact on the implementation of the science operation models that make extensive use of queue schedulino or service observing. Several 8-10 m class telescopes currently implement one of the two approaches:
1. Application of a single administered queue mode observing system (as for ESO)
2. Application of a partner queue mode observing system (as for the Large Binocular Telescope, LBT).
It is evident that the selection of the strategy is widely influenced by organisational issues (the single European agency in the former case, a consortium of a few institutes in the latter) and for this reason the absolute efficiency of a telescope is not the only criterion in selecting a given strategy. However, it is certainly useful to quantify these efficiencies so as to be aware of what might be lost or gained through alternative solutions. On this topic Fernando Comeron noted that, currently at the VLT, a sizeable fraction of observations ($\sim$20%) have to be repeated because conditions strayed outside constraints during the execution; this is an important, hidden source of inefficiency. It is therefore obvious that a tool for the prediction of the state of the atmosphere would defineitely be a major step towards increasing the efficiency of the service mode at the VLT. Thus, it appears evident that the goal of the OT prediction on the time scale of a few hours in advance remains an important objective for observatory operations. The closing discussion session evinced the success of this first step towards a productive collaboration between astronomers and meteorologists. The most evident feature of such a constructive interaction was the decision, promoted by the Principal Investigator (PI) of the E-ELT (Roberto Gilmozzi) to prepare a detailed document outlining the main steps necessary to prepare an

efficient site-testing campaign, benchmark test expressly conceived for the validation of mesoscale atmospheric models for applications to the astronomy. We are all confident that this document will represent the first step on a path that this conference has definitively and unequivocally traced.

The Local Organizing Committee, the Scientific Programme Committee deserve thanks in supporting such a productive meeting, with an exciting programme of talks and poster presentations.

E. Masciadri & M. Sarazin
(Editors of the International Conference OTAM08)

Alghero, Italy
15 September 2008

CONFERENCE GROUP PICTURE

Conference group picture taken in front of Capo Caccia, Porto Conte, Alghero, Sardegna, (Italy) beside a robotic DIMM mount (P. Aniol).

Conference promoted by

E. Masciadri – Istituto Nazionale di Astrofisica (INAF), Florence, Italy

Scientific Program Committee

P. Bougeault – European Center for Medium Weather Forecast (ECMWF), Reading, United Kingdom
M. Chun – Insitute for Astronomy (IfA), Hawaii, United States
J-M. Conan – ONERA, Paris, France
S. Esposito – Istituto Nazionale di Astrofisica (INAF), Florence, Italy
E. Masciadri – Istituto Nazionale di Astrofisica (INAF), Florence, Italy
R. Racine – University of Montreal, Montreal, Canada
M. Sarazin – European Southern Observatory (ESO), Garching, Germany
J. Storey – New South Wales University, Sydney, Australia

Local Organizing Committee

L. Fini – Istituto Nazionale di Astrofisica (INAF), Florence, Italy
E. Masciadri – Istituto Nazionale di Astrofisica (INAF), Florence, Italy
E. Masini – Istituto Nazionale di Astrofisica (INAF), Florence, Italy

Sponsored by

- ForOT, *Marie Curie Excellence Grant*, MEXT-CT-2005-023878, FP6 Program
- European Southern Observatory (ESO), Garching, Munich, Germany
- Istituto Nazionale di Astrofisica (INAF), Italy

IN MEMORY OF ANDRE ERASMUS

Dr. Erasmus passed away on 11 March, 2007 after years of courageous fight against a progressively disabling disease.

Dr. Erasmus was based Cape Town and was affiliated with the South African Astronomical Observatory and the Department of Earth Sciences at the University of Northern Colorado where he was on the faculty from 1991 to 1997. Between 1980 and 1989 Dr. Erasmus worked at the University of Hawaii at Manoa where he obtained the Ph.D. degree in meteorology in 1984. He then held a postdoctoral fellowship and a faculty position in the Institute for Astronomy where he conducted meteorological site characterization and comparison studies for many of the world's largest telescopes. Prior to Dr. Erasmus' departure from South Africa, he worked for the South African Weather Bureau and attended the University of Cape Town where he obtained the B.Sc.(Hons.), 1977 and M.Sc.(cum laude), 1980 degrees. He matriculated from Rondebosch Boys' High School. He has been a guest speaker at the California Institute for Technology, National Center for Atmospheric Research (U.S.A.), National Optical Astronomy Observatories (U.S.A.), European Southern Observatories, Kiepenheuer-Institut, Max Planck Institut, University of Nice, Arcetri Observatory, Tokyo Astronomical Observatory, Kyoto University, Universidad de Chile and the University of Cape Town among others.

SCIENTIFIC PROGRAM

Session 1: Optical Turbulence and Astroclimatic Parameters

J. Beckers	Turbulence in high angular resolution techniques in astronomy
R. Racine	An astronomer's view of the optical turbulence

Session 2: Measurements: Instruments - Techniques - Sites surveys (I)
'Profilers'

J. Beckers	Using scintillation of extended objects to probe the lower atmosphere
P. Hickson	Optical turbulence profiles at CTIO from a 12-elements lunar scintillometer
R. Avila	LOLAS results at Mauna Kea: extreme altitude resolution monitoring of turbulence-profiles in the boundary layer
E. Masciadri	HVR-GS at Mt. Graham: characterization of the optical turbulence at standard and high vertical resolution
J. Borgnino	Profilometry for the lower terrestrial atmosphere
R. Wilson	SLODAR site monitors for site characterization and real-time support of adaptive optics
T. Butterley	High resolution SLODAR measurements on Mauna Kea
B. Garcia-Lorenzo	Vertical turbulence profiles at Canary Islands Astronomical Sites
C. Robert	C_N^2 profiles measurements from Shack-Hartmann data
J. Storey	Modern instrumentation for measurements of atmospheric turbulence profiles

Session 3: Measurements: Instruments - Techniques - Sites surveys (II)
'Surveys - Outer Scale'

R. Riddle	TMT site selection survey: Instruments, methods and operations
M. Schoeck	Site selection survey: calibration and results
M. Sarazin	Wavefront characterization campaign at Paranal: concurrent measurements with DIMM, MASS, GSM, MOSP, LuSci and Scidar
A. Ziad	Turbulence outer scale for high angular resolution techniques
C. Arcidiacono	Retrieving high layer turbulence statistics on E-ELTs scales
J. Melnick	Detailed analysis of the turbulence in the surface layer above Paranal

Session 4: Mesoscale Models and Assimilation Data

C. Lac	The mesoscale meteorological models Meso-Nh and AROME
P. Brousseau	Data Assimilation and mesoscale modelization for optical turbulence forecast
P. Le Moigne	A new Meso-Nh surface scheme optimized for polar conditions: first validation tests for astronomical applications
J. Powers	Mesoscale NWP over Antarctica: AMPS and support for ground-based astronomy

Session 5: Optical and Dynamic Turbulence Modelling

S. Businger	The development of Mauna Kea Center: forecasting planetary-scale to turbulence-scale circulations for astronomy
E. Masciadri	ForOT: a new approach to the optical turbulence forecasts studies for the ground-based astronomy
T. Cherubini	Modeling Optical Turbulence using WRF: sensitivity analysis to horizontal and vertical resolution

F. Lascaux	Numerical simulations of the wintertime optical turbulence in Antarctica with the mesoscale model Meso-Nh
O. Cuevas	Meteorological study of an event of turbulence in TMT sites: case 13 August 2006

Session 6: Optical Turbulence and Meteorological Parameters: Studies

S. Els	The ground layer as seen by MASS and DIMM and its relation to micrometeorology
G. Lombardi	Astroclimatological analysis of ground-based observatory
S. Hagelin	Comparison of the atmosphere above the South Pole, Dome C and Dome A: first attempt
T. Travouillon	Understanding the ground layer turbulence of the TMT sites
T. Varela	Climatology at the Roque de los Muchachos Observatory
C. Bonner	SNODAR: a new instrument to measure the height of the atmospheric boundary layer on the Antarctic plateau
E. Carrasco	Sierra Negra Characterization: an update

Session 7: AO and Interferometry versus Optical Turbulence

P. Wizinovich	AO and interferometry in the international astronomical community: present and future systems and impacts on turbulence
E. Gendron	A new differentiation wavefront sensor for open-loop laser tomography on ELTs
O. Lai	IMAKA: Imaging from Mauna Kea
P. Hinz	Preparing for Interferometry with the LBT
R. Ragazzoni	Dealing with turbulence: MCAO experience and beyond
J. Beckers	Pupil Slicing Adaptive Optics: making extremely large telescopes diffraction limited at short wavelenghts

Session 8: Science Operation for Ground Based Astronomy

F. Comeron	Why is the VLT very efficient?
R. Green	Future Look-Like Science Operation for the LBT
F. Comeron	Planning Operations for the European Extremely Large Telescope

Session 9: Posters

C. Bonner	SNODAR: Measuring the atmospheric boundary layer on the high Antarctic plateau
A. Chachon	Evaluation of GFS and MM5 meteorological models for Paranal, Pachon and Macon zones
A. Eff-Darwich	The impact of seismicity on high angular resolution astronomy: the case of Canaries Observatories
J. Castro-Almazan	Turbulence structure and meteorological conditions at Teide and Roque de los Muchachos observatories (Canary Islands)
B. Garcia-Lorenzo	Studying the relationship between the average velocity of the turbulence (v0) and the high altitude winds (v200) at the Teide Observatory
F. Lascaux	The Antarctica Polar Vortex: study of winter 2005
J. Osborne	Slodar profiling of the surface layer of optical turbulence
J. Stoesz	Reducing optical turbulence profiler data to a representative set for GLAO simulations
J. Thomas-Osip	Giant Magellan Telescope site testing and characterization at Las Campanas Observatory

OPTICAL TURBULENCE IN HIGH ANGULAR RESOLUTION TECHNIQUES IN ASTRONOMY

JACQUES M. BECKERS
Unaffiliated
Scottsdale, AZ 85255, USA
jbeckers@cox.net

All astronomical observations are done best from space where the absorption by and turbulence in the Earth atmosphere are absent. One has access to the entire electro-magnetic radiation spectrum and the absence of seeing allows unlimited angular resolution. However, the cost of facilities in space and their operation is 3 orders of magnitude of similar sized facilities on Earth. Experimental astrophysicists have therefore in the past decades pursued the development of techniques to overcome the seeing limitations by the atmosphere. So far they have been very successful at this and much more is almost certain to come. Adaptive Optics (AO) will make very large (8 - 10-meters diameter) and extremely large (30 – 42 meters diameter) telescopes diffraction limited first at infrared wavelengths and eventually at visible wavelengths. The development of fast optical turbulence/seeing wavefront sensing techniques using artificial sources (Laser Beacons) will enable doing that over the entire sky. Atmospheric Tomography (AT) needed for Multi-Conjugate Adaptive Optics (MCAO) will give 3D maps of the rapidly variable atmospheric turbulence. Large interferometers with baselines of hundreds of meters will further enhance the angular resolution using fringe tracking for both co-phasing and coherent operation. Ground-based astronomy is therefore entering a new era in which milli-arcsecond observations and better are foreseen even of objects at the edge of the universe. The astronomical techniques will result in information of atmospheric optical turbulence which is likely to be of interest for meteorologists.

1. Introduction

When I presented in 1964 my Doctor's thesis on the structures in the solar chromosphere to the Utrecht University it included among others the following the following sub-thesis (translated from Dutch):

"The study of temperature fluctuations on a small scale (~ 5 cm) in the Earth' atmosphere is of great importance for astronomical observations. More interest by the Meteorologists for such research would be most desirable."

It is now 44 years later, and here we are having a conference between astronomers and meteorologists which is responsive exactly to this statement. Much has happened in those years in astronomy.

The detrimental effects of the atmosphere on optical astronomical observations fall in seven categories:

1. *Clouds* which obstructs any observations at optical wavelengths.
2. *Atmospheric Absorption* which limits the spectral region that can be observed from the ground to visible wavelengths (0.3 – 1.0 μm), a number of "windows" in the infrared region of the spectrum (1.0 – 24 μm), some windows in the submillimeter radio spectrum and the radio spectrum up to about 20 m wavelength.
3. *Atmospheric Scattering* due to molecules and aerosols are of special concern for observations of the solar corona and observations at night in so-called "bright time" when the illuminated moon is present.
4. *Atmospheric Emission* in the thermal infrared from aerosols and spectral emissions from atoms and molecules. Also the many spectral emission lines from OH in the near-infrared are a problem.
5. *Atmospheric Pollution* resulting, for example, from cirrus and contrails. This limits astronomical photometry.
6. *Atmospheric Scintillation* resulting from high-altitude optical turbulence limits photometry especially in cases where high time resolution is required.
7. *Atmospheric Seeing* which is the term used to describe the image deterioration resulting from optical turbulence in the atmosphere, the topic of this conference. I will review it in the second section of this paper. Sections 3 and 4 will discuss ways to overcome the seeing effects.

Because of the topic of this meeting my paper will focus on the last item (and to a lesser extent on the sixth item), I will describe the properties of atmospheric seeing and ways that have been devised to correct for its effects either by after image detection techniques (speckle interferometry) or pre-detection techniques by adaptive optics and fringe tracking in interferometry. The paper is directed largely to the meteorological attendees of the audience rather than to the astronomical audience. It is not intended to be a review paper. I will therefore not attempt to list the full literature on the topic.

At this point one might ask why astronomers do not abandon ground-based astronomical observations altogether since all effects mentioned above are absent in observations made from satellite-borne telescopes. Indeed space observatories are of course essential at wavelengths where the radiation is absorbed by the atmosphere like most sub-millimeter and infrared wavelengths and at all UV (< 300 nm), XUV, X-ray, and γ-ray wavelengths. Even at optical wavelengths where the atmosphere is transparent the observations with the 2.4 meter Hubble Space Telescope (HST) with its 0.1 arcsec image quality over a large field-of-view are unsurpassed by larger ground-based telescopes except where the larger light gathering power of current 8 - 10-meter and future 30 – 42-meter telescopes are concerned. The 6.5-meter infrared James Webb Space Telescope (JWST) to be launched in the middle of the next decade will have a thermal telescope/atmosphere emission about a million times less than that of earth-based telescopes, will have access to the full infrared spectrum and will be diffraction limited. Nonetheless ground-based telescopes are put to good use because many science programs do not require the outstanding properties of these space facilities, because their construction and operations costs are less by two to three orders of magnitudes and because the astronomer's demand for telescope time far exceeds the time available on HST and JWST.

The realization of adaptive optics (AO) techniques in the last two decades is now correcting for atmospheric seeing. With it image qualities comparable to that of the HST (and in the future JWST) are already obtained. Further developments will improve image quality even further and extend the field-of-view for diffraction limited imaging for ground based telescopes. The 30 – 42-meter ground-based Extremely Large Telescopes ELTs) which are to be constructed in the near future include plans to incorporate multi-conjugate adaptive optics (MCAO) to give unequalled sharp images over substantial field-of-views. Initially this will be done for near-infrared wavelengths but ultimately, as technologies improve, they will include visible wavelengths as well. It is in fact this higher angular resolution that provides the main scientific justification for these ELTs. These ground-based telescopes will exceed in angular resolution capability anything that can realistically be realized in space.

2. Refractive Properties of the Turbulent Atmosphere

An excellent reference for a description of the optical properties of the turbulent atmosphere is the paper by Roddier[2] titled "The Effects of Atmospheric Turbulence in Optical Astronomy". I refer to it for many of the details.

2.1. *Average Atmosphere*

Figure 1 shows the variation of the refractive index n of the atmospheric gas as a function of optical wavelength and relative humidity for sea level and for a mountain site. For wavelengths above 0.5 microns the variation is quite small and almost independent of the relative humidity. The latter is not the case for radio wavelength where n is very humidity dependent. Some of that humidity dependence is thought to persist at mid-infrared wavelengths (> 10 μm).

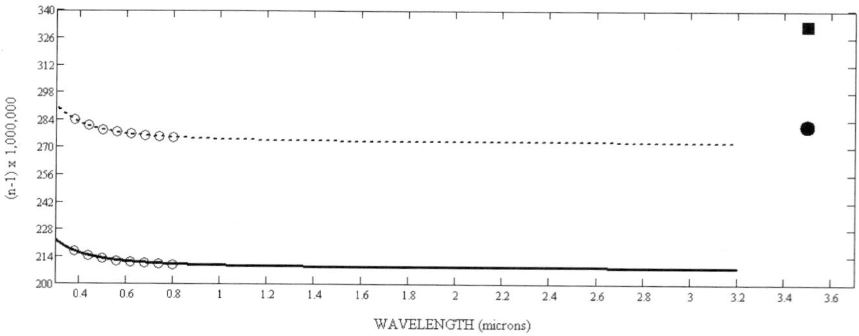

Figure 1 Refractive index variation n for a sea level site (dashed line) and for Cerro Paranal (height 2635 meters) as a function of wavelength. The curves are for zero relative humidity (RH), the open circles for RH = 1. For comparison the refractive index for radio wavelengths (sea level) is also shown in filled symbols (circle for RH = 0, square for RH = 1).

Figure 1 refers to wavelengths where the atmosphere is transparent. In the vicinity of the strong infrared absorption bands anomalous dispersion effects cause significant variations in n as shown in figure 2 taken from Mather[1].

2.2. *The Turbulent Atmosphere*

Turbulence in the Earth' atmosphere results in both temperature and density variations coupled by the pressure equilibrium. Refractive index variations with wavelength follow the n variation with wavelength shown in figure 1. The optical path length variations due to turbulence are therefore also fairly achromatic at the longer wavelengths. (> 0.5 μm) when measured in linear dimensions. So are wavefront tilts and curvatures, a behavior used in astronomical wavefront sensing (section 3.1). When expressed in waves the optical path length variations decrease of course with increasing wavelengths.

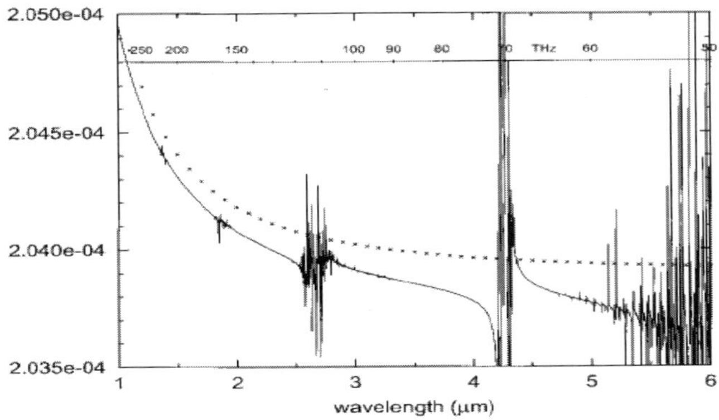

Figure 2 Effect of anomalous dispersion on the atmospheric refractive index (from Mather[11]).

For Kolmogorov turbulence the so-called Structure Function for n equals

$$D_n(\rho) = <|n(r+\rho)-n(r)|^2> \equiv C_n^2 |\rho|^{2/3}, \tag{1}$$

where r and ρ are the distance between two points and C_n^2 is the refractive index structure constant. Figure 3 shows a typical $C_n^2(h)$ variation for a sea level and mountain site (Cerro Paranal).

A frequently used parameter for expressing the total seeing caused by the atmospheric optical turbulence is the so-called Fried Parameter r_0. It equals:

$$r_0(\lambda, \zeta) = 0.185\,\lambda^{6/5} \cos^{3/5}\zeta\,(\int C_n^2 dh)^{-3/5} \tag{2}$$

where ζ is the zenith distance Within a circle of r_0 diameter the wavefront RMS distortion equals about 1/6 wave. A telescope with that (sub)aperture therefore give a diffraction limited image (Airy disk). Other dependencies on $C_n^2(h)$ for $\zeta = 0$ and telescope aperture diameter D are (integral is over the atmosphere):

Image Size (FWHM) $\approx \lambda/r_0$ (radians) $= 5.42\,\lambda^{-1/5}(\int C_n^2 dh)^{3/5}$ (3)

MS Image Motion $\approx 5.8\,D^{-1/3}\int C_n^2 dh$ (4)

Scintillation Stars $\equiv (\Delta I_{RMS}/I)^2 = 19.1\,\lambda^{-7/6}\int h^{5/6} C_n^2 dh$ (5)

Scintillation Sun/Moon $\equiv (\Delta I_{RMS}/I)^2 \approx 1.1*10^6 \int h^{-1/3} C_n^2 dh$ (6)

Time Constant $\tau \approx r_0/<V_{wind}>_{Cn2}$ (7)

Radius of "Isoplanatic Patch" $\varphi_0 \approx 0.3\ r_0/\langle h \rangle_{Cn2}$ (8)

Number of Speckles $\approx (D/r_0)^2$ (9)

Speckle Size = λ/D (radians) (10)

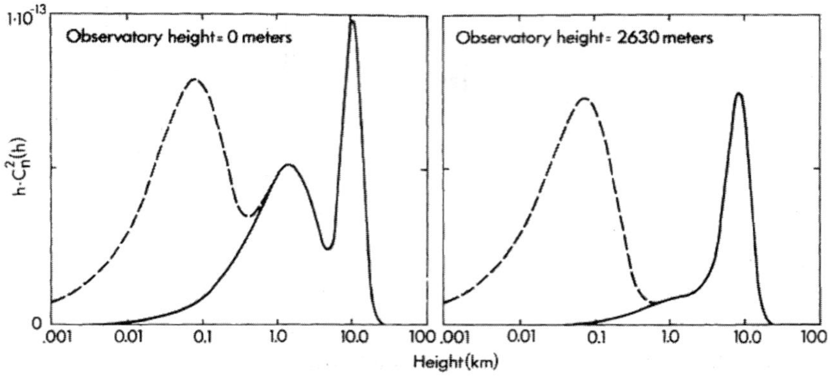

Figure 3 Variation of C_n^2 with height for the Hufnagel-Valley model. The $h{\cdot}C_n^2(h)$ vs. log(h) scaling was chosen to give a realistic representation of the contributions of the different height intervals. The dashed and full lines refer respectively to the boundary and free atmosphere layer contributions. The boundary layer refers to nighttime conditions. Daytime conditions are worse.

Figure 4 shows the spatial structure of the refractive index variations. It follows the dependence predicted by Kolmogorov turbulence except for large and short spatial scales beyond the so-called outer scale of turbulence L_0 and the inner scale of turbulence l_0. The outer scale L_0 is the scale of the "input" motions which drive the turbulence through inertial decay. The inner scale l_0 is the scale at which viscous dissipation sets in (a few mm).

The inner scale l_0 is only of interest for laser propagation where the beam widths are small. For astronomical purposes the outer scale of turbulence L_0 is of primary interest. Experimental results obtained with telescopes (Hershell/La Palma, Keck/Mauna Kea, VLT/Paranal, OHP/France) and interferometers (Sydney USI, Palomar TI) show values for L_0 ranging from 2 through 50 meters with a medium of 15 meters. L_0 is thought by some to correspond to the thickness of the narrow turbulent layers in the atmosphere, others believe it to result from local vertical wind patterns and others have postulated that L_0 varies linearly with height. It is quite likely that there is a height, time and location dependence of L_0, but I am not aware of any observations that establish that. Also when turbulence in the telescope and enclosure is an issue their size becomes important. A small L_0 improves the performance of adaptive optics in large telescopes ($D \geq L_0$) and decreases the Optical Path Differences (OPDs) between the different arms of an interferometer.

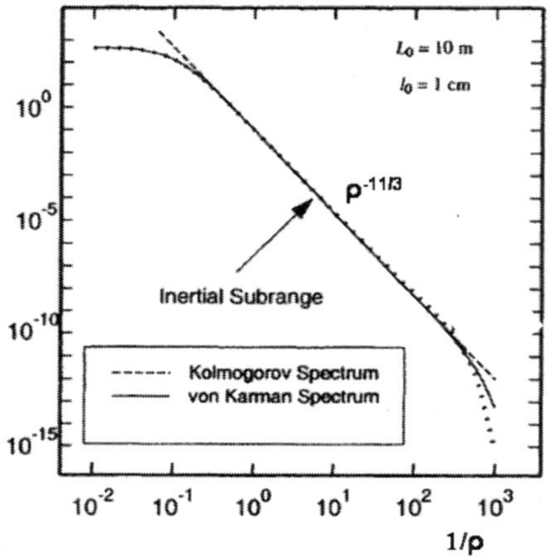

Figure 4 Spatial power spectrum of the refractive index variations. ρ is the distance between points; L_0 and l_0 are the outer- and inner-scale of turbulence (adapted from Andrews and Phillips[3]).

2.3. *Optical Turbulence under Special Conditions*

The $C_n^2(h)$ profiles given in figure 3 represent a reasonable approximation for many nighttime conditions. Because of solar heating the boundary layer in daytime is generally worse especially in the latter half of the day. There are, however, major variations in this behavior. In particular I want to point out two cases: (i) at lake sites, like the Big Bear Solar Observatory, daytime heating is minor so that the boundary layer does not change much between day and night. (ii) At Dome C in Antarctica, the best seeing in summer time occurs in the mid-afternoon and not during the night. Beckers and Travouillon[4] interpret this to be the result of a balancing between the radiation cooling of the snow surface and the solar heating. Without solar heating the temperature difference between the surface and the air is larger thus causing larger temperature differences in the wind-driven turbulence.

3. Adaptive Optics

Figure 5 is a schematic of a classical adaptive optics (AO) system for astronomical telescopes following Babcock's[5] suggestion. For more detailed general descriptions I refer to Beckers[6] and Roddier[7].

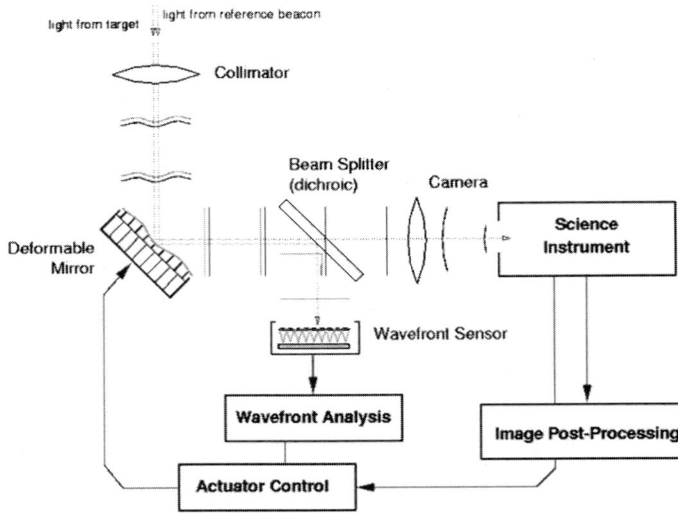

Figure 5 Schematic diagram of an astronomical Adaptive Optics (AO) system. The collimator forms an image of the telescope aperture on the deformable mirror (DM) and on the Shack-Hartmann (S-H) wavefront sensor. Equation (9) gives the approximate number of sub-segments of the DM and S-H. A rapid processor closes the fast (equation 7) complex servo loop that couples both to flatten out the distorted wavefront.

In his talk during this conference Wizinovitch[8] also discusses various AO systems and shows a number of interesting results of the improved imaging that can be obtained with them.

3.1. *Wavefront Sensing*

There are three ways that are frequently used to sense the wavefront of real or artificial stars. They are: (i) Shack-Hartmann (S-H) sensors shown already in figure 5. It measures the wavefront tilts across the telescope aperture, (ii) Roddier-Beckers[9] (R-B) curvature sensors which uses out-of-focus images to measure the wavefront curvature, and (iii) knife edge sensors using the Foucoult optics test. It also measures the wavefront tilts. The Pyramid Wavefront sensor by Ragazzoni[10] used the latter in a sophisticated way. From the wavefront tilts and curvatures the wavefront itself can be determined but in some AO system the tilts and curvatures are used directly for tilt and curvature actuated DMs.

Most AO systems use the S-H sensors combined with modal control using Zernike polynomial fitting to recover the wavefront. Figure 6 shows it in more detail.

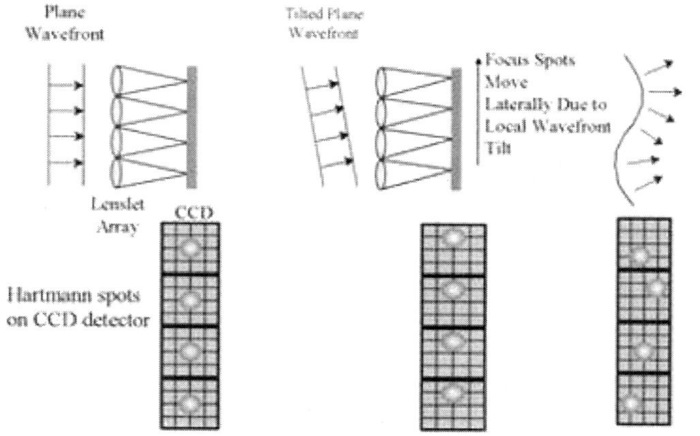

Figure 6 Schematic of a Shack-Hartmann sensor. A 2-D lenslet array placed in an image of the pupil conjugate forms multiple images of the stellar object. General or local tilts of the wavefront move these images around by an amount directly proportional to the wavefront tilts. A CCD array measures the displacements (from Boston Micromachine Corporation website; David Williams).

The S-H sensor works not only with stellar images but also with images of extended objects like the Sun. Figure 7 shows an example of a solar S-H sensor using a sunspot region.

3.2. *Laser Beacons or Guide Stars*

Normally there are, especially at short wavelengths, not enough natural stars available to apply AO to any object in the sky. Foy and Labeyrie[11] therefore suggested the use of artificial stars created by laser scattering on very high atmospheric layers (similar to those used by LIDAR) for measuring wavefront distortions. These can be placed anywhere in the sky. These stars are commonly referred to as Laser Beacons or Laser Guide Stars (LGSs).

Most frequently used are Sodium LGSs where the laser light is scattered off the neutral Sodium layer in the mesosphere at an altitude of about 90 km. This layer is about 10 km thick and time variable so that Na-LGSs vary in brightness with time. By transmitting the laser from the center of the telescope the LGS looks like a star in the center of the aperture. Near the edge one has to correct for the so-called perspective elongation resulting from the Na layer thickness[12]. Because of the finite distance of the LGS it is necessary to correct for the so-called "cone effect". This is done by Atmospheric Tomography[13,14] (occasionally also referred to as "Laser Tomography"). Occasionally Rayleigh scattering with fast shuttering on atmospheric molecules lower in the atmosphere is used. It creates brighter LGSs but suffers from a larger cone effect and possibly more perspective elongation.

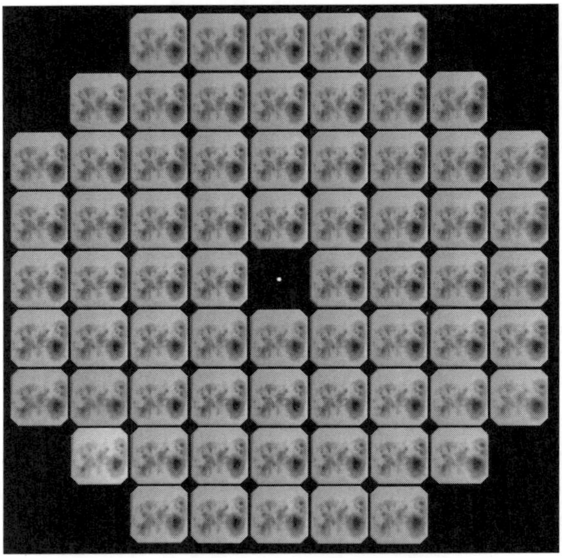

Figure 7 Shack-Hartmann wavefront sensor used to measure the wavefront distortion of a solar sunspot region. It results from a 9 x 9 lenslet array placed across an image of the 76 cm diameter entrance aperture of the Dunn Solar Telescope at the National Solar Observatory at Sacramento Peak, New Mexico. Each image covers a 127 x 127 arcsecond area on the Sun. This large area was needed for an atmospheric tomography/ multi-conjugate adaptive optics experiment.

3.3. *Multi-Conjugate Adaptive Optics*

One of the major limitations of classical AO is the small angular area on the sky over which the atmospheric seeing is corrected. It is called the "Isoplanatic Patch" (equation 8). I[13,14] suggested the use of Multi-Conjugate Adaptive Optics (MCAO) to increase that area. The concept is illustrated in figure 8.

Atmospheric Tomography benefits from the knowledge of the vertical structure of $C_n^2(h)$. A number of papers in this conference and in the literature describe methods to obtain this height profile. I refer to those.

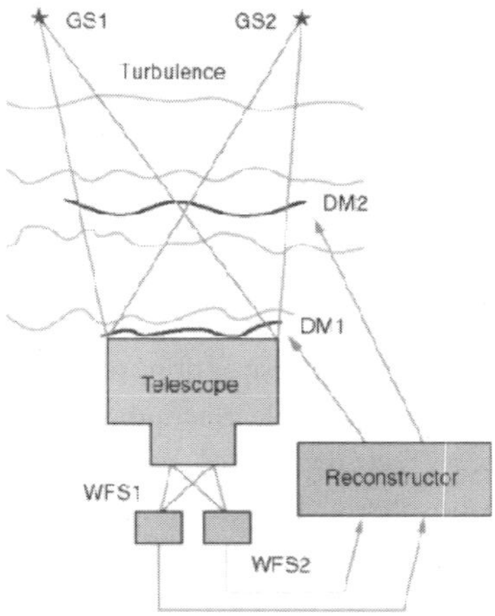

Figure 8 Concept of Multi-Conjugate Adaptive Optics. In it a number of natural stars (NGSs) and/or LGSs are used to measure the atmospheric wavefront disturbances in 3-D using Atmospheric Tomography (AT). A number of DMs are then used at optimum conjugates to correct these wave fronts in 3-D.

There are a number of variants of the MCAO concept like Dual Conjugate Adaptive Optics (DCAO), Ground Layer Adaptive Optics (GLAO), Layer Oriented Adaptive Optics (LOAO) and Multi-Object Adaptive Optics (MOAO). Nice recent descriptions of MCAO can be found in Esposito[15] and other papers in the same issue of that journal. Recently spectacular results have been obtained both in solar and nighttime astronomy with the first generation of MCAO facilities[16]. The era of MCAO is just beginning. The plan is to incorporate MCAO in all major future extremely large solar and nighttime telescopes.

4. Astronomical Interferometry

In interferometry an array of telescopes is used to achieve extremely high angular resolution. This resolution equals the maximum telescope separation divided by the wavelength used. For maximum sensitivity each of the telescopes should be outfitted with AO. Because of the optical turbulence the optical path differences (OPDs) between the telescopes fluctuate. Sometimes it is possible to "fringe track" on a field star. In general triple correlation between triads of telescopes can be used to achieve phase closure and do images provided the OPDs are within the so-called coherence length[17]. I refer to Wizinovich paper[8] in this conference for results obtained with current interferometers.

5. Conclusion

In the last few decades it has become clear that the correction for optical turbulence in astronomical imaging is possible. This open up huge new capabilities for ground-based astronomical telescopes and interferometers. The path to achieve this will be expensive and time consuming. It took 30 years to realize Babcock's AO concept. It took two decades to implement MCAO. These first realizations are so far limited. Future progress requires: (i) extension of AO to shorter wavelengths, (ii) the construction of DMs with many actuators, (iii) the implementation of Deformable Secondary Mirrors, (iv) increasing the brightness of Na-LGSs, (v) correction of perspective elongation techniques, (vi) optimization of Atmospheric Tomography, and (vii) full exploitation of interferometers including optimized imaging algorithms and the use of long base lines. The next decades in ground-based astronomy will be very exciting!

References

1. R.J. Mather, *Applied Optics* **43**, 928 (2004)
2. F. Roddier, *Progress in Optics* (Ed. E. Wolf) **XIX**, 281 (1981)
3. L.C. Andrews and R.L. Phillips, "Laser Beam Propagation through Random Media", *SPIE Optical Engineering Press* (1998)
4. J.M. Beckers and T. Travouillon, *Proceedings of the ARENA-2 Potsdam Meeting*, EDP Sciences, in press (2008)
5. H.W. Babcock, *Publ. Astr. Soc. Pacific* **65**, 229 (1953)
6. J.M. Beckers, *Ann. Rev. Astron. and Astrophys.* **31**, 13 (1993)
7. F. Roddier (Editor), *Adaptive Optics in Astronomy*, Cambridge University Press (1999)
8. P. Wizinovich, proceedings this conference (2008)
9. J.M. Beckers, *J. Opt. Soc. America* **A11**, 425 (1994)
10. R. Ragazzoni, *Modern Optics* **43**, 289 (1996)
11. R. Foy and A. Labeyrie, *Astron. and Astrophys.* **152**, L29 (1985)
12. J.M. Beckers, M. Owner-Petersen and T. Andersen, 2003, *Proceedings SPIE* **5169**, 123 (2003)
13. J.M. Beckers, *ESO Conference and Workshop Proceedings* **30**, 693 (1988)
14. J.M. Beckers, *Proceedings SPIE* **1114**, 215 (1989)
15. S. Esposito, *Comptes Rendus Physique* **6**, 1039 (2005)
16. E. Marchetti and co-authors, *The ESO Messenger* 129, 8 (2007)
17. J.M. Beckers, *Experimental Astronomy* **2**, 57 (1991)

AN ASTRONOMER'S VIEW OF OPTICAL TURBULENCE

RENÉ RACINE
Département de physique, Université de Montréal
and
Association of Canadian Universities for Research in Astronomy
CP 6128, Succursale Centre-Ville, Montréal, Qc, H3C 3J7 Canada

Optical turbulence is a key determinant of an astronomical telescope science performance. Forecasting optical turbulence at altitude is thus of paramount importance for astronomy. And so are the understanding and management of ground layer and locally induced turbulence. This paper presents an astronomer's view of both the futility and the utility of various site characterization activities, of the challenges they pose and of dreams one has about understanding and forecasting site qualities. It is pointed out that, for likely suitable sites, the average integrated optical turbulence can be calculated to good accuracy by a simple model and that turbulence profiles differ less between sites that between nights or times at a given site. The challenges therefore stem from the temporal and spatial variability of the turbulence. This variability is illustrated and briefly discussed. Collaboration between astronomers and atmospheric physicists must hold the key to the astronomers' dream of knowing in detail what the optical turbulence at the site will be tomorrow night, or next week, and which science program will then make best use of the facility.

1. Introduction

The characterization, selection and use of astronomical sites are challenging affairs. This Conference deals with the challenges posed by the characterization and forecasting of optical turbulence at a site. Much will be said about present and future turbulence measurement techniques, campaign results, and interpretations. Turbulence modeling, now-casting and forecasting will be discussed at length. This paper addresses some of these topics, as they are of great importance for astronomy.

Figure 1 shows how strongly the radiometric signal-to-noise ratio (SNR) of an astronomical image depends on the integrated optical turbulence. The "seeing", or full width at half maximum intensity (FWHM) of the point spread function (PSF), is used as a proxy for turbulence. SNR increases as the reciprocal of FWHM. In excellent natural seeing, say 0.5 arc second ("), the

SNR is three times larger than with mediocre 1.5 arc second PSFs. Or, in a more operational sense, one achieves the same science output nine times faster since SNR is proportional to the square root of the integration time. With the adaptive optics (AO) compensation of turbulence, the benefit of good seeing is spectacularly greater. Once the AO system becomes able to compensate wave front patches of size comparable to the wave front coherence length, which occurs in excellent seeing beyond 500 nm on 8-m telescopes with the AO removal of about 150 Zernike modes, images become diffraction limited. Observing time is then 1000 times more rewarding.

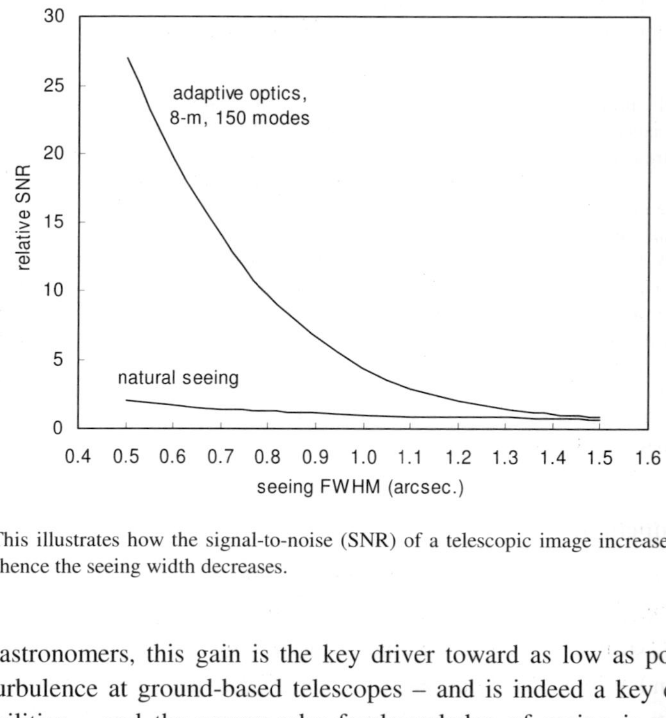

Figure 1. This illustrates how the signal-to-noise (SNR) of a telescopic image increases as optical turbulence hence the seeing width decreases.

For astronomers, this gain is the key driver toward as low as possible an optical turbulence at ground-based telescopes – and is indeed a key driver for space facilities – and the reason why foreknowledge of seeing is crucial for observing run planning. There are two approaches to realizing this gain. The first, which I will contend is futile and wasteful, is to invest efforts in the characterization and subsequent selection of sites that show the best average characteristics of turbulence: median *FWHM*, turbulence profile, etc. The second, still in its infancy but of immense future utility, is to *forecast* seeing and turbulence profiles at the site.

2. Futility

The average integrated turbulence and turbulence profile above a site are highly predictable. Isaac Newton[1] already knew in 1730 that the basic rule was to install telescopes "…on the tops of the highest Mountains, above the grosser Clouds". Racine[2] has shown that a simple model where the median FWHM is calculated from the altitude of a site and the elevation of the seeing sensor above the ground successfully represents the data from all 44 pre-2005 seeing campaigns at 23 sites to an accuracy of ±0.04". In addition, the FWHM distributions were found in all cases to be undistinguishable from log-normal distributions with median-to-quartiles ratios of 5/4 and 4/5. Figure 2 illustrates these and more recent results. It also emphasizes the deleterious effect of ground layer turbulence.

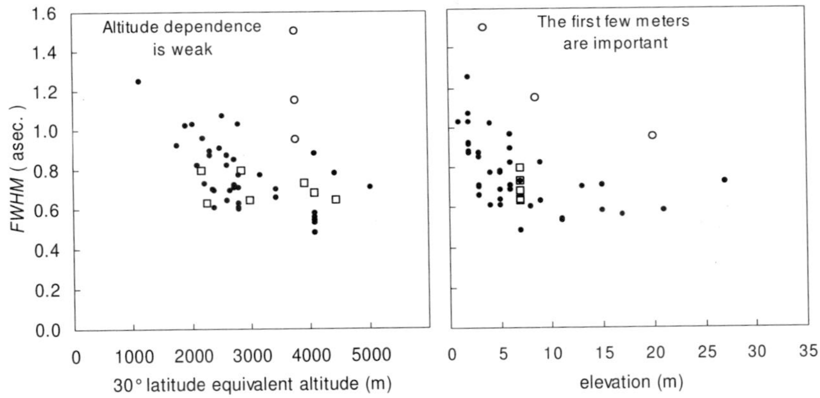

Figure 2. The dependence of the median seeing FWHM on site altitude (left) and instrument elevation above ground (right) are shown for 44 pre-2005 night-time campaigns at 23 sites (dots), for the precision 2005-2007 TMT campaigns at 7 sites (squares) and for Dome C in Antarctica (open circles). Results from multi-campaign sites reveal that much of the scatter arises from sampling uncertainties. The strongest seeing determinant is the elevation. This is particularly striking at Dome C. See Racine[2] for details.

Figure 3 compares the results of model calculations to the results of the recent Thirty Meter Telescope (TMT) precision campaigns. The root mean squares difference in FWHM is 0.06" while the accuracy of the model is estimated to be 0.04". It is difficult to believe that such small differences can be decisive in the selection of a telescope site.

Figure 4 shows that the *variability* with time of turbulence profiles above astronomical sites is much stronger from minutes to minutes than from night to night or, especially, from site to site. The SCIDAR data are from Roddier et al.[3], Vernin and Munoz-Tunon[4], Racine and Ellerbroek[5] who used an unpublished

Mauna Kea profile data base by Vernin and Azouit, Avila et al.[6], Vernin at al.[7] Masciadri et al.[8], Avila et al.[9], Tokovinin et al.[10] and Egner et al.[11]. They include 7 sites, 83 nights and over 1000 profiles. Here again, the differences between sites do not appear to be a strong selection criterion

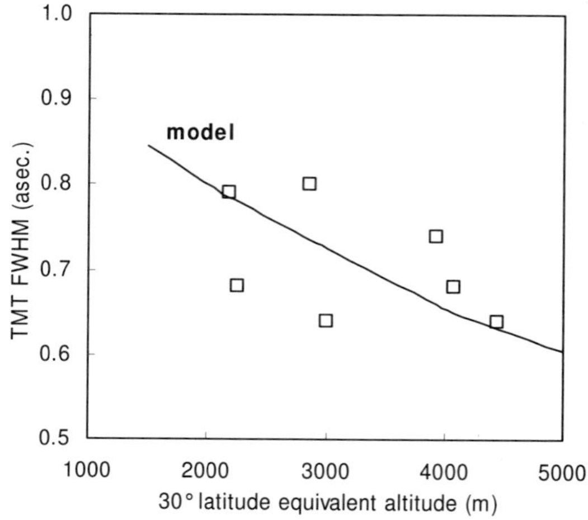

Figure 3. The results of precision TMT seeing campaigns are compared to the calculations from the Racine[2] seeing model as a function of site altitude.

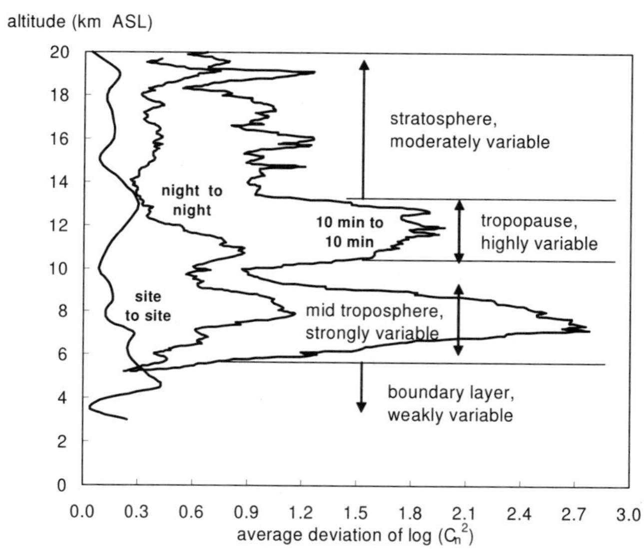

Figure 4. Turbulence profiles are highly variable but their averages differ less between sites than between nights or times.

The facts presented above indicate that extensive site testing for the purpose of site selection based on the resulting averaged turbulence statistics is a rather futile activity. The differences between promising sites are either not significant enough to be decisive or can be calculated to sufficient accuracy from simple models.

In addition, there are other considerations, some of a more geo-political and often decisive nature, for the selection of an observatory site. The average clear sky fraction is of course primordial. It is readily accessible from weather satellite data bases (Erasmus and Sarazin[12]). So is the level of precipitable water vapor, which must correlate with altitude. Funding scenarios, costs and possible delays caused by various cultural or political debates surrounding the site can be strong determinants. However adamant telescope project managers and site testers can be about testing sites, they should keep these considerations in mind.

3. Utility and Challenges

From the preceding declarations, it is clear that what is relevant to astronomers is what the turbulence profile *is* or *will be*. Realistic modeling, now-casting and forecasting are immensely important and, although still in infancy, are finally becoming realities (Masciadri[13], Masciadri and Jabouille[14], Masciadri[15], Ogasawara et al.[16]). In time they will allow adjusting science programs and instrumental configurations – such as the conjugate altitude of an AO system – to the soon-to-prevail atmospheric conditions in order to maximize the productivity of flexible and queue scheduling.

3.1. *Large and Rapid Variability with Time*

The large amplitude and rapid variability of optical turbulence makes the forecasting challenges staggering. Figure 5 shows six months of DIMM seeing data at a TMT prospective site. The AO SNR fluctuates by a factor of ~10 in an hour and of 100 from day to day and from week to week. There are longish periods of better and worse seeing (dashed ellipses in Fig. 5) and it would already be valuable to forecast those. But the general variability of the phenomenon will require immense efforts in meteorological data gathering and turbulence modeling for the forecast to be reasonably well mastered in detail.

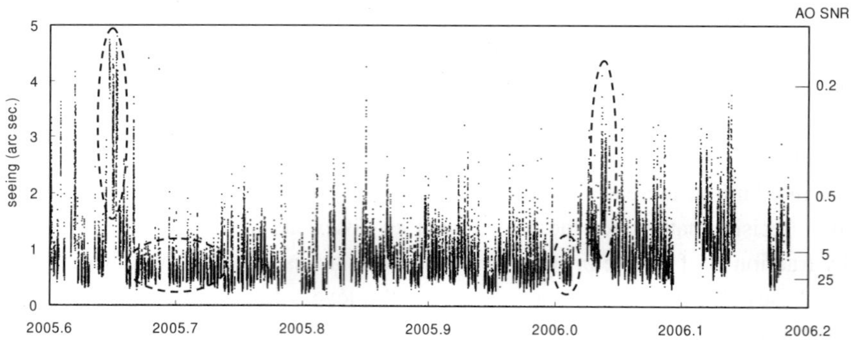

Figure 5. Six months of DIMM seeing data at a TMT site show large and rapid variability with time.

3.2. *Difference between Sub-Sites*

A second challenge has to do with very significant differences in the seeing measured strictly simultaneously at sub-sites only a few hundred meters apart. The phenomenon is illustrated by data provided by the TMT project office in Fig. 6. The seeing differences between the two sub-sites can reach 3" and are

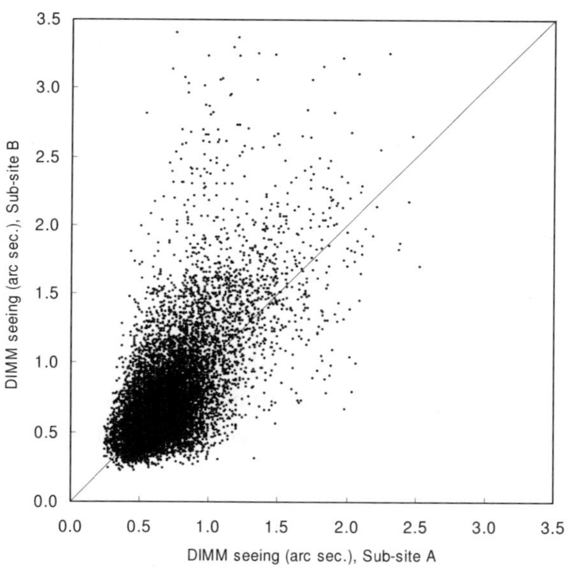

Figure 6. Simultaneous DIMM seeing measurements (8 m elevation, 2 minute samples) from two sub-sites at nearly the same altitude separated horizontally by 500 m.

frequently so large that turbulence forecasts for that summit would be largely irrelevant unless local effects are suitably modeled. The characteristic decay time of the seeing differences is a few hours. An interpretation that fits the data is that each sub-site is affected by uncorrelated sporadic enhancements of their local turbulence, of probability of occurrence 50% and with an average seeing component of 0.15" at sub-site A, and of 25% at sub-site B with a 0.65" seeing component. Analysis shows that differences in the DIMMs' line of sight (different stars) through the planetary boundary layer and the free atmosphere contribute a relative seeing dispersion of only ~15%.

3.3. *"Facility Seeing"*

Turbulence generated in the telescope and in its enclosure has historically been the dominant cause of telescopic image quality (IQ) degradation. Until ca. 1980, "good seeing" at large telescopes was anything better than 1.5" and sub-arc second seeing was unheard of. Improved thermal management in large domes has now made things much better. A ~0.7" IQ has become the norm. Figure 7 shows the historical evolution of IQ at the Canada-France-Hawai'i Telescope.

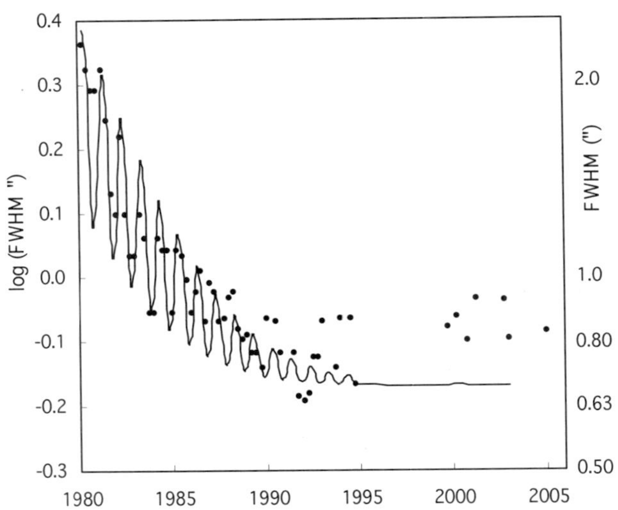

Figure 7. The quarterly averaged image quality at the Canada-France-Hawai'i Telescope on Mauna Kea has steadily improved after its commissioning in September 1979. The wiggly curve of decaying mean and oscillation amplitude is a model that peaks in mid-winters, when the outside air temperature is colder. It follows early year data rather well.

The decay of the local optical turbulence between 1980 and ~1995 displayed by these data can be attributed to increasingly stringent and successful, but totally empirical, seeing management measures. The commissioning of increasingly sophisticated, optically complex and power dissipating instruments that started in the mid 1990's seems to have worsened the IQ somewhat.

The important point here is that astronomers and observatory managers do not have a quantitative physical understanding of the relation between IQ degradation and thermal differences between the primary mirror and ambient air or between the air inside and outside the dome. How small a ΔT (°C) is good enough? All this local stuff may not be exciting to experts in optical turbulence and atmospheric physics. But it is of paramount importance to astronomers. Astronomers can take advantage of the research developed so far in optical turbulence. Collaboration between astronomers and optical turbulence experts can shed much needed light on these "local" issues.

4. Dreams

Astronomers dream of clear skies, superb seeing through very weak optical turbulence, trustworthy turbulence forecasts, including the range of dominant layers, and mitigation of the troublesome ground layer and facility turbulence. Weak local winds are also important for the next generation of extremely large, hence easily buffeted, telescopes. Finally, low levels of precipitable water vapor are essential for work in the mid infrared spectral domain.

To make these dreams come true, astronomers call upon a number of time-proven but mostly empirical strategies. Because all the dreams rest on the management of variable astro-climatic parameters, the astronomers' task can be greatly facilitated by collaborations with climatologists and atmospheric physicists.

Major telescopes are now generally installed "on the top of the highest Mountains, above the grosser Clouds", preferably near the tropics, to enjoy as much clear sky as possible. Yet, reliable weather forecasts are necessary to optimize the use of these sites.

High altitude sites also afford lower integrated turbulence, hence better seeing. But, as Fig. 3 shows, the median seeing improves rather little between a 2000 m (FWHM ~ 0.80") and a 4000 m (FWHM ~ 0.65") site, the pressure scale height of the atmosphere being some 8500 m. Here again, reliable forecasts of integrated turbulence can much facilitate the science strategy. High altitude or near-polar sites are also advantageous because they offer much lower levels of precipitable water vapor and a darker sky in the thermal infrared.

Ground layer (elevation < 30 m) turbulence can be dominant at all sites, is poorly understood and never forecasted. Its deleterious effects can be mitigated by elevating the telescopes on tall structures. This is being proposed for Antarctica sites (see Tonry, this Conference) but can be a costly proposition. A clear physical understanding of ground layer turbulence would be tremendously precious, even if only to help selecting sub-sites around an otherwise promising mountaintop. Forecasting of ground layer turbulence deserves more attention from aerodynamicists. One could expect that the very local nature of the phenomenon would make it more tractable than mesospheric climatology.

5. Confronting Reality

The collaboration between climatologists and astronomers should naturally lead to joint projects where the models and forecasts are confronted to the reality of the observations in double-blind tests. The first and, to astronomers, most important observable to test is the integrated turbulence or seeing. It is very difficult to find published results of such tests. Consequently, it is hardly possible to evaluate the reliability, nay, the credibility and usefulness of current efforts in turbulence modeling and forecasting. Perhaps the field is still too young.

A test could be constructed as follows. A DIMM is installed at a site. A team of turbulence experts inspect and instrument the site and its environment with meteorological sensors and uses the data to *calculate* the zenith seeing. Meanwhile, an independent DIMM observer *measures* the zenith seeing. Turbulence experts and DIMM observer do not communicate. They transmit their time-flagged results to an independent compiler who reports the observed *minus* calculated differences. These are used to improve the calculations until some target, say <0.05", is consistently met. Then the turbulence experts forecast the seeing for the next night and the process continues. The challenge may be demanding. But Mother Nature is the only judge of a model validity and accuracy.

Astronomers are all aware that their dreams of optimizing science management and productivity at the telescope can only become reality through close collaboration with climatologists and atmospheric physicists.

Acknowledgments

The presentation of this work was supported by the Association of Canadian Universities for Research in Astronomy (ACURA) and by the OTAM 2008 organization through an EC Marie Curie Excellence grant to the ForOT project.

References

1. I. Newton, *Optiks,* 4th ed; London: R. Society (1730)
2. R. Racine, *P.A.S.P.* **117**, 401 (2005)
3. F. Roddier, L. Cowie, J. E. Graves et al., *S.P.I.E.* **1236**, 485 (1990)
4. J. Vernin and C. Munoz-Tunon, A&A **284**, 311 (1994)
5. R. Racine and B. Ellerbroek, *S.P.I.E.* **2544**, 248 (1995)
6. R. Avila, J. Vernin, M. R. Chun et al., *S.P.I.E.* **4007**, 721 (2000)
7. J. Vernin, A. Agabi, R. Avila at al., *Gemini Report* **AO-G0094** (2000)
8. E. Masciadri, R. Avila, L. J. Sanchez et al., RevMexA&A, **19**, 63 (2003)
9. Avila, E. Masciadri, J. Vernin & Sanchez, L. J., *P.A.S.P.* **116**, 682 (2004)
10. A. Tokovinin, J. Vernin, A. Ziad and M. Chun, *P.A.S.P.* **117**, 39 (2005)
11. S. E. Egner, E. Masciadri, D. McKenna and T. M. Herbst, *S.P.I.E.* **6272**, 165 (2006)
12. D. A. Erasmus and M. Sarazin, *A.S.P. Conf.* **266**, 310 (2002)
13. E. Masciadri et al., *Astroph. & Space Sci.* **220**, 389 (1997)
14. E. Masciadri and P. Jabouille, *A&A* 376, **1063** (2001)
15. E. Masciadri, *S.P.I.E.* **6267**, 44 (2006)
16. R. Ogasawara, A. Miyashita, K. Kosugi et al., *S.P.I.E.* **4844**, 493O (2002)

USING THE SCINTILLATION OF EXTENDED OBJECTS TO PROBE THE LOWER ATMOSPHERE

JACQUES M. BECKERS
Unaffiliated
Scottsdale, AZ 85255, USA
jbeckers@cox.net

The scintillation of point-like objects is primarily caused by thermal fluctuations in the upper atmosphere. For it the scintillation index (σ_I^2) is proportional to the height integral of $C_n^2(h)$ weighted by a height dependent function $F(h) = h^\alpha$ where $\alpha = +5/3$. For extended objects like the Sun or the Moon the height contribution to the (much smaller) scintillation is quite different. Because of their size the effects of the optical turbulence is averaged over an ever increasing area as the distance to the detector increases. Assuming vertical viewing, the area diameter increases like $h*\Omega$ where Ω is the angular diameter of the Sun or Moon. For Kolmogorov turbulence the function F(h) then still has the same shape, but with $\alpha = -1/3$ so that the lower layers contribute more to scintillation. This makes it a good tool for the probing of the lower atmospheric layers. Unlike the σ_I^2 for stellar scintillation, the σ_I^2 for the Sun and the Moon is wavelength independent. Using an array of scintillometers one can probe the $C_n^2(h)$ distribution of those lower layers in a technique called SHABAR. SHABARs have been used in site testing for lower atmosphere probing for solar and nighttime telescopes. The aim is to establish the height to place telescopes, like the Advanced Technology Solar Telescope (ATST), to minimize boundary layer seeing effects. SHABAR site tests using the Moon are planned both for Arctic sites (Hickson's contribution to this meeting) and Antarctic sites (Storey's contribution to this meeting) where boundary layer heights are very site dependant reaching sometimes very small values. In my contribution I described some of the solar results related to the ATST site testing. The scintillation of planets have an F(h) function different in shape from that of the Sun or Moon. For low heights, where their beams still are narrow, F(h) has an α of +5/3 (as for stars); for large heights it is -1/3 (as for the Sun & Moon). For Mars the height contributions F(h) for seeing and scintillation are similar.

Synopsis

In a paper in 1993 E. Seykora[1] describes how he measured the scintillation of integrated sunlight at the National Solar Observatory, Sacramento Peak site and found it to be exceedingly well correlated with the atmospheric seeing. The amount of scintillation is much smaller than that of stars (0.1% - 0.01% vs. about 10%) but clearly detectable and separated from atmospheric transparency changes. Beckers[2] showed how the solar scintillation index σ_I^2 relates to the atmospheric refractive index structure $C_n^2(h)$ as proportional to $\int h^{-1/3} \cdot C_n^2(h)dh$, quite different from that for stars where the weighting function is $h^{+5/3}$. The solar scintillation is therefore dominated by ground/boundary layer turbulence. The good correlation between it and atmospheric seeing is the result of the fact that in daytime ground layer seeing dominates at Sacramento Peak.

Beckers[3] used the properties of solar scintillation to develop a device to sense the height profile of $C_n^2(h)$ in the ground layer by using an array of solar scintillometers called the SHABAR (= ***SHA***dow ***BA***nd ***R***anger). It was used for the site testing for the 4-meter Advanced Technology Solar Telescope (ATST)[4]. The inversion of the scintillometer array signals to $C_n^2(h)$ has been described by Socas-Navarro[5]. The full report of the ATST site testing effort can be found on the web site of the National Solar Observatory[6].

The figure below is taken from Beckers[3] and shows the amount of scintillation resulting from 1 arcsec seeing layers located at different altitudes for a variety of extended objects.

SCINTILLATION INDEX σ_I^2 FOR VARIOUS OBJECTS
(FOR 1 ARCSEC SEEING LAYER AT HEIGHT H)

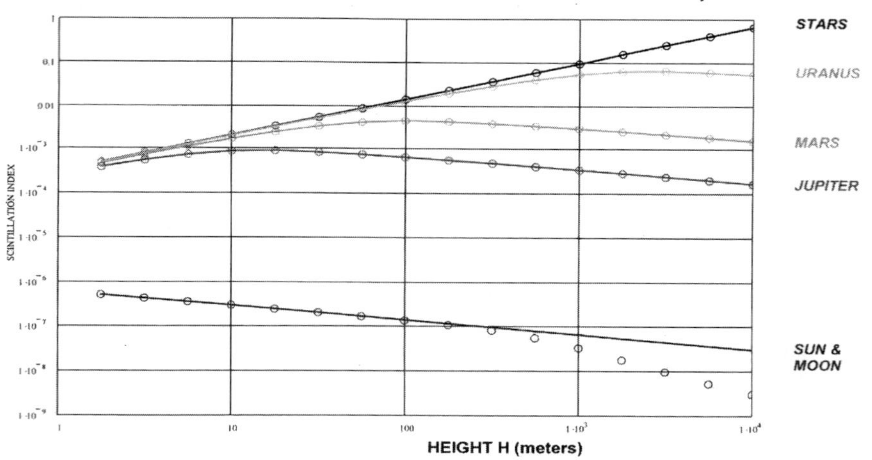

NOTE: FULL DRAWN LINES ARE FOR OUTER SCALE L_0 = INFINITE; CIRCLES ARE FOR OUTER SCALE L_0 = 20 METERS

Because of its approximately equal size the lunar and solar scintillation indices are approximately equal. For the planets the cone size at low elevations is small, so the index follows the $h^{+5/6}$ of stars. At higher elevations the cone size becomes large enough to follow the Sun/Moon $h^{-1/3}$ rule.

The figure also indicates the effect of the finite outer scale of turbulence on the scintillation index. It is only of interest for scintillation arising at high elevations.

References

1. E.J. Seykora, "Solar scintillation and the monitoring of solar seeing", *Solar Physics* **145**, 389 (1993)
2. J.M. Beckers, "On the Relation between Scintillation and Seeing Observations of Extended Objects", *Solar Physics* **145**, 399 (1993)
3. J.M. Beckers, "Seeing Monitor for Solar and other Extended Object Observations", *Experimental Astronomy* **12**, 1 (2001)
4. F. Hill, J. Beckers and co-authors, "Solar site testing for the Advanced Technology Solar Telescope", *Proceedings SPIE* **5489**, 122 (2004)
5. H. Socas-Navarro, J.M. Beckers and co-authors, "Solar Site Survey for the Advanced Technology Solar Telescope. I. Analysis of the Seeing Data", *Pub. Astron. Soc. Pacific* **117**, 1296 (2005)
6. "ATST Site Survey Working Group Final Report", *http://atst.nso edu/site/finalreport*, (2008)

OPTICAL TURBULENCE PROFILES AT CTIO FROM A 12-ELEMENT LUNAR SCINTILLOMETER

PAUL HICKSON, THOMAS PFROMMER†

Department of Physics, University of British Columbia, 6224 Agricultural Road Vancouver, BC V6T1Z1, Canada

ARLIN P. CROTTS

Department of Astronomy, Columbia University, 550 West 120th Street, New York, NY 10027, USA

We report initial results from a program of observations of boundary-layer turbulence at Cerro Tololo, Chile, using a 12-element lunar scintillometer. Observations were conducted on 75 clear nights over a period from July 2007 to June 2008, from a location near the northern end of the summit area. The median seeing FWHM at a height of 2.5 m above ground was found to be 0.60 arcsec. In median conditions, 50% of the contribution of the ground layer to the turbulence integral originates in the first 400 m of the atmosphere.

1. Introduction

In recent years increasing attention has been focused on the characterization of the planetary boundary layer for astronomical applications. Turbulence in the first km or so of the atmosphere usually provides the dominant contribution to astronomical seeing – the blurring of images due to phase distortion of propagating light. The vertical profile of turbulence in this region is important for assessing the quality of astronomical sites, locating telescopes, and optimizing the performance of adaptive optics systems.

The choice of instruments that can provide turbulence profiles in this region is limited. Generalized SCIDAR and SLODAR systems provide good resolution, but employ relatively large telescopes [1 – 4]. SODAR systems have been used successfully with careful calibration [5]. Techniques based on the scintillation of extended sources, such as SHABAR [6, 7] have proven valuable for daytime site testing [8], and have been further developed for night-time

† Work partially supported by the Natural Sciences and Engineering Research Council of Canada.

observations using the moon as a source [9, 10]. These systems have the advantage of relative simplicity. They need only a baffle to protect against stray light and means to ensure that the sensors track the sun or moon.

In order to further develop the technique, and to study the turbulence profile at a well-characterized astronomical site, we have deployed a lunar scintillometer at Cerro Tololo Inter-American Observatory (CTIO). This instrument uses a linear array of 12 large-area silicon photodiodes to measure correlations in intensity fluctuations of moonlight. Since June of 2007, the instrument has operated autonomously, acquiring and tracking the moon on clear nights.

This paper briefly describes the instrument and our analysis technique, and presents some initial results from the first year of observations. A more extensive discussion will appear in a forthcoming publication.

2. Observations

Our lunar scintillometer is located on the Cerro Tololo summit area, near the north end of the summit, approximately 30 m south of the CTIO MASS/DIMM tower. It is housed in a small enclosure to protect it from the weather. The roof, and side panels, are opened and closed manually on clear nights by CTIO staff. The instrument is attached to a Losmandy equatorial tracking mount and controlled by a small computer running Linux. The mount is shared by a 25 cm telescope, equipped with a CCD imaging camera, which is used as a monitor for transient lunar phenomena [11]. Image centroid data from the CCD camera, updated three times per minute, ensures precise tracking of the moon.

The scintillometer employs 12 large-area silicon photodiodes, arranged in a linear array, with positions determined by an optimal Golomb rule in order to minimize degeneracy in the resulting baselines. The 66 baselines, which correspond to the distances separating any two diodes, range from 0.018 m to 1.530 m. The orientation of the array is perpendicular to the polar axis, so baselines are along the direction of right ascension.

The signal processing electronics are the same design as was used for our earlier Mark I scintillometer [9]. The signal from each sensor enters a two-stage amplifier and is then digitized with 16-bit resolution at a rate of 100 samples per second. The sampling is synchronized so that the signals from all 12 diodes are captured simultaneously. The gain of the first (DC) amplification stage is programmable and is set automatically, according to the phase of the moon, to maximize the dynamic range.

The results presented here are based on observations obtained over the period 1 July 2007 – 30 June 2008. Data were recorded when all of the following conditions were met:

- The Moon's elongation was between 90° and 270°
- The Moon's altitude exceeded 30°
- The Sun's altitude was less than -12°
- The weather was suitable for observations (clear sky, low humidity)

A further cut was made during the data analysis, to remove periods of time for which conditions were not photometric. This resulted in 75 nights during which at total of 291 hrs of photometric data were recorded.

3. Analysis

The primary observable measured by the lunar scintillometer is the covariance of intensity,

$$B_{ik} = \left\langle \left[I_i / \langle I_i \rangle - 1 \right] \left[I_k / \langle I_k \rangle - 1 \right] \right\rangle, \tag{1}$$

where $I_i = I(\mathbf{r}_i)$ is the intensity recorded by diode i ($i = 1, \cdots, n$ where $n = 12$) and $\langle\ \rangle$ denotes a time average. This average is taken over a time that is long compared to the atmospheric coherence time, in order to accurately reflect the statistical properties of the turbulence. The quantity of interest is the index of refraction structure constant $C_N^2(h)$ as a function of height h. These two quantities are related by a Fredholm integral equation [9]

$$B_{ik} = \int_0^\infty C_N^2(z \sec \zeta) K_{ik}(z) dz. \tag{2}$$

The integration is performed along the line of sight to the moon, which is at zenith angle ζ. $K_{ik}(z)$ is a function of the baseline $\mathbf{r}_i - \mathbf{r}_k$, the lunar phase, and the turbulence spectrum. To compute values of $K_{ik}(z)$ for times other than full moon, we adopt the elliptical-moon approximation introduced by Tokovinin and Rajagopal [12]. They have shown this approximation to be accurate to better than 10% within 7 days of the full moon. The turbulence spectrum was assumed to be Kolmogorov. This is a good approximation in the boundary layer because in this region the width of the cone subtended by the moon is small compared to the outer scale. However, the outer scale becomes important for high altitude turbulence. We accounted for this by applying a correction, described later.

The aim of the analysis is to determine the turbulence profile, $C_N^2(h)$, given B_{ik}. However, because the data invariably contain noise, the solution is not unique. There are an infinite number of profiles that are consistent with the data.

Clearly some can be rejected as unphysical. By definition, C_N^2 is positive definite, so profiles containing negative values are unphysical. But, of all the possible profiles, how does one determine which is the best?

Our approach is to determine the *most probable* profile, in a Baysian sense [13]. This is the profile that maximizes the posterior probability

$$P(M,D) = P(M)P(D,M). \tag{3}$$

Here $P(M)$ is the prior probability of the model $M = C_N^2(h)$, given our prior knowledge, and $P(D,M)$ is the probability of obtaining the data $D = B_{ik}$, given the model M (also called the likelihood function). As a simple, yet plausible, model we adopted a double exponential profile for the boundary layer, plus a thin high-altitude component,

$$C_N^2(h) = a_1 \exp(-h/a_2) + a_3 \exp(-h/a_4) + a_5 \delta(h - h_0). \tag{4}$$

Here h_0 is the height of the high-altitude component of turbulence, which we take to be 10 km. The exact value matters little as the covariance on metre-scale baselines has very low sensitivity to the height above a few km. The constants a_r, $r = 1,\cdots, 5$ are model parameters, to be determined by the analysis. The form of Eq. (4) follows the parameterization employed for a recent study of optical turbulence at Cerro Pachon [14].

The recorded data were grouped into blocks of 12000 samples, corresponding to an interval of 2 minutes. For each block, the intensity covariance was computed for every baseline. In addition, we computed the covariance matrix

$$\mathrm{Cov}(B_{ij}B_{kl}) = \frac{1}{T}\int_{-T/2}^{T/2}\left[B_{ik}(\tau)B_{jl}(\tau) + B_{il}(\tau)B_{jk}(\tau)\right]d\tau, \tag{5}$$

where $B_{ik}(\tau)$ is the temporal cross correlation if the intensity fluctuations of diodes *i* and *k*,

$$B_{ik}(\tau) = \frac{1}{T}\int_{-T/2}^{T/2}\left[I_i(t)/\langle I_i\rangle - 1\right]\left[I_k(t+\tau)/\langle I_k\rangle - 1\right]d\tau. \tag{6}$$

(A derivation of this formula is given by Tokovinin [15].) By ordering the baselines, the resulting covariances can be written as the components of a vector B_α, where $\alpha = 1,\cdots, m$ and $m = n(n-1)/2 = 66$. The covariance matrix can now be written as $C_{\alpha\beta} = \mathrm{Cov}(B_\alpha B_\beta)$.

The likelihood function can now be written as

$$P(B_\alpha, a_r) = \exp\left\{-\frac{1}{2}\sum_{\alpha,\beta=1}^{m}\left[B_\alpha - \hat{B}_\alpha(a_r)\right]C_{\alpha\beta}^{-1}\left[B_\beta - \hat{B}_\beta(a_r)\right]\right\}, \tag{7}$$

where $\hat{B}_\alpha(a_r)$ is the predicted covariance vector for a given set of model parameters a_r, calculated from Eqn. (2). The parameters a_r were assumed to have a lognormal probability distribution, consistent with the Cerro Pachon turbulence model. Thus, the prior probability can be written as

$$P(a_r) = \exp\left\{-\sum_{r=1}^{5}\frac{(\ln a_r - m_r)^2}{2\sigma_r^2}\right\}, \tag{8}$$

where m_r and σ_r are the mean and standard deviation of $\ln a_r$, estimated from the data reported in [14].

The aim now is to find the set of parameters a_r that maximize the posterior probability, and to determine confidence intervals for these parameters. This was done using a Markov-chain Monte-Carlo optimization code employing the Metropolis-Hastings algorithm [16, 17]. This technique has the advantage of converging quickly, while at the same time avoiding trapping by local maxima. Its output is probability distribution functions for all of the model parameters. From this one can determine the mean or median values, and confidence intervals and standard errors for all functions of these parameters (such as the Fried parameter). Because the outer scale reduces the intensity fluctuations produced by high-altitude turbulence, the parameter a_5 is underestimated. To correct for this, we multiplied this parameter by a constant multiplicative scale factor of 2.43, chosen to make the mean value (for the entire year) equal to the mean free-atmosphere turbulence integral of the Cerro Pachon data. This is justified because Cerro Pachon and Cerro Tololo are separated by only a few km and should therefore see essentially the same average high-altitude turbulence.

For wavelength λ and zenith angle ζ, the Fried parameter at height h above ground $r_0(h)$ is related to the vertical turbulence integral

$$J(h) = \int_h^\infty C_n^2(x)\,dx \tag{9}$$

by the equation

$$r_0(h) \simeq \left[16.6\lambda^{-2}\sec\zeta J(h)\right]^{5/3}, \tag{10}$$

and the FWHM of the long-exposure PSF is given by $w(h) = 0.976\lambda / r_0(h)$ [18].

4. Results

Fig. 1 shows the time variation of the seeing FWHM (at the zenith and a wavelength of 0.5 um), at seen from heights of 1, 10, 200 and 1000 m above the scintillometer (add 2.5 m to get the height above ground), for one night (April 10, 2008). The variations are much larger than the noise in the measurement and reflect temporal changes in the turbulence structure during the night. In Fig. 2 we show turbulence integrals as a function of height for the boundary layer only. From this it can be seen that, in median to poor conditions, half of the turbulence in the boundary layer comes from the lower 400 m. Under conditions of good seeing, this thick boundary layer is mostly absent, with most of the low level turbulence coming from the first 40 m.

Cumulative probability distributions of the seeing FWHM are shown in Fig. 3, for a range of heights (this includes the high-altitude turbulence). The median value of the seeing is 0.6 arcsec at a height of 3.5 m above ground. Fig. 4 shows the FWHM as a function of height for the median, 25th percentile and 75th percentile seeing (at 3.5 m height).

Acknowledgments

This work was supported by grants from the Natural Sciences and Engineering Research Council of Canada, and by the ALPACA telescope consortium.

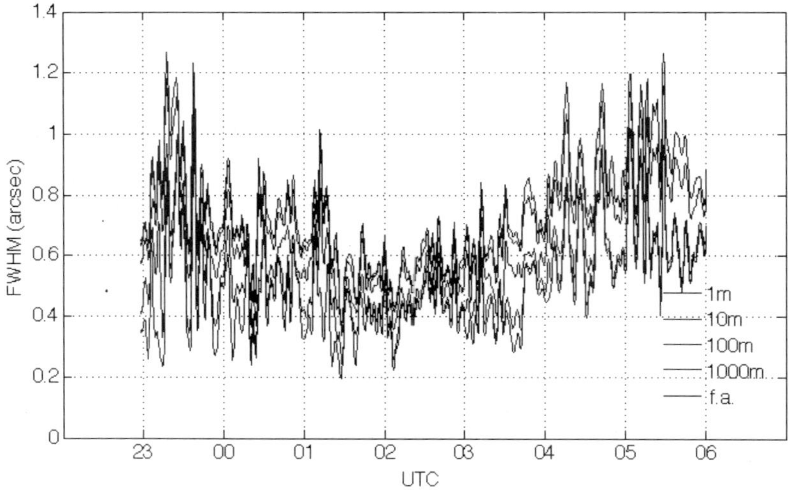

Figure 1. Seeing FWHM, as would be observed by a telescope located at the specified heights, looking towards the zenith at a wavelength of 0.5 um. The data were recorded on the night of April 10, 2008. The lowest curve shows the contribution of high-altitude turbulence alone.

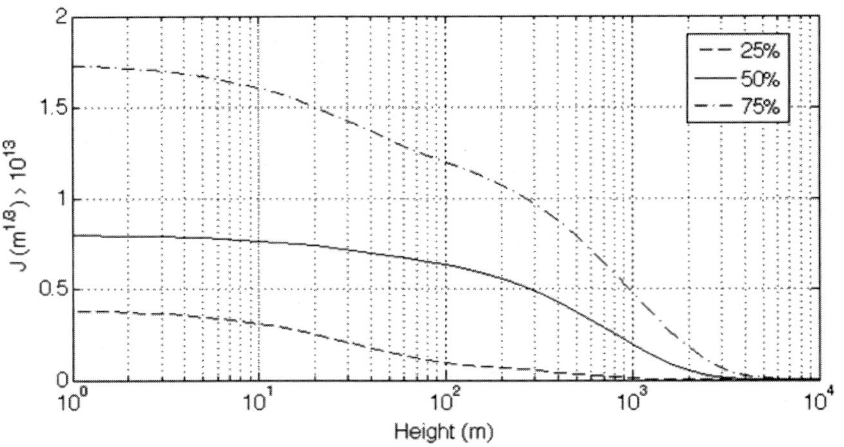

Figure 2. The turbulence integral for the boundary layer, as a function of height, for 25, 50 and 75 percentile values (of the turbulence integral at 1 m) from the entire data set.

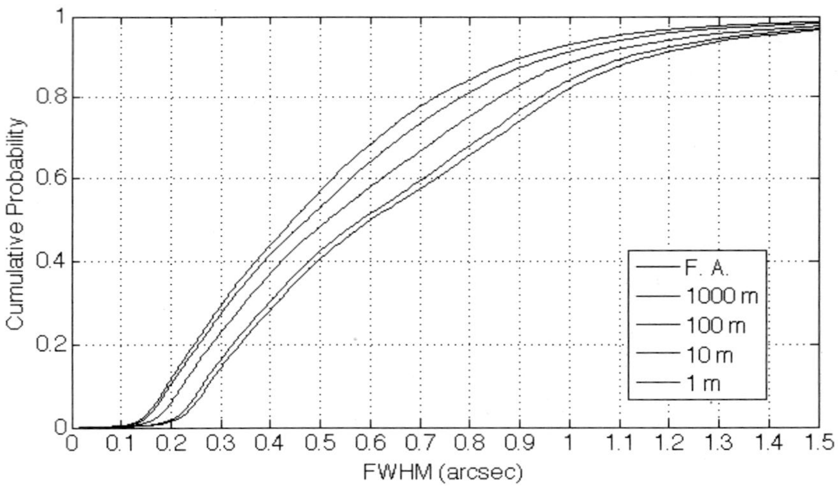

Figure 3. Cumulative probability distributions of FWHM for four heights. The top curve shows the contribution from the free atmosphere (taken to be a thin turbulent layer at a height of 10 km).

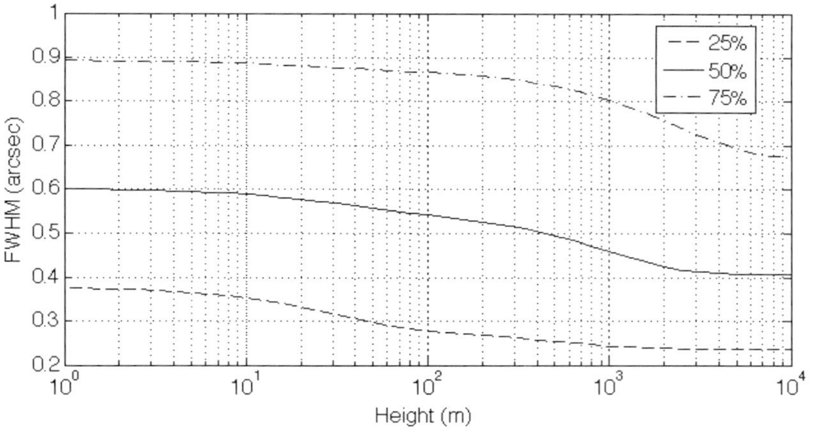

Figure 4. FWHM vs. height for turbulence profiles of the 25, 50 and 75 percentile 1-m seeing.

References

1. A. Rocca, F., Roddier and J. Vernin, *J. Opt. Soc. Am.* **64**, 1000 (1974).
2. S. Egner and E. Masciadri, *Pub. Astron. Soc. Pacific* **119**, 1441 (2007).
3. R. Wilson, *Mon. Not. Roy. Astron. Soc.* **337**, 103 (2002).
4. R. Wilson, *et al.*, *Proc. Soc. Photo-Opt. Instr. Eng.* **7015**, 70154K (2008).
5. T. Travouillon, *Proc. Soc Photo-Opt. Instr. Eng.* **6267**, 626720 (2006).
6. J. Beckers, *Solar Phys.* **145**, 399 (1993).
7. J. Beckers, *Exp. Astron.* **12**, 1 (2001).
8. F. Hill, *et al.*, *Proc. Soc. Photo-Opt. Instr. Eng.* **6267**, 62671T (2006).
9. P. Hickson and K. Lanzetta, *Pub. Astron. Soc. Pacific* **116**, 1143 (2004).
10. J. Rajagopal, A. Tokovinin, E. Bustos and J. Sebag, *Proc. Soc. Photo-Opt. Instr. Eng.* **7013**, 70131P-1 (2008).
11. A. Crotts, *et al.*, *Lunar Plan. Sci.* **39**, 2430 (2008).
12. A. Tokovinin and J. Rajagopal, unpublished technical note, *Modelling the effect of the Moon's phases on scintillation* (2008).
13. P. Gregory, *Baysian Logical Data Analysis for the Physical Sciences*, Cambridge Univ. Press, Cambridge (2005).
14. A. Tokovinin and T. Travouillon, *Mon. Not. Roy. Astron. Soc.* **365**, 1235 (2006).
15. A. Tokovinin, unpublished technical note, *Random measurement errors in LuSci*, http://www.ctio.noao.edu/~atokovin/profiler/errcalc.pdf (2008).
16. N. Metropolis, *et al.*, *J. Chem. Phys.* **21**, 1087 (1953).
17. W. Hastings, *Biometrika* **57**, 97 (1970).
18. F. Roddier, *Prog. Optics* **19**, 281 (1981).

HVR-GS AT MT. GRAHAM: OPTICAL TURBULENCE VERTICAL DISTRIBUTION WITH STANDARD AND HIGH RESOLUTION

E. MASCIADRI*, J. STOESZ, F. LASCAUX, S. HAGELIN

INAF, Osservatorio Astrofisico di Arcetri, Firenze, Italy
** E-mail: masciadri@arcetri.astro.it*

Since a few years measurements of the optical turbulence vertical distribution have been done at Mt. Graham with a Generalized Scidar (GS) located at the focus of the 1.75 m Vatican Advance Technological Telescope (VATT). Such a telescope is placed on the summit of Mt. Graham (Arizona) at around 250 m far away from the Large Binocular Telescope (LBT). Thanks to a new technique[1] that we recently proposed based on the GS observations of wide-binaries (30-35 arcsec), measurements of C_N^2 profiles characterized by a vertical resolution as high as 20-30 m in the first 600 m have been also collected. The statistic sample of measurements consists, at present, of 43 nights distributed in different periods of the year. In this contribution we present the main scientific motivations as well as the analysis of this extended survey and new insights into the turbulence characterization achieved so far by this on-going activity.

Keywords: Turbulence; site testing; generalized scidar.

1. Optical Turbulence Characterization at Mt. Graham: Scientific Motivations

Measurements of the C_N^2 profiles have been obtained with a Generalized Scidar (GS) placed at the focus of the Vatican Advanced Technology Telescope (VATT, D =1.75 m) on the Mt.Graham summit, around 250 m far away from the Large Binocular Telescope (LBT, D = 8 m). As shown in Fig. 4,[1] the primary mirror of the LBT is located $\sim$ 35 m above the dome of the VATT.

The scientific motivations for such a long-term site testing campaign are: **(1)** To collect an as rich as possible statistical sample of optical turbulence vertical distribution (C_N^2 profiles) to be compared with simulations obtained with mesoscale atmospheric models.[2] This is a key milestone for the ForOT project[a] whose final goal is to predict the optical turbulence above astronomical sites.[3] **(2)** To provide a characterization of all the most important integrated astroclimatic parameters above the site of the Large Binocular Telescope (LBT). **(3)** To provide an as rich as possible statistical sample of optical turbulence high resolution vertical distribution in the first hundreds of meters to support the feasibility studies of new generation instruments for LBT such as ARGOS, the LBT-LGS system[4] that in its first baseline will work in a GLAO configuration whose efficiency strongly depends on our ability in quantifying the turbulence near the ground.

2. GS and HVR-GS Techniques

The GS technique[5] is based on the observation of binaries with a typical separation θ within (3-10) arcsec, m_1,$m_2 \leq 5$ mag and $\Delta(m_{1,2}) \leq 1$ mag. When two plane wavefronts propagating from a binary meets a turbulence layer located at a height h from the ground, they produce on the detector plan, optically placed below the ground at about $h_{gs} = 3$ km, two scintillation maps made by a set of characteristic shadows appearing in couple separtated by a distance d that is geometrically related to the position of the turbulence layer as $d = \theta \cdot |h - h_{gs}|$ (Ref. 7-slide 5). The calculation of the auto-correlation (AC) of the scintillation map produces the so called 'triplet'. The central peak is located in the centre of the AC, the lateral peaks are located at a symmetric distance d from the centre. The height of the lateral peaks is proportional to the strength of the turbulence of the layer located at the height h weighted by the scintillation that such a layer produces on the detector.

In a multilayer atmosphere different turbulent layers produce triplets located at a different distance d^* from the center of the AC frame. To monitor the whole troposphere the pupil size needs to be enough large (D $\geq$ 1.5 m) to avoid that one of the shadows that form a couple (or both of them) fall outside the pupil map of the telescope. In the AC frame the triplets are all placed along the direction of conjunction of the binary. To retrieve the C_N^2 profile it is necessary to eliminate the central peak of each triplet in which are present sources of noise such as the photon noise. Finally the C_N^2 is obtained inverting the Fredholm equation (Ref. 7-slide 6) using the conjugated gradient[8] algorithm. We refer the reader to Ref. 8 for a detailed description of the procedure. The vertical resolution $\Delta H(h)$ of the GS technique depends on our ability in discriminating two different

[a] *http://forot.arcetri.astro.it*

later peaks and it is proportional to the Fresnel Zone (FZ):

$$\Delta H(h) = \frac{0.78 \cdot \sqrt{\lambda |h - h_{gs}|}}{\theta} = \frac{0.78 \cdot FZ}{\theta} \tag{1}$$

For typical values of the observable binaries with a standard GS the typical vertical resolution is $\Delta H(0) \sim 1$ km.

The HVR-GS technique has been introduced recently[1] and aims to measure the C_N^2 profiles with a higher vertical resolution in the first kilometer above the ground. If we abandon the idea to monitor the whole 20 km we can easily increase the vertical resolution up to a factor 4 using the standard GS technique based on the calculation of the AC obtain from the of scintillation maps. If we use simultaneously the AC and the cross-correlation maps (CC) taken with 20-40 msec, the triplets in the CC frames are not more aligned on the same direction identified by the binary separation but the central peak of each triplet is located on the direction of the wind speed of each turbulent layer and the distance from the centre of the CC frame is proportional to the wind speed of the same turbulent layer. Under this assumption it is possible to prove[1] that the vertical resolution is finally limited by the accuracy with which we can estimate the position of the lateral peaks in the CC frames that is smaller than the pixel size projected on the pupil (Δ x_{pix}):

$$\Delta h_{pix} = 0.56 \frac{\Delta x_{pix}}{\theta} \tag{2}$$

. For $\theta = 30$ arcsec and Δ $x_{pix} = 7$ mm we retrieve a typical vertical resolution Δ $h_{pix} \sim$ 25-30 m. In conclusion, the main HVR-GS concept consists on: **(1)** taking a wide binary of the order of 30-35 arcsec monitoring the first 1 km and **(2)** treating simultaneously the AC and CC frames. From a practical point of view, the C_N^2 retrieved from the AC frames is characterized by an energy that is redistributed in a set of thinner layers within the first kilometer whose vertical resolution is of the order of 25 m. The AC frames are fundamental to renormalize the total energy in the first kilometer. The final high vertical resolution C_N^2 profile is retrieved from:

$$\int_{-\Delta h_{\max}(0)/2}^{h_{\max}} C_{N,AC}^2(h) \cdot dh = f_{scale} \cdot \sum_i C_{N,CC}^2(h_i) \tag{3}$$

f_{scale} is a factor that takes into account the decorrelation of the central peak of the triplets with $V \neq 0$ with respect to the zero velocity triplet.[1] We measured that f_{scale} in the range $\Delta H = 1$ km can be considered the same for all the thin turbulent layers. Two slightly different approaches have been proposed recently to increase the vertical resolution near the ground: the HVR-GS[1] and LOLAS.[9] HVR-GS is based on the employment of the same instrument (GS) but it uses a different technics. LOLAS is a different instrument from the GS but it uses the same GS technics.

3. Astroclimatic Parameters: Main Results

We collected and analyzed so far observations related to 43 nights (Ref. 7 - slide 11) i.e. 31750 frames correspondent to roughly 245 hours. The code used to identify the typology of data-set is: 'GS' for the standard GS measurements extended on the whole troposphere (43 nights), 'WB' for the C_N^2 retrieved from the AC frames (ΔH $\sim$ 200-250 m) (15 nights) and 'HVR-GS' for the C_N^2 with the high vertical resolution (ΔH $\sim$ 25-30 m) (15 nights). The samples of the three categories are differently rich because the HVR-GS technique is more recent and the HVR-GS has a higher rate of rejection of measurements. The binaries observed are reported in Table 2 (Ref. 10). Table 1, in this paper, summarizes the median (50th), first (25th) and third (75th) quartiles for the three main integrated astroclimatic parameters calculated for the following groups: the all 43 nights, the summer and the winter time. The summer time range includes [April-June] period and winter [October-March]. We refer to Ref. 10-Fig. 1 for the cumulative

Table 1. Median, first and third quartiles values of the main integrated astroclimatic parameters above Mt. Graham (43 nights).

	Total			Summer			Winter		
Parameter	25th	50th	75th	25th	50th	75th	25th	50th	75th
ε (arcsec)	0.65	0.95	1.34	0.53	0.61	0.72	0.89	1.19	1.50
θ_0 (arcsec)	1.6	2.5	3.6	3.1	3.8	4.5	1.4	2.0	2.7
τ_0 (msec)	2.7	4.8	8.7	6.4	10.1	14.6	2.5	3.8	6.2

distribution calculated for the total 43 nights and for the separated seasons: summer and winter. In May 2005 and May 2007 it has been observed, in both cases, 10-15 days of extremely good seeing and large isoplanatic angle. For the τ_0 it has been used a composite wind speed profile. Below 2 km, the wind speed retrieved from the GS has been used, above 2 km, the wind speed profile retrieved from the European Center for Medium-range Weather Forecast (ECMWF) analyses extracted in the nearest grid point (32.75°N, 110.00°W) to the Mt. Graham ($\sim$ 20 km northwest of the summit) has been used. From the total seeing (ε) it has been subtracted the contribution from the dome (ε_d), calculated with the method described in Ref. [8], to quantify the contribution of the seeing provided only by the atmosphere. The method consists in discriminating the triplets located at the ground with a velocity equal to zero from those located at the same height and having a velocity different from zero. The velocity resolution of our system is 0.8 to 0.2 m/s per pixel with time lags within the range (10 to 40 ms). The median value of the 'dome seeing' is ε_d= 0.52 arcsec. For the first time it has been observed an interesting seasonal trend in the dome contribution; such a feature will be certainly carefully investigated in the future. **Knowing that the median seeing in the whole atmosphere (included the dome seeing) is ε=0.95 arcsec and that ε_d= 0.52 arcsec, it follows that the median seeing related to the whole atmosphere without the dome contribution for the richest statistic we collected so far (43 nights) is ε_{tot}= 0.72 arcsec.**

4. AO Applications: LBT-LGS System ARGOS

The *'composite profiles'* are an efficient and practic way to represent, in statistical terms, the vertical distribution of the optical turbulence in a discretized way suitable for AO simulations. The method consists in identifying a finite number of vertical slabs covering the whole troposphere ($\sim$ 20 km) with their correspondent value of seeing ($\Delta\varepsilon_i$ or ΔJ_i) so that the total turbulence integrated on the troposphere is conserved (see Ref. 10). Ref. 7 (slide 17) reports the ΔJ_i values calculated at different heights and obtained from the C_N^2 associated to a r_0 related to the 70-80 % of the cumulative distribution ('good' case), the r_0 related to the 45-55 % ('typical' case) and the r_0 related to the 20-30 % ('bad' case). The turbulence in the free atmosphere (h>1 km) and in the boundary layer (h$\leq$1 km) is treated, as a first approximation, in an independent way to permit to study different combinations of probabilities for the OT vertical distribution. It can be calculated[6] that for ARGOS[4] (field of view θ=4 arcmin), we do not need to sample the C_N^2 with a vertical resolution better than 200 m in the first unit cell above the ground. It is interesting to quantify with great accuracy the portion of turbulence developed below 200 m because the H_{min} of the 'gray zone' rarely falls below 200 m. Composite profiles done with the C_N^2 profiles retrieved from the 'WB' sample (AC frames and Δh = 200 m) are therefore suitable for this purpose and it is not necessary a higher vertical resolution. To take into account the different statistic obtained with the standard 'GS' and wide binary sample 'WB', each number of the table in

Ref. 7 (slide 17, h$\leq$1 km) has to be multiplied by the correction factor $f_{gl}=(\varepsilon_o/\varepsilon'_o)^{5/3}$ (Ref. 7, same slide). The numerator is the seeing measured in the first kilometer from the ground (from the 'GS' sample). The denominator is the seeing measured in the first kilometer from the ground (from the 'WB' sample).

5. HVR-GS Vertical Distribution for h $\leq$ 1 km

The OT vertical distribution at high resolution (20-30 m) is obtained with the method presented in Ref. 1. The HVR-GR data-set is obtained taking the integral of the C_N^2 profiles retrieved from the AC frames redistributing the energy in the first 1 km according to the detectable triplets in the CC frames. Fig. 1 shows the mean of the C_N^2 profiles (Δh $\sim$ 25-30 m) calculated for 15 nights (asterisks) extracted from the sample 'HVR-GS'. This sample covers all the nights for which measurements with HVR-GS are available and reliable. To retrieve the typical scale height B of the exponential decay we fitted the measurements done below 125 m with an exponential law (Eq. (4)):

$$y = A \cdot e^{(-h/B)} \tag{4}$$

where A and B are free parameters. The calculation has been performed in the case in which the dome is included and taking C_N^2 =10^{-16} where there is no signal. This threshold corresponds to the minimum C_N^2 value detected in the CC frames and associated to an equivalent J=$2.5{\cdot}10^{-15}$. It corresponds also to the sensitivity threshold of our system. [An inferior limit can be calculated taking C_N^2=0 where there is not signal - to be discussed in a forthcoming paper]. The fit gives A=$4.72{\cdot}10^{-15}$ and B=37.5 m (Fig. 1-left). These results definitely indicate that the HVR-GS technique is able to prove that the turbulence decays above typical astronomical sites in stable night time conditions in a much more sharper way than what believed so far. Indeed, the Hufnagel model in proximity of the surface[11] in night conditions states that the turbulence scales as $h^{-2/3}$. In that case (Fig. 1-right) the C_N^2 decreases of one order of magnitude within 1 km while our results indicate that the C_N^2 scales of one order of magnitude within $\sim$ 25 m (see zoom in Ref. 7-slide 19).

Figures 2 show the percentage of the turbulence located below 30 m (height of the LBT primary mirror from the ground) with respect to the turbulence in the whole 20 km for all the 15 nights analyzed in this paper. The dome contribution is included. Both cases (normalized and not

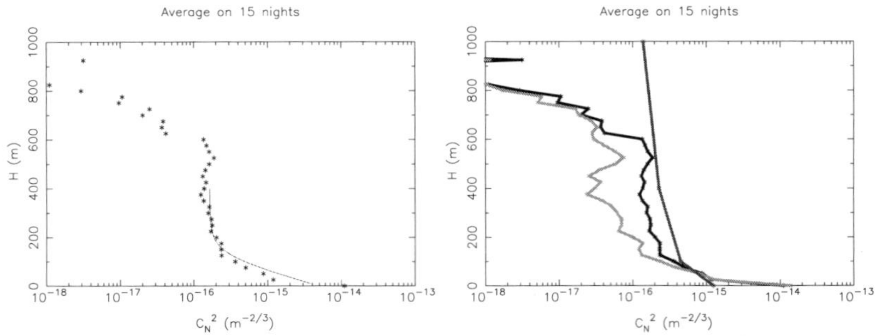

Fig. 1. (color on-line) **Left:** Mean C_N^2 profile (Δh~20-30 m) calculated with the method described in Ref. 1 with the sample 'HVR-GS' - dome included (black asterisks). Full red line: analytical fit. **Right:** measured 'HVR-GS' C_N^2 profiles (black line) versus Hufnagel model (red line). The green line is the average C_N^2 profile obtained when the threshold is not taken into account (C_N^2 =0).

for the f_{gl} factor) are considered. We observe that a consistent percentage of turbulence (~49-54 %) develops in the first 30 m from the ground. A sharp vertical distribution is certainly a promising scenario for GLAO-AO correction system.

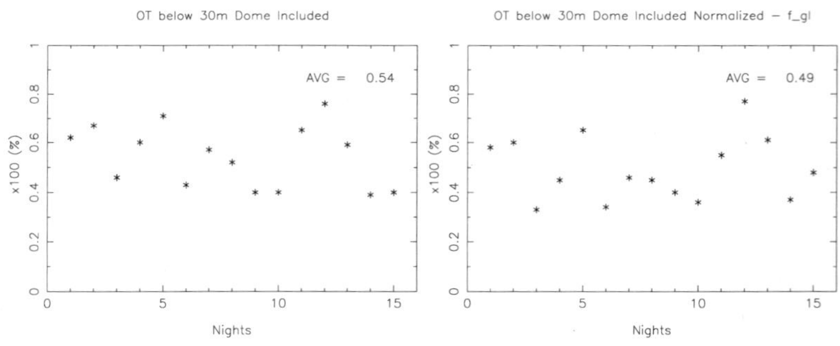

Fig. 2. **Left:** percentage of turbulence developed below 30 m for all the 15 nights. **Right:** values of the panel on the left are normalized by the f_{gl} factor to improve the statistic reliability.

6. Conclusions

In this paper we present the status report of an on-going study aiming to characterize the optical turbulence at Mt. Graham. We present a general overview of the statistics (43 nights) of all the main integrated astroclimatic parameters and their seasonal trends. With a median seeing ε= 0.95 arc-

sec (ε= 0.72 arcsec without dome contribution), isoplanatic angle θ_0= 2.5 arcsec and a wavefront coherence time τ_0=4.8 msec, Mt. Graham confirms its good quality in terms of turbulence characteristics typical of the best astronomical sites in the world. The ground layer is characterized for the first time with a high statistical reliability with a resolution of 200-250 m and 20-30 m. A composite distribution of the turbulence on the whole 20 km is calculated to be used as input of AO simulations of ARGOS. The main conclusions we achieved are:
(1) The turbulence exponentially decays above Mt. Graham with a much sharper profile than what supposed so far. It is the first time that such a trend has been put in evidence above a mid-latitude site.
(2) The percentage of turbulence developed in the first 30 m above the ground (below the primary mirror of the LBT) is mostly 49-54 %. This trend, joint with the favorable large θ_0 in the spring/summer time are conditions particularly favorable for astronomical observations assisted by a LGS/GLAO systems such as ARGOS.[4]
(3) For the first time, we observed a seasonal trend of the dome seeing. This is certainly a topic that is worth of a more careful investigation in the future.

Acknowledgments

This work is funded by a Marie Curie Excellence Grant (ForOT), MEXT-CT-2005-023878 - FP6 Program.

References

1. Egner, S. & Masciadri, E., 2007, PASP, 119, 1441
2. Masciadri, E., Avila, R., Sánchez, L. J., 2004, RMxAA, 40, 3
3. Masciadri, E., 2006, SPIE, 6267, 62671C
4. Rabien, S. et al., 2008, SPIE 2008, 7015, 701515
5. Vernin, J. & Azouit, M., 1983, J. Optics, 14, 131
6. Masciadri, E., Stoesz, J., Lascaux, F., Hagelin, S., 2009, ARGOS-Phase 2 Report
7. Masciadri, E., Stoesz, J., Lascaux, F., Hagelin, S., 2008, on-line presentation: *http://forot.arcetri.astro.it/otam* – 08 - Link: Program
8. Egner, S., Masciadri, E., McKenna, D., 2007, PASP, 119, 669
9. Avila, R., Aviles, J-L., Wilson, R., Chun, M., Butterley, T., Carrasco, E., 2008, MNRAS, 387, 4, 15-11
10. Stoesz, J., Masciadri, E., Lascaux, F., Hagelin, S., 2008, SPIE, 7012, 70121W
11. Roddier, F., Progress in Optics, XVIII, 1981

PROFILOMETRY FOR THE LOWER TERRESTRIAL ATMOSPHERE

J. BORGNINO*, A. BERDJA, A. ZIAD and J. MAIRE

UMR 6525 H. Fizeau, Université de Nice Sophia Antipolis,
Centre National de la Recherche Scientifique,
Observatoire de la Côte d'Azur,
Campus Valrose, F-06108 Nice Cedex, FRANCE
**E-mail: Julien.Borgnino@unice.fr*

The principle of a differential Optical Turbulence Profiler based on Angle-of-Arrival statistics as the SLODAR is presented. This instrument is well-adapted to study the terrestrial atmosphere boundary-layer in daytime and nighttime conditions.

Keywords: Optical Turbulence; Angle of Arrival; Profiler; Atmospheric boundary-layer.

1. Introduction

The implementation of modern techniques used for High Angular Resolution astronomical observations with new generation large telescopes needs the knowledge of the altitude distribution of terrestrial atmospheric turbulence as well as its statistical properties. A particular attention must be focused on the boundary layer and more particularly on the surface layer which extends typically on the first 100 meters. Dedicated experiments have been built allowing, from star scintillation observation, measurements of the turbulence vertical distribution with an altitude resolution more or less high (Multi-Aperture Scintillation Sensor (MASS)[1,2], Generalized - SCIntillation Detection And Ranging (G-SCIDAR)[3,4], and more recently Low Layer SCIDAR (LOLAS)[5,6]. Other experiments founded on the observation of angle-of-arrival (AA) fluctuations have been used to deduce the coherence parameters associated to the perturbed wavefronts arriving on the ground-based telescope entrance pupil (Differential Image Motion Monitor (DIMM)[7], Generalized Seeing Monitor (GSM)[8,9]). Recently, instruments have also been used to obtain profiles of atmospheric turbulence energy

$(C_n^2(h))$ (SLOpe Detection And Ranging (SLODAR)[10,11]) and of the outer scale $(L_0(h))$(Monitor for Outer Scale Profile (MOSP)[12,13]). Finally, for monitoring parameters of optical turbulence, a combination of MASS and DIMM, which implies simultaneous observations of star scintillation and AA fluctuations, is now currently used[14,15].

In this paper is presented the principle of an optical turbulence profiler based on the observation of AA fluctuations shown pointing the solar limb. These fluctuations are transformed in intensity fluctuations using a technique similar to the Foucault test performed to evaluate optical surface quality[16].

2. Wavefront Analysis : Observation of AA Fluctuations in the Pupil Plane[17]

An incoherent extended object (solar limb) is observed using a telescope with an entrance pupil described by a complex amplitude transmission $P(\mathbf{r})$, in which $\mathbf{r}$ is a vector in planes perpendicular to the optical axis, and a focal length f_T. The image is formed in the focal plane where is placed a thin slit, of complex amplitude transmission $t(\mathbf{r})$, and a lens of focal length f_L. The width of the slit is of the order of a few arcseconds. The height is typically equal to dozens arcseconds. On Fig. 1. (section 4) is presented the case of 2 slits.

The intensity distribution in the pupil image formed by the lens $(I_0(\mathbf{r}))$ may be obtained performing a Fourier optics calculation. As a first approximation, it is assumed that the limb profile is proportional to the angular direction α (linear model for the solar limb)[17]. This approximation may be validated from different models proposed to describe the solar-limb darkening. In this case, intensity fluctuations (Solar Flying Shadows) observed in the pupil plane image are expressed as a function of $\beta(-\mathbf{r})$, component of AA fluctuations considered in the direction perpendicular to the limb, under the form of a linear relation :

$$\widetilde{I_0}(\mathbf{r}) = \mathbf{C}\frac{f_T}{f_L}[\frac{\lambda f_T}{f_L}]^2|\hat{t}\frac{\mathbf{r}}{\lambda f_L}|^2 * P(-\mathbf{r}\frac{f_T}{f_L})\beta(-\mathbf{r}\frac{f_T}{f_L}) \tag{1}$$

where, $\hat{}$ notes a Fourier transform and $*$ symbolizes a convolution product. λ is the wavelength and $\mathbf{C}$ is a proportionality factor in the linear model. Scintillation and edge effects (derivatives of $P(\mathbf{r})$) have been neglected. In conclusion, if the solar-limb darkening may be assumed to be linear, the intensity fluctuations observed in the pupil image are proportional to the AA fluctuations.

The thin slit performs two types of filtering. The first one is due to diffraction and the second one is due to an angular integration varying linearly with altitude[17].

3. AA Statistics

3.1. *Transverse Spatial Covariance of AA Fluctuations*

It can be shown[18] that this covariance for the component β of AA fluctuations may be expressed as :

$$\boldsymbol{C}_{\beta,0}(b) = \pi\lambda^2 \int_0^{+\infty} df\, f^3 \mathbf{W}_\varphi(f)[\mathbf{J}_0(2\pi bf) + \mathbf{J}_2(2\pi bf)][\frac{2\mathbf{J}_1(\pi D_T f)}{\pi D_T f}]^2 \quad (2)$$

where f is a spatial frequency, b the baseline and D_T, in the general case, the telescope diameter. $\mathbf{J}_0$, $\mathbf{J}_1$ and $\mathbf{J}_2$ are Bessel functions of first kind. One can note that Fresnel diffraction has been neglected because we are mainly interested by the boundary-layer (typically for an altitude h lower than 1km).

In the case of the von Kármán model[19], if the terrestrial atmosphere is constituted of several turbulent layers, the power spectrum of phase fluctuations is given by

$$\mathbf{W}_\varphi(\mathbf{f}) = 0.38\lambda^{-2} \sum_j \mathbf{C}_n^2(h_j)\delta h_j (f^2 + L_0(h_j)^{-2})^{-11/6} e^{-1.126 f^2 l_0(h_j)^2} \quad (3)$$

where L_0 and l_0 are respectively outer and inner scales. Introducing Eq. (3) in Eq. (2) allows to obtain the transverse spatial covariance of AA fluctuations.

3.2. *Transverse Spatio-Angular Covariance of AA Fluctuations*

This covariance is obtained considering AA fluctuations, at ground level, observed in 2 angular directions separated by θ. In the case of a multi-layer atmosphere, using Eq. (3) and assuming that the inner scale is equal to 0, Eq. (2) becomes:

$$\begin{aligned} \boldsymbol{C}_{\beta,\theta}(b) = 1.19 \sum_j \mathbf{C}_n^2(h_j)\delta h_j \int_0^{+\infty} df\, f^3 (f^2 + L_0(h_j)^{-2})^{-11/6} \\ [\mathbf{J}_0(2\pi(b-\theta h_j)f) + \mathbf{J}_2(2\pi(b-\theta h_j)f)][\frac{2\mathbf{J}_1(\pi D_T f)}{\pi D_T f}]^2 \end{aligned} \quad (4)$$

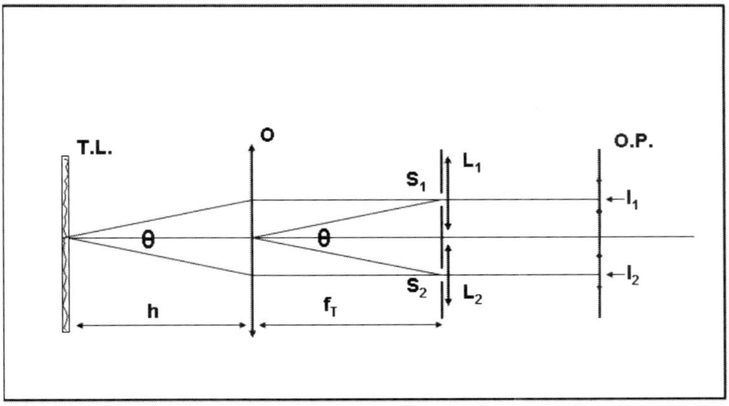

Fig. 1. Principle of the Optical Turbulence Profiler.

4. Optical Turbulence Profiler

4.1. *Principle*

This principle is schematized on Fig. 1.

T.L. is a turbulent layer localized at the altitude h. The telescope pupil (O) is observed through 2 slits (S_1 and S_2) placed in the telescope focal plane on the solar limb image with an angular separation equal to θ. Two images (I_1 and I_2) are formed using 2 lenses (L_1 and L_2) in the observation plane O.P. In each angular direction is obtained a map of $\beta(x, y)$, angle-of-arrival component considered in the direction perpendicular to the solar limb. One can see in the following that the estimation of the 1D-covariance according to the direction parallel to the limb leads to the vertical energy of the optical turbulence $C_n^2(h)$.

Thus, a triangulation method, based on the observation of AA fluctuations simultaneously in 2 angular directions, may be developed. Let us assume for an illustration that the first kilometer of the terrestrial atmosphere has 4 turbulent layers localized at the altitudes $h_1 = 0$, $h_2 = 100m$, $h_3 = 500m$ and $h_4 = 800m$ with respective weights equal to 0.6, 0.25, 0.1 and 0.05. One assumes also that these layers represent 75 percent of the whole turbulent energy. The wavelength is $\lambda = 468nm$. The Fried parameter r_0 for the whole atmosphere is 6cm. With the above assumptions this parameter is equal to 7.13cm for the turbulence localized in the first kilometer. The outer scale is equal to 10m and is assumed constant with the altitude. The width of the slit is 3 arcseconds that corresponds to a size of the smaller structure which may be observed in the pupil image equal to 3cm. The filtering imposes by diffraction is predominant for h lower than 1km.

Using Eq. (4) and the above assumptions, one presents on Fig. 2 the spatio-angular covariance when the angular separation θ is equal to 200 arcseconds (a.s. on the figure).

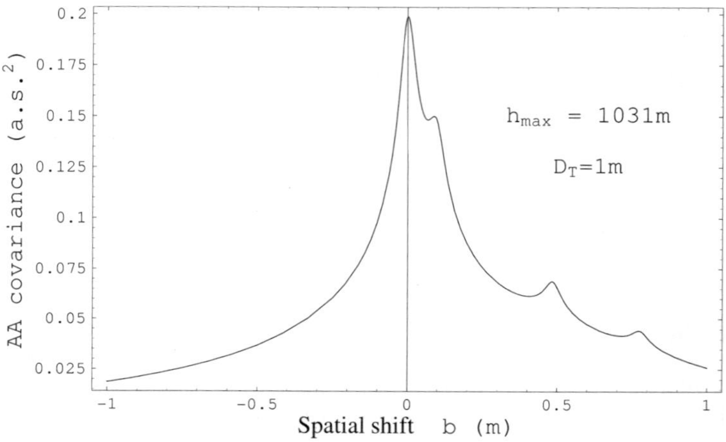

Fig. 2. AA transverse spatio-angular covariance function $\boldsymbol{C}_{\beta,\theta}(b)$.

If D_T is the telescope diameter, the maximum sensing altitude is given by $h_{max} = \frac{D_T}{\theta} = 1031m$.

To estimate the altitude resolution one can use the ratio between the Full Width at Half Maximum ΔC of the spatial covariance and the angular separation between the slits. Here, this resolution is $R_A = \frac{\Delta C}{\theta} = 140m$.

4.2. *A Differential Profiler*

The profiler could be affected by telescope vibrations which may have different origins : tracking, wind,... Studies of the effects of these eventual vibrations have been already performed related to High Angular Resolution techniques[20,21,22]. Wilson et al.[23] have more particularly studied the effects of vibrations due to the tracking system of the 4.2m William Herschel Telescope at La Palma.

A solution to this vibration problem is to implement a differential method. Let us consider 2 images I_1 and I_2 obtained simultaneously with 2 slits separated by θ. On the images, one notes x-direction and y-direction respectively parallel and perpendicular to the limb. Recalling the angular directions, AA fluctuations are respectively $\beta_1(x, y, -\theta/2)$ and $\beta_2(x, y, +\theta/2)$. The differential spatio-angular covariance writes as :

$$\mathbf{CC}_{\beta,\theta}(b) =< [\beta_1(x, y, -\theta/2) - \beta_2(x, y, +\theta/2)][\beta_1(x + b, y, -\theta/2) - \beta_2(x + b, y, +\theta/2)] > \tag{5}$$

which leads to

$$\mathbf{CC}_{\beta,\theta}(b) = \mathbf{C}_{\beta,0}(b) + \mathbf{C}_{\beta,0}(b) - \mathbf{C}_{\beta,0}(b + \theta h) - \mathbf{C}_{\beta,0}(b - \theta h) \tag{6}$$

where <> notes an ensemble average, $\mathbf{C}_{\beta,0}$ is the 1D-spatial covariance, b is a spatial shift on the pupil image and h the altitude of the turbulent layer.

On Fig. 3 is shown $\mathbf{C}_{\beta,0}(b+\theta h) + \mathbf{C}_{\beta,0}(b - \theta h) = \mathbf{C}_{\beta,-\theta}(b) + \mathbf{C}_{\beta,+\theta}(b)$, which has been obtained doing $2\mathbf{C}_{\beta,0}(b) - \mathbf{CC}_{\beta,\theta}(b)$.

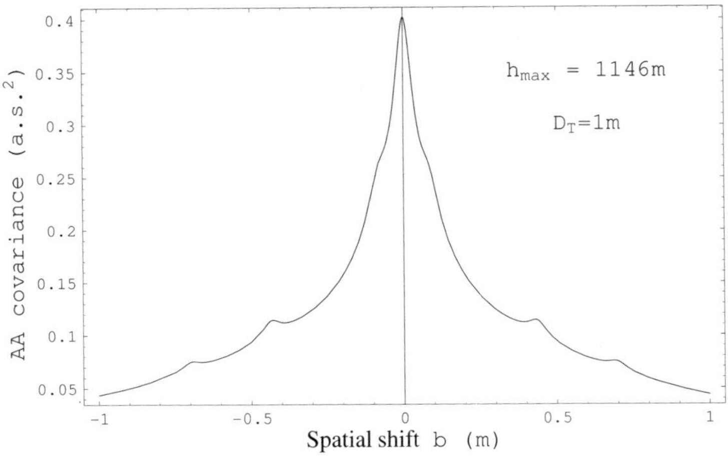

Fig. 3. AA transverse spatio-angular covariance function $\mathbf{C}_{\beta,-\theta}(b) + \mathbf{C}_{\beta,+\theta}(b)$.

The turbulence in the first kilometer has been assumed to be identical to the one presented above. The separation between the 2 slits has been taken equal to 180 arcseconds and the telescope diameter is 1m, which gives a maximum sensing altitude $\mathbf{h}_{max} = 1146m$. One can note here that, using Eq. (4), the function $\mathbf{C}_{\beta,-\theta}(b) + \mathbf{C}_{\beta,+\theta}(b)$ which is shown in Fig. 3 may be rewritten as :

$$\mathbf{C}_{\beta,-\theta}(b) + \mathbf{C}_{\beta,+\theta}(b) = \sum_j \mathbf{C}_n^2(h_j)\delta h_j F(b, L_0(h_j), \theta, D_T) \qquad (7)$$

Retrieving $C_n^2(h)$ (and eventually $L_0(h)$) from the above function is a non-linear inverse problem. As that is performed in the case of the MOSP[12,13], one can use simulated annealing algorithm for minimizing a cost function.

5. Conclusion

The method may also be used in the case of nighttime conditions (observation of the lunar limb). In this case, the profiler will be coupled with the Monitor for Outer Scale Profile (MOSP) in a differential version.

One can also perform a spatio-temporal analysis (estimation of the spatio-temporal covariances) in order to study the motions of turbulent layers localized at different altitudes. The method is similar to the one which allowed to study the motions of turbulent layers observing star scintillation. One follows the motions of the correlation peaks increasing the temporal difference. One can thus estimate the speed and the direction of the wind (horizontal component).

Using 4 slits in non-redundant positions one can obtain, in a direction parallel to the limb, 6 angular baselines (multi-resolution). One can so obtain the profile $\mathbf{C}_n^2(h)$ (and possibly $\mathbf{L}_0(h)$) simultaneously on 6 altitude ranges overlaping but more or less extended, with variable altitude resolution R_A(for example from 560 to 93m when the angular separation between the slits varies from 50 up to 300 arcseconds). This is similar to the use of the SLODAR observing multiple sources (natural or laser guide stars)[24]. Having 6 baselines instead of one is advantageous and allows to improve the altitude explorer, the resolution and so the global precision of the estimations.

A study (theoretical with numerical simulations) of the influence of scintillation due to high turbulent layers will be developed, with the angular separation between the slits as parameter.

References

1. V. Kornilov, A.A. Tokovinin, O. Vozyakova, A. Zaitsev, N. Shatsky, S.F. Potanin and M. Sarazin, *Proc. SPIE* **4839**, 837 (2003).
2. A.A. Tokovinin, V. Kornilov, N. Shatsky and O. Vozyakova, *Mon. Not. R. Astron. Soc.* **343**, 891 (2003).
3. R. Avila, J. Vernin and E. Masciadri, *Appl. Opt.* **36**, 7898 (1997).
4. R.A. Johnston, C. Dainty, N.J. Wooder and R.G. Lane, *Appl. Opt.* **41**, 6768 (2002).
5. R. Avila and M.R. Chun, *Proc. SPIE* **5490**, 742 (2004).
6. R. Avila, R. Wilson, M. Chun, S. Cuevas and A. Farah, *Proc. of "Symposium on Seeing"* held in Kona, Hawaii, 20-22 March 2007.
7. M. Sarazin and F. Roddier, *Astron. Astrophys.* **227**, 294 (1990).
8. R. Avila, A. Ziad, J. Borgnino, F. Martin, A. Agabi and A. Tokovinin, *J. Opt. Soc. Am. A* **14**, 3070 (1997).
9. A. Ziad, R. Conan, A. Tokovinin, F. Martin and J. Borgnino, *Appl. Opt.* **39**, 5415 (2000).
10. R.W. Wilson, *Mon. Not. R. Astron. Soc.* **337**, 103 (2002).
11. T. Butterley, R.W. Wilson and M. Sarazin, *Mon. Not. R. Astron. Soc.* **369**, 835 (2006).
12. J. Maire, A. Ziad, J. Borgnino and F. Martin, *Mon. Not. R. Astron. Soc.* **377**, 1236 (2007).
13. A. Ziad, J. Maire and J. Borgnino, *Proc. of "Symposium on Seeing"* held in Kona, Hawaii, 20-22 March 2007.
14. A. Tokovinin and V. Kornilov, *Mon. Not. R. Astron. Soc.* **381**, 1179 (2007).
15. V. Kornilov, A. Tokovinin, N. Shatsky, O. Voziakova, S. Potanin and B. Safonov, *Mon. Not. R. Astron. Soc.* **382**, 1268 (2007).
16. J. Borgnino and F. Martin, **in French**, *J. Optics (Paris)* **8**, 319 (1977).
17. J. Borgnino, A. Berdja, A. Ziad and J. Maire, *Proc. of "Symposium on Seeing"* held in Kona, Hawaii, 20-22 March 2007, *http : //mkwc.ifa.hawaii.edu/symposium/* - Program and Events.
18. R. Conan, J. Borgnino, A. Ziad and F. Martin, *J. Opt. Soc. Am. A* **17**, 1807 (2000).
19. V. V. Voitsekhovich, *J. Opt. Soc. Am. A* **12**, 1346 (1995).
20. A. Tokovinin, *Mon. Not. R. Astron. Soc.* **316**, 637 (2000).
21. S. Altarac, P. Berlioz-Arthaud, E. Thiébaut, R. Foy, Y.Y. Balega, J.C. Dainty and J.J. Fuensalida, *Mon. Not. R. Astron. Soc.* **322**, 141 (2001).
22. Y. Clénet, M. Kasper, N. Ageorges, C. Lidman, T. Fusco, O.P. Marco, M. Hartung, D. Mouillet, B. Koehler, G. Rousset and N. Hubin, *SF2A-2004 : Semaine de l'Astrophysique Francaise*, Paris, France, June 14-18 2004, Edited by F. Combes, D. Barret, T. Contini, F. Meynadier and L. Pagani, EDP Sciences Conferences Series, p. 179, (2004).
23. R.W. Wilson, N. O'Mahony, C. Packman and M. Azzaro, *Mon. Not. R. Astron. Soc.* **309**, 379 (1999).
24. L. Wang, M. Schöck and G. Chanan, *Appl. Opt.* **47**, 1880 (2008).

SLODAR TURBULENCE MONITORS

R. W. WILSON, T. BUTTERLEY and J. OSBORN

Department of Physics, University of Durham,
South Road, Durham, DH13LE, UK

SLODAR turbulence monitors have been installed and operated at the Cerro Paranal, Mauna Kea and SAAO Sutherland observatories. The instruments, developed at Durham University, provide real-time measurements of the atmospheric turbulence strength, altitude and velocity, for site characterization and for real-time support of adaptive optics for astronomy. We present sample results and compare contemporaneous data obtained with SLODAR, MASS and DIMM monitors at the ESO Paranal observatory.

Keywords: Turbulence profiling; Site Characterization; SLODAR.

1. Introduction

SLODAR[1,2] is an optical turbulence profiling method based on Shack-Hartmann wavefront sensing, using double star targets. The turbulence profile $C_n^2(h)$ is recovered from the time-averaged cross-covariance of the wavefront sensor (WFS) centroids for the two stars.[3] The advantages of the SLODAR sensor are that it is sensitive to turbulence at all altitudes, including the ground level, with a response function that is only a weak function of altitude. It is straightforward to calibrate in terms of the absolute turbulence strength, and need not be calibrated for the effects of photon/shot noise, since the WFS cross-covariance is not biased by shot noise. Turbulent layer translational (wind) velocities can also be measured.

SLODAR can be implemented on large telescopes for detailed profiling using a high-order WFS[1,4] or on small telescopes as a facility site monitor.[5] The number of sampling elements of the turbulence profile is equal to the number of WFS sub-apertures subtended across the telescope pupil. The altitude sampling resolution for SLODAR can be tuned by selecting double star targets with appropriate angular separation: for small angular separations the whole profile, from the ground to high altitude, is mapped with lower resolution. For wider double stars the lower part of the atmosphere can be studied in detail. Hence the system can be configured to operate as a general-purpose profiler or to study the ground layer. In either case the total turbulence strength for the whole atmosphere is also measured.

2. Current SLODAR Profilers

Self-contained turbulence profilers, based on the SLODAR technique applied to small telescopes, have been developed at Durham CfAI. A prototype portable SLODAR system was commissioned by ESO for statistical characterization of the ground layer (GL) of atmospheric turbulence at the Cerro Paranal observatory, in relation to the development and application of adaptive optical instrumentation at ESO.[5] A second, similar, monitor was installed at the Mauna Kea observatory as part a study of GL turbulence for the Gemini observatory in 2006.[6] A third SLODAR system was deployed at the site of the South African Large Telescope (SALT) in 2008.

Details of the Paranal and Mauna Kea systems and their operation have been described elsewhere, and the system at SAAO is similar. Each of these instruments comprises a 40cm telescope equipped with a WFS subtending 8x8 sub-apertures. The key to the implementation of the portable SLODAR profilers is the use of electron-multiplication CCD (EM-CCD) cameras to overcome the effects of read-out noise for wavefront sensing with small apertures and short exposure times. The use of EM-CCDs permits effective operation of the systems for targets as faint as 7th magnitude. There are then sufficient targets available for continuous operation.

The data products of the SLODAR profilers are measures of the total integrated seeing angle, the integrated turbulence strength in each of the 8 directly sampled profiling bins, and of the integrated turbulence strength for all altitudes above the highest altitude for direct sampling. Turbulent layer velocities are not measured automatically but can be recovered by off-line analysis of the spatio-temporal cross-covariance of the centroid data.

The SLODAR system at Cerro Paranal has an altitude resolution of approx. 150m in GL profiling mode, limited by the field of view of the WFS camera. The systems at Mauna Kea and SAAO employ larger format cameras, so that targets with wider separations can be observed, and the vertical resolution is reduced to approx. 60m (the maximum altitude for profiling is then reduced to approx. 450m). The limiting sensitivity of the systems, in terms of the minimum detectable integrated turbulence strength per bin per sample, is approx. $3 \times 10^{-14} m^{1/3}$.

Fig. 1. The SLODAR turbulence profiler at the Cerro Paranal observatory.

3. Sample Results

3.1. *Cerro Paranal*

Figure 2 shows an example data sequence recorded by the SLODAR profiler for one night at Cerro Paranal. In this example we see that there is always significant turbulence strength in the first sampling bin centered at zero altitude, whereas turbulence at intermediate altitudes is intermittent. Frequently the surface layer of turbulence is not resolved by the system - i.e. there is little or no signal in the second bin. This indicates that the scale height of the surface layer is much smaller than the altitude range spanned by the first bin. The first bin is half-height, since it is centered at the ground, so that the surface layer thickness is then much less than 80m.

Figure 3 shows the mean turbulence profile measured by the SLODAR at Cerro Paranal from February 2005 to January 2006. Here the profiles have been re-binned to the lowest common resolution (largest bin width)

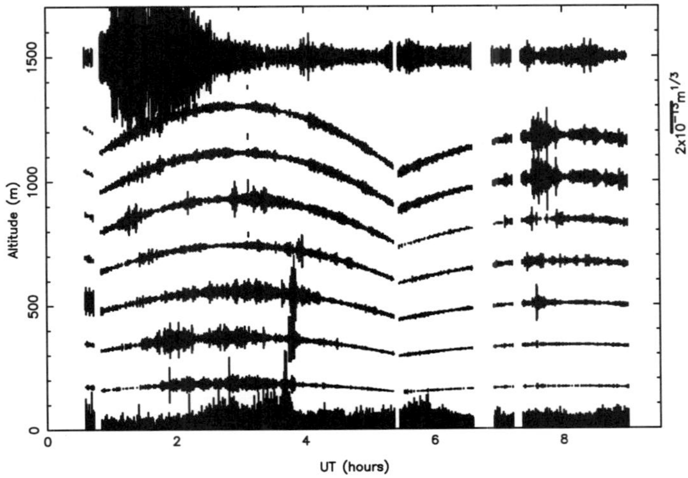

Fig. 2. Example SLODAR results for a single night (9 January 2006) at Cerro Paranal. The bottom trace is centered at the ground. The top trace is the integrated turbulence strength for all altitudes above 1300m.

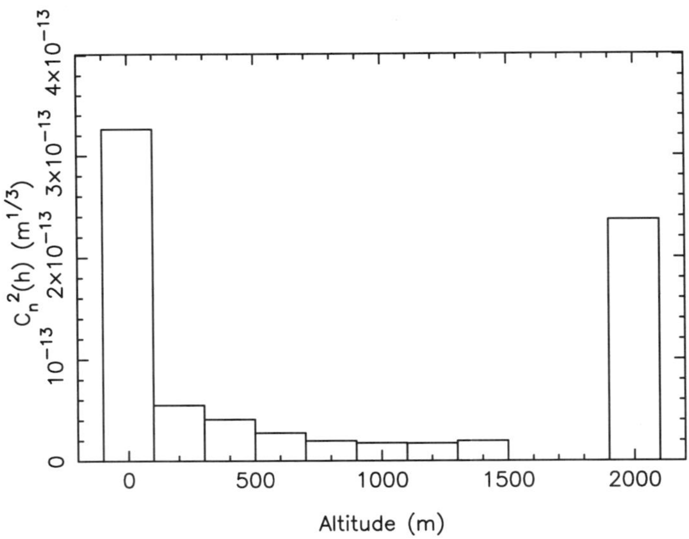

Fig. 3. Mean SLODAR profile measured at Cerro Paranal Observatory, February 2005 to January 2006, including observations from 70 nights. The bin centered at 2000m is the total turbulence strength for all altitudes above 1500m.

to allow for the effects of the variable bin width in the data set. The mean profile clearly shows a strong (unresolved) signal in the lowest altitude bin.

3.2. *Mauna Kea*

The SLODAR system at Mauna Kea forms part of a combined SLODAR-LOLAS monitor[7] to study the GL at the site, for the Gemini observatory. Figure 4 shows example data for two nights for the SLODAR-LOLAS system at Mauna Kea. Figure 4 (top) is an example of a night in which there was strong turbulence within the first 500m altitude, with a significant change in the altitude distribution of the turbulence during the night. However, figure 4 (bottom) shows a more representative example for the Mauna Kea data set, with little optical turbulence detected at low altitudes, above the lowest resolution element. The statistical results of the campaign and their implications for GLAO at the site are discussed elsewhere.[6]

4. Comparison with MASS and DIMM

The SLODAR system at the Paranal observatory is sited within 20m of the MASS and DIMM instruments of the automated Site Monitor (ASM), so that useful cross-comparisons of the instruments can be made. For the DIMM, we compare the integrated seeing value to that for the SLODAR. We compare the SLODAR and MASS integrals for altitudes of 1km and above, as a measure of the 'free atmosphere' seeing for the two instruments. For MASS we use the sum of the profile excluding the lowest (0.5km) bin. For SLODAR we use the sum of the bins for altitudes of 1km or higher, with weighting functions to match the response function of the MASS.

The MASS and DIMM instruments require a very bright single-star target, whereas SLODAR requires double stars. Hence the two instruments do not normally observe in the same line of site. Furthermore the MASS and DIMM are mounted on a tower at altitude 6m, whereas the SLODAR is at ground level. Both instruments produce measurements at intervals of less than one minute, but the measurements are not synchronized. Hence we may expect a significant scatter of the individual comparison measurements due to the spatial and temporal non-stationarity of the turbulence profile.

Figure 5 (top) shows the scatter of the contemporaneous SLODAR and DIMM data acquired between February 2005 and January 2006. Data are averaged over five minute intervals and we use only those data for which there were at least 3 SLODAR and DIMM measurements in each five-minute bin. We find a small but significant skew of the scatter. There are a

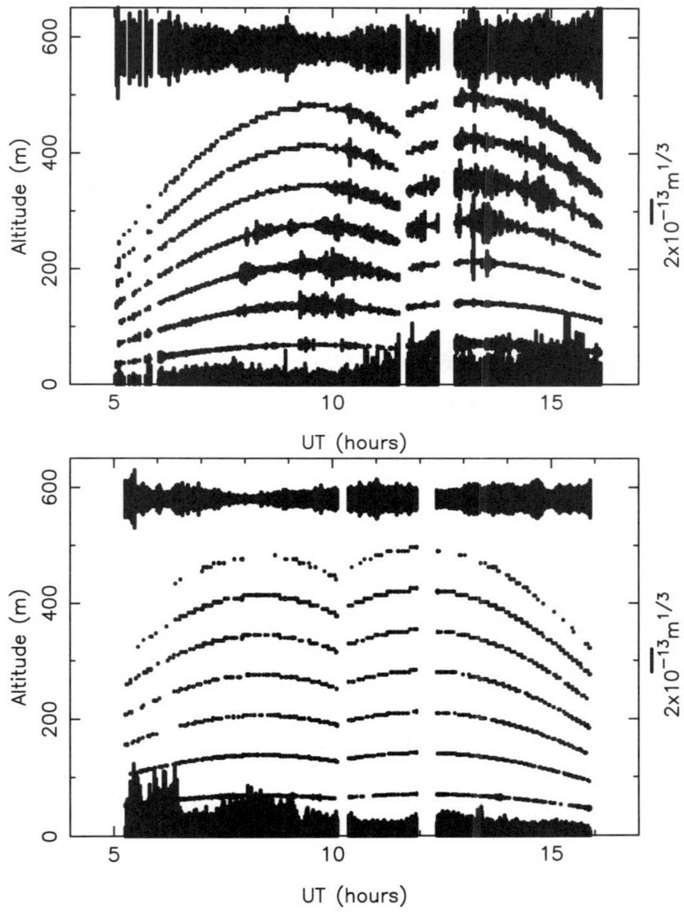

Fig. 4. Example SLODAR data sequences for SLODAR at Mauna Kea. Top: 2 February 2007. Bottom: 3 March 2007.

significant number of nights on which the SLODAR recorded consistently higher seeing values than the DIMM. This is most obvious in periods of very good seeing (~0.5arcsec). During these periods we typically find that the turbulence above the lowest level (SLODAR bins 2-8) is very weak, and that the SLODAR and MASS monitors agree on the strength of the free atmosphere turbulence. Hence we associate the excess turbulence (mean strength approx. $7 \times 10^{-14} m^{1/3}$) with the first SLODAR bin. Given the strong concentration of turbulence within a few tens of meters of the ground, this suggests that the excess results from the difference in altitude of the

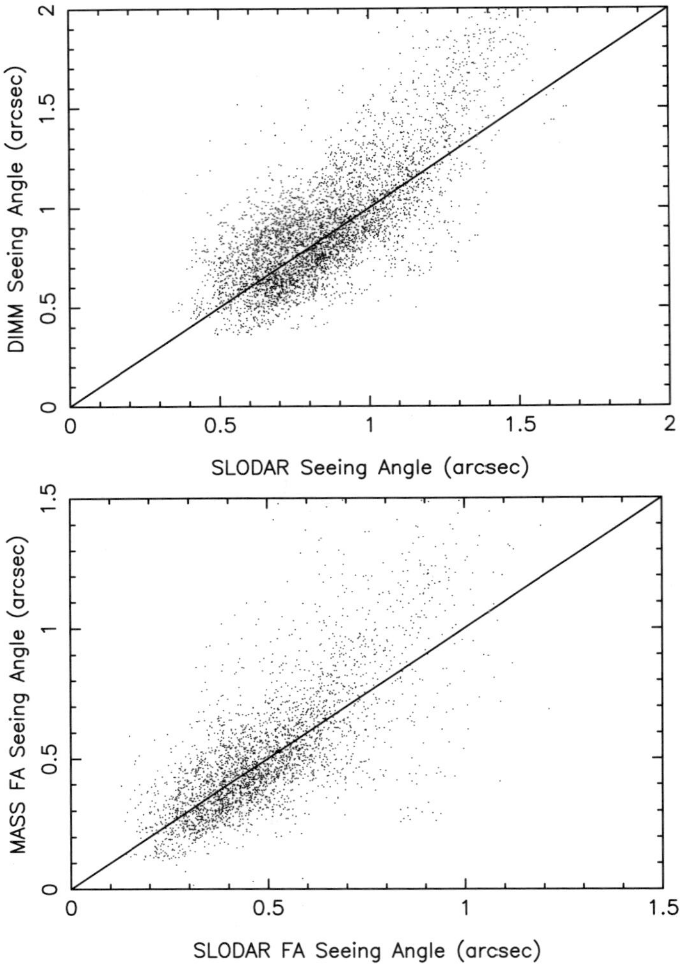

Fig. 5. Cross-comparison of SLODAR and the MASS-DIMM monitor at Paranal (70 nights, February 2005 to January 2006). Top: DIMM versus SLODAR seeing angle for the full atmosphere. Bottom: MASS versus SLODAR seeing angle for the free atmosphere.

DIMM and SLODAR monitors. Profiling of the surface layer turbulence with higher vertical resolution is required to confirm that this is the case.

In rare periods of very poor seeing (>1.25arcsec), we see that the SLODAR seeing value is consistently lower than for the DIMM. This results from clipping of large centroid excursions for the SLODAR: the use of a small format CCD camera limited the allowable inter-spot distance for the

WFS, and hence the window size for centroiding. For larger format detectors this problem is overcome since the spot separation can be increased.

Figure 5 (bottom) is a scatter plot of the contemporaneous MASS and SLODAR values for the free atmosphere seeing. We find a good general agreement between MASS and SLODAR for the integrated free-atmosphere seeing. However again (for the same reason) the SLODAR gives significantly lower seeing values that the MASS in periods of very poor seeing.

5. Summary and the Future

SLODAR turbulence monitors have been implemented at three observatories, for statistical characterization of the optical turbulence profile. Significant data sets characterizing the ground-layer turbulence have been obtained at Cerro Paranal and Mauna Kea. Current and planned developments include a modified implementation permitting profiling of the surface-layer turbulence with a resolution of 10m,[8] and greater automation of the existing systems. Future applications of the monitors may usefully include verification of meso-scale turbulence model forecasts for astronomy. As well as the turbulence profile, SLODAR would provide direct verification of the wind velocity prediction for the altitude of any turbulent layers.

Acknowledgments

We are grateful for assistance from the staff of the SAAO, the ESO Paranal and Gemini observatories and the University of Hawaii Institute for Astronomy. We acknowledge financial assistance from ESO and the UK STFC. Data from the automated site monitor of the VLT were courtesy of ESO.

References

1. R. W. Wilson, *Mon. Not. R. Astron. Soc.* **337**, 103 (2002).
2. M. Goodwin, C. Jenkins and A. Lambert, *Optics Express* **15**, 14844 (2007).
3. T. Butterley, R. W. Wilson and M. Sarazin, *Mon. Not. R. Astron. Soc.* **369**, 835 (2006).
4. L. Wang, M. Schock and G. Chanan, *Applied Optics* **47**, 1880 (2008).
5. R. W. Wilson, J. Bate, J. C. Guerra, N. Hubin, M. Sarazin and C. D. Saunter, *Proc. SPIE* **5490**, 758 (2004).
6. M. R. Chun, R. W. Wilson, R. Avila, T. Butterley, J. L. Aviles, D. Weir and S. Benigni, submitted to Mon. Not. R. Astron. Soc..
7. R. Avila, J. L. Aviles, R. W. Wilson, M. Chun, T. Butterley and E. Carrasco, *Mon. Not. R. Astron. Soc.* **387**, 1511 (2008).
8. J. Osborn, R. Wilson, T. Butterley, these proceedings.

HIGH RESOLUTION SLODAR MEASUREMENTS ON MAUNA KEA

T. BUTTERLEY and R. W. WILSON
University of Durham, United Kingdom

M. R. CHUN
University of Hawai'i at Hilo, USA

R. AVILA and J.-L. AVILES
Universidad Nacional Autónoma de México, Mexico

We have made high order (32×32 subaperture) Shack-Hartmann wavefront sensor observations of binary stars with separations of approximately 20 arcseconds using the University of Hawaii 2.2 m telescope. We present preliminary results of a Slope Detection and Ranging (SLODAR) analysis of the data yielding measurements of turbulence strength, wind velocity and velocity dispersion as a function of altitude, with approximately 500 m vertical resolution. The aim of the investigation is to explore the validity of the Taylor frozen flow approximation and the implications for layer-oriented predictive AO reconstruction algorithms.

Keywords: SLODAR; turbulence profiling.

1. Introduction

Slope Detection and Ranging (SLODAR) has been demonstrated as an effective turbulence profiling technique on large telescopes[1,2] and on small telescopes as a dedicated site monitor.[3] SLODAR is a crossed beams technique, based on the use of Shack-Hartmann wavefront sensors (WFS) observing double star targets.[4] The turbulence profile is recovered from the cross-covariance of the WFS measurements for the two stars.[5] The velocity of turbulent layers is also measured, from the spatio-temporal cross-covariance of the WFS data.[6] This provides a direct measurement of the spatio-temporal properties of the wavefront aberrations associated with each turbulent layer.

The wavefront sensors employed for SLODAR are essentially identical to those used in typical adaptive optical systems, so that the data directly characterize the wavefront measurements that are input to AO control algorithms. Hence, if sufficient spatial and temporal resolution can be achieved, SLODAR characterization of the turbulence can permit realistic and accurate modelling of predictive control for advanced AO systems. The use of prediction in AO control can potentially improve correction performance.[7] For a single conjugate AO system the usefulness of prediction is limited, since there are typically multiple layers with different wind speeds and directions. However, for advanced AO systems, for example employing MCAO or GLAO correction, correcting elements address individual turbulent layers. Hence predictive control may be effective for future advanced AO systems. The performance of predictive control will depend on the validity of Taylor's frozen flow hypothesis for individual turbulent layers at each altitude. Realistic modeling therefore requires detailed characterization of the temporal properties of turbulence as a function of altitude, for the best sites.

For SLODAR, the turbulence strength is measured in a number of resolution elements of equal extent in altitude. The number of resolution elements is equal to the number of sub-apertures of the WFS subtended across the aperture of the telescope.[5] Here we describe SLODAR measurements made at the University of Hawaii (UH) 2.2 m telescope using a high order WFS, with 32 by 32 subapertures across the aperture of the telescope. This system order is a factor of two higher than has been employed previously for SLODAR systems.[1,2] The high resolution measurements permit detailed characterization of the strength, altitude and temporal properties of the turbulence at all altitudes, from the ground to ~ 18 km. The observations were made in the context of the Gemini North ground-layer turbulence characterization study.[8]

We give a brief overview of the SLODAR instrument and analysis used – further details have been described previously.[9] We describe the observations made and discuss example data obtained, in the context of modelling requirements for AO predictive control.

2. SLODAR Instrument

The SLODAR instrument was a 32×32 subaperture Shack-Hartmann wavefront sensor mounted at the Cassegrain focus of the UH 2.2 m telescope. The camera used was an Andor Technology iXon DV885 electron multiplication CCD (EMCCD) with on-chip charge multiplication resulting in effectively zero read noise. Exposure times in the range 2-4 ms were used, and the frame rate was typically 50 Hz.

The optical arrangement did not allow the Shack-Hartmann patterns for the two stars to be fully separated on the detector. Instead, they were interleaved as shown in Fig. 1. The optical arrangement was mounted on a rotator to allow the separation angle of the two S-H patterns on the detector to be oriented such that the wavefront sensor spots for the two target stars did not overlap.

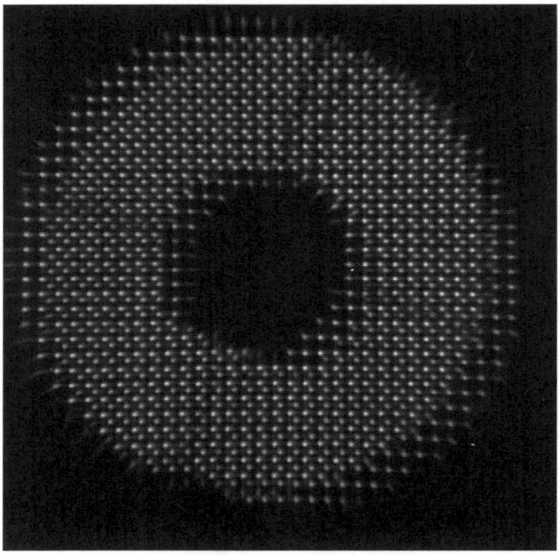

Fig. 1. Time-averaged wavefront sensor image showing the two interleaved Shack-Hartmann patterns. Note that the shadow of the telescope secondary mirror support is clearly visible in the pattern.

3. Data Analysis

The procedure for analysing the data has been described previously[5,9] and is summarized as follows: the Shack-Hartmann spots for the two stars are located and their centroids (relative to their mean positions) are measured using a standard centre-of-gravity algorithm. A high-pass temporal filter is optionally applied to the centroids to remove slow wavefront changes characteristic of telescope tube seeing. For each star, common centroid motion is subtracted – thus removing wavefront tip-tilt and telescope shake from the data. The total seeing is measured via a fit to the centroid spatial auto-covariance for the brighter star. The turbulence profile is determined from the spatial cross-covariance of the centroids for the two stars; a set of (theoretical) functions describing the response of the system to a layer at a given altitude are fitted to the cross-covariance yielding the turbulence strength in each altitude bin. The velocities of turbulent layers are extracted from the spatio-temporal cross-covariance (the cross-covariance calculated with a range of temporal offsets between the centroids for the two stars) – the peak in the cross-covariance corresponding to a particular layer moves with the same velocity as the layer.

4. Results

Here we present sample results demonstrating the effectiveness of the instrument. The sample data were collected on a target with a separation of 21.6" at around 06:45 UT on 23 September 2007. The target elevation was 64°, so the altitude bins were 590 m in size and the maximum sensing altitude was ~18 km.

Figure 2 shows the spatial cross-covariance of the centroid measurements. The top plot shows the full 2-dimensional cross-covariance. The solid black line shows the direction corresponding to the separation of the two stars. Peaks along this line indicate turbulent layers, with the centre of the plot corresponding to the ground and increasing altitude towards the edge. The lower plot shows a cut along the solid line, with a cut along the dotted line at 90° to the star separation direction shown for reference.

Figure 3 shows the preliminary C_n^2 profile extracted from the data in Figure 2. Five distinct turbulent layers are resolved which are clearly visible in both the cross-covariance and the recovered profile. There is an apparent signal at ~17 km but this is likely to be a noise artefact – the signal to noise ratio is poor at the largest spatial offsets, as can clearly be seen in Figure 2. An additional strong signal is present at 3 km in the recovered profile that is not evident in the cross-covariance. This is most likely an artefact resulting from the effects of cross-contamination of centroid measurements due to the close proximity of the Shack-Hartmann spots on the detector. We are currently addressing this problem via a different choice of centroiding method. Previously we relied on thresholding rather than windowing[10] but we have concluded that windowing is more appropriate when the spots are close together. The initial results with the updated algorithm (not included here) indicate a substantial reduction in the amount of cross-contamination in the centroids.

Figure 4 shows a series of cross-covariances with increasing temporal offsets between the two stars' centroids. The most striking feature of this

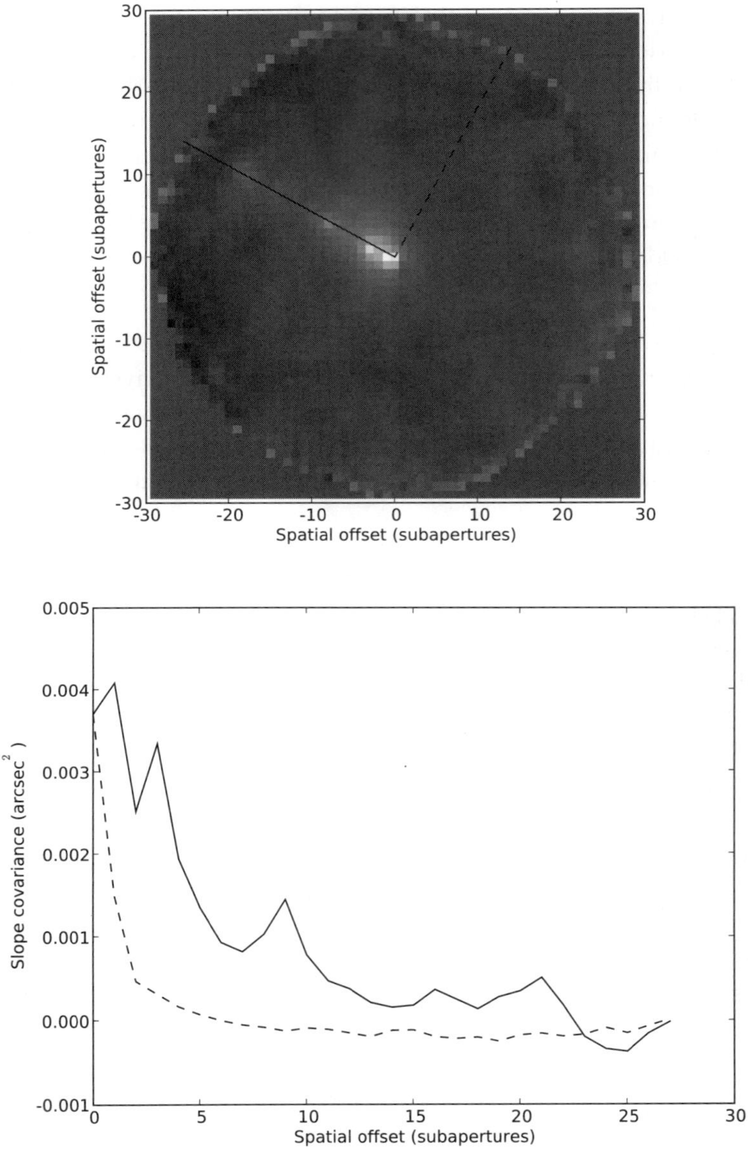

Fig. 2. *Top:* Average of x- and y-centroid cross-covariance for data taken at 06:45 UT on 23 September 2007. Averaged over 60 packets of 50 frames (53 seconds) of data. Higher values are lighter and lower values are darker. The black lines correspond to the cuts in the bottom plot. *Bottom:* Radial cut through the cross-covariance; the solid line is a cut through the direction of the profile information and the dotted line is a cut at 90° to this.

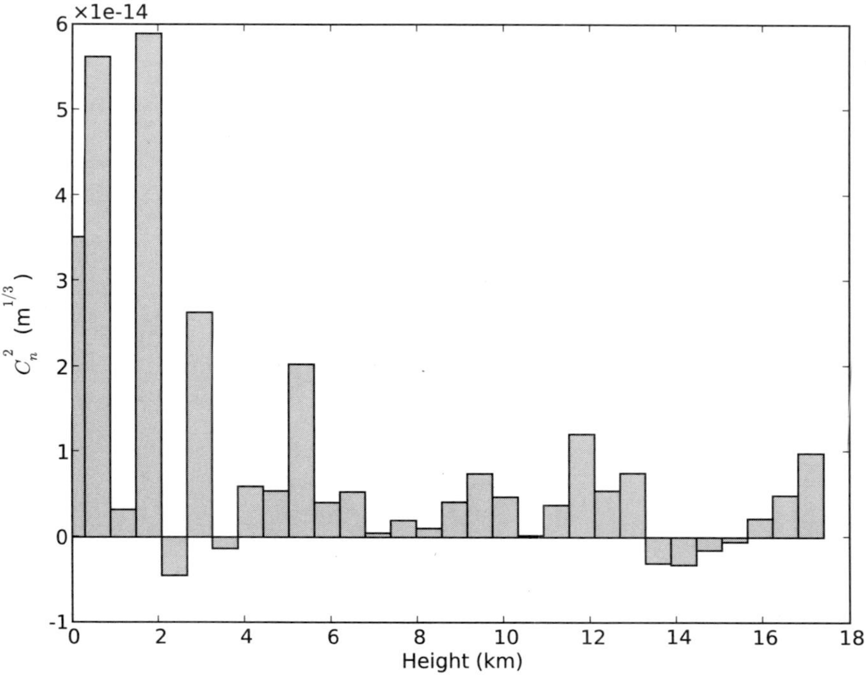

Fig. 3. Preliminary C_n^2 profile extracted from the cross-covariance in Fig. 2. r_0=17.5 cm. Note that the lowest-altitude bin effectively has half the thickness of the other bins because it is centred at the ground.

figure is the considerable velocity dispersion in the lower altitude turbulence. Multiple peaks are clearly visible at larger time offsets, all originating in the first ~2 km, but with a spread of speeds and directions. This implies the benefits of predictive AO reconstruction algorithms would be limited in these conditions, at least for reconstructing lower-altitude layers. Obviously this is only a single example, and statistical data would need to be accumulated in order to assess the potential of predictive reconstruction for ground layer AO at the site.

The profile fit will be improved further by carrying out an in-depth analysis involving fitting 2D theoretical covariance functions to the full spatio-temporal cross-covariance to measure accurately the velocity and velocity dispersion of each layer.[11] The motion of the peaks corresponding to different layers should help to resolve ambiguities in the strength and altitude of layers that are not well separated in altitude. We also aim to

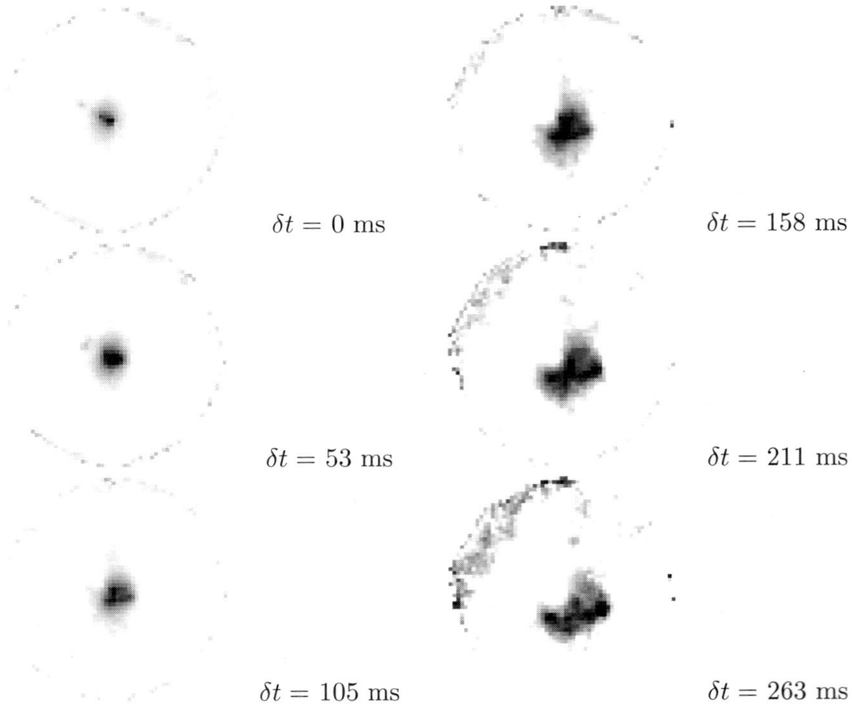

Fig. 4. Cross-covariances with temporal offsets between the measurements from the two stars. For clarity, a threshold is applied and each cross-covariance is normalised separately. Higher values are darker and lower values are lighter.

measure the decay of the covariance associated with each layer peak with increasing time offset, as a measure of the validity of Taylor's frozen flow hypothesis.

5. Conclusion

A 32×32 subaperture Shack-Hartmann WFS has been deployed on the UH 2.2 m telescope and used to make SLODAR turbulence profile measurements. The instrument provides high resolution profiling of the atmospheric turbulence with a vertical resolution of 400-700 m. Detailed measurements of the velocities of turbulent layers, including velocity dispersion within layers, were possible. Improvements to the data analysis algorithms are required to fully recover the complete turbulence characteristics. The high resolution data provide information on the validity of the frozen flow approximation for individual layers. Statistical data over many nights would be required to reliably model the performance of AO systems with predictive reconstruction. Future high resolution SLODAR instruments will have a larger Shack-Hartmann spot separation on the CCD to avoid cross-contamination between adjacent centroids.

Acknowledgements

The authors are grateful to the University of Hawai'i for the use of the 2.2 m telescope.

References

1. R. W. Wilson, *Mon. Not. R. Astron. Soc.* **337**, 103 (2002).
2. M. Goodwin, C. Jenkins and A. Lambert, *Optics Express* **15**, 14844 (2007).
3. R. W. Wilson, T. Butterley, M. S. Sarazin, G. Lombardi, M. R. Chun, S. Benigni, D. Weir, R. Avila and J. L. Aviles, SLODAR turbulence monitors for real-time support of astronomical adaptive optics, in *Proceedings of the SPIE 7015*, 2008.
4. A. Tokovinin, M. Le Louarn, E. Viard, N. Hubin and R. Conan, *A&A* **378**, 710 (2001).
5. T. Butterley, R. W. Wilson and M. Sarazin, *Mon. Not. R. Astron. Soc.* **369**, 835 (2006).
6. R. W. Wilson, J. Bate, J. C. Guerra, N. N. Hubin, M. Sarazin and C. D. Saunter, Development of a portable SLODAR turbulence profiler, in *Proceedings of the SPIE*, 2004.
7. L. A. Poyneer, B. A. Macintosh and J.-P. Véran, *Journal of the Optical Society of America A* **24**, 2645 (2007).
8. M. R. Chun, R. W. Wilson, R. Avila, T. Butterley, J. L. Aviles, D. Weir and S. Benigni, Mauna Kea ground-layer characterisation campaign *submitted to Mon. Not. R. Astron. Soc.* 2008.
9. T. Butterley, R. W. Wilson, M. R. Chun, R. Avila and J.-L. Aviles, High-resolution SLODAR measurements on Mauna Kea, in *Proceedings of the SPIE 7015*, 2008.
10. A. Tokovinin, *PASP* **114**, 1156 (2002).
11. L. Wang, M. Schöck and G. Chanan, *Applied Optics* **47**, 1880 (2008).

VERTICAL TURBULENCE PROFILES AT CANARY ISLANDS ASTRONOMICAL SITES

B. GARCIA-LORENZO*, J.J. FUENSALIDA, J. CASTRO-ALMAZAN and M.A.C. RODRIGUEZ-HERNANDEZ

Instituto de Astrofísica de Canarias
C/ Via Lactea S/N, 3800-La Laguna, Tenerife, Spain
**E-mail: bgarcia@iac.es, jjf@iac.es, jcastro@iac.es, ahernan@iac.es*

We present the statistical results of the monitoring of the optical turbulence profiles at the Roque de los Muchachos (ORM) and Teide observatories (OT), Spain. The data has been obtained using the Generalized-SCIDAR technique at the 1m Jacobous Kaptein telescope at the ORM and 1.5m Carlos Sánchez telescope at the OT. Statistical profiles are based on 150 nights of measurements recorded at the ORM from February 2004 to July 2008 and 140 nights at the OT obtained from November 2002 to July 2008. The monthly statistical profiles for different years present a similar structure for different years, suggesting a stable seasonal evolution of the turbulence at the Canary Islands sites. Statistically, the turbulence structure at ORM and OT is similar and present a similar evolution along the seasons. The consistence of the behaviour of the turbulence in different time scales is crucial for AO/MCAO systems implementation.

Keywords: Atmospheric Effects – Turbulence – Instrumentation; Adaptive Optics – Site Testing – Telescopes.

1. Introduction

Atmospheric optics has gained importance in modern astronomy due to the excellent image quality requirements of current projects on large telescopes. The proper knowledge of the turbulence is crucial for optimizing the efficiency of adaptive optic (AO) systems. In particular, the input parameters for the design of AO systems working with several deformable mirrors are determined through the statistical behaviour of turbulence layers. SCIDAR (from SCIntillation Detection and Ranging) is the most contrasted and efficient remote sensing technique to study the optical atmospheric structure. This work presents the statistical profiles derived from the largest database of vertical turbulence profiles obtained from G-SCIDAR measurements at any astronomical site, the ones at the Canary Island.

There are two astronomical sites in the Canaries, The Teide Observatory (OT hereafter), on the island of Tenerife, and the Roque de los Muchachos Observatory (ORM), on the island of La Palma, both being separated about 160 km. Tenerife is the largest island of the Canarian Archipelago and it is located between latitudes 28-29 deg N and longitudes 16-17deg W, 280 km distant from the African coast. Both astronomical sites are at about 2400 meters in altitude above sea level. More information about these astronomical sites and their sky quality can be found in the web page: http://www.iac.es/site-testing/.

The Instituto de Astrofísica de Canarias (IAC, http://www.iac.es) is a research centre based on the island of Tenerife. The IAC has a research and instrumental team, GARE (from the spanish "Grupo de Alta Resolución Espacial"), involved in the characterization of the vertical atmospheric structure with statistical significance for multi-conjugated adaptive optics systems for large telescopes. With this purpose, GARE team is working in different tasks: (a) they are involved in the development of instruments with common user facilities for the characterization of the atmospheric turbulence. In fact, GARE has built three completely automated SCIDAR instruments for ORM (Fuensalida et al. 2004[1]), OT (Rodríguez-Hernández et al. 2007[2]), and Paranal (Vázquez-Ramió et al. 2008[3]) sites; (b) GARE has also developed a data reduction software to obtain the turbulence profiles quasi-on-line, with a delay of about 5 minutes. At the moment, this software is installed at the SCIDAR in Paranal. This software includes the on-line dome seeing correction (Fuensalida, García-Lorenzo, & Hoegemann 2008[4]); (c) GARE has also developed an algorithm to obtain the velocity of turbulence layers from SCIDAR data without human supervision, that is not obtained on-line, but can be run after observations; (d) This team has been also studying the applicability of data from climate diagnostic archives to astronomical site evaluation (García-Lorenzo et al. 2005[5]); (e) GARE has a program for the monitoring of the vertical turbulence structure at the both Canary Islands astronomical observatories, ORM and OT.

2. The Instruments

GARE developed the first automatically controlled SCIDAR instrument, the so-called Cute-SCIDAR (Fuensalida et al. 2004[1]), to carry out the systematic monitoring of the turbulence structure at the ORM on the island of La Palma. The Cute-SCIDAR is in operation at the Jacobus Kaptein Telescope (1m-JKT), a 1m telescope operated by the Isaac Newton Group of telescopes. Each detector pixel covers a square of side 1.94 cm on the 1m-JKT pupil. A friendly graphical interface was developed to control all the instrument mechanisms as well as the data acquisition system. This graphical interface also has a preview of the autocorrelation and cross-correlation of the scintillation patterns in real time.

In order to monitor the turbulence structure at the OT, GARE have also developed a hybrid turbulence profiling instrument (Rodríguez-Hernández et al. 2007[2]). The hybrid instrument is installed at the 1.5m Carlos Sánchez Telescope (1.5m-TCS). The hybrid instrument is completely automated and it is formed by two arms. In the vertical arm there is a clone of the Cute-SCIDAR installed in La Palma, while the horizontal arms host the wave front sensor, a Shack-Hartmann. The instrument includes a beam splitter to observe with the two arms simultaneously. SCIDAR is the normal mode of observations and Shack-Hartmann data are sporadically obtained. Each detector pixel covers a square of side 2.81 cm on the 1.5m-TCS pupil.

For the moment, data are processed after observations although the implementation of the software to derive the turbulence profile on-line (already working at the SCIDAR instrument developed by GARE for Paranal site) is pending. In both instruments, we use 1000 images of scintillation patterns to derive the average normalized autocovariance and five frames that are measurements of the 2D spatiotemporal cross-correlation functions separated by a lapse of Δt.

Figure 1 shows the Cute-SCIDAR and the hybrid instruments installed at the 1m-JKT and 1.5m-TCS, respectively. Therefore, the SCIDAR technique is being used to monitor the turbulence structure at the ORM and OT with the previous described instruments.

3. Observations and Data

The systematic campaigns at La Palma began in February 2004. The frequency of the observational runs was one-week per month during the first year. Between February 2005 to January 2008 the number of nights was reduced to four per month. In 2008, runs of eight consecutive nights per month were carried out. At the moment, more than 150 useful nights of G-SCIDAR measurements at the ORM are available.

At the OT, the systematic monitoring began in November 2002 with a manual G-SCIDAR prototype and in July 2005 the hybrid instrument was installed. The frequency of the observational runs has been always about four or six nights per month. There is useful data for around 140 nights at this site.

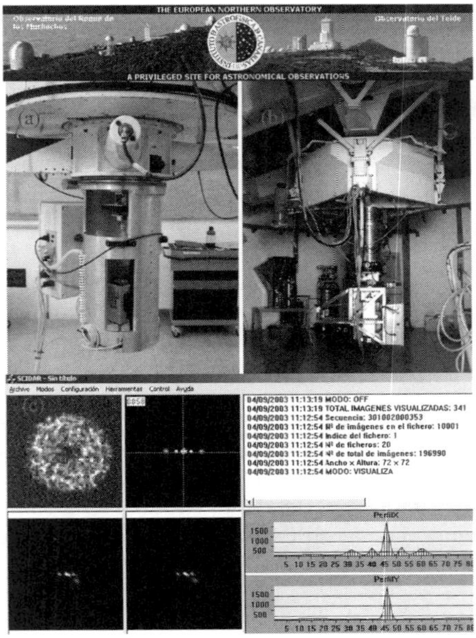

Fig. 1. (a) The Cute-SCIDAR instrument installed at the 1m-JKT and a view of the Roque the los Muchachos observatory; (b) The hybrid instrument installed at the 1.5m-TCS above of a view of the Teide observatory; (d) User interface window showing an example of the on-line data evaluator.

As an example, figure 2 presents the evolution of the turbulence along the same night and for the same period of time at the ORM and OT. The most intense turbulence is concentrated at the observatory level. The dome contribution was removed from the turbulence profiles following a procedure based on the evenness properties with Fourier analysis (Fuensalida, García-Lorenzo & Hoegemann 2008[4]). The turbulence structure presents a similar behaviour in both sites, although turbulence layer present different evolution: the layer at around 14 km is clearly formed after 2 UT at the ORM, whereas it is present during all the night at the OT.

4. Statistical View of the Seasonal Evolution of the Turbulence

From February 2004 to July 2008, 150 nights of useful SCIDAR data have been recorded at La Palma site. The total number of useful profiles obtained at this site is 145231, distributed along the months and years from 2004 to 2008, except for 2007, for which do not have any data. This was due to a combination of technical problems with the instrument, with the telescope

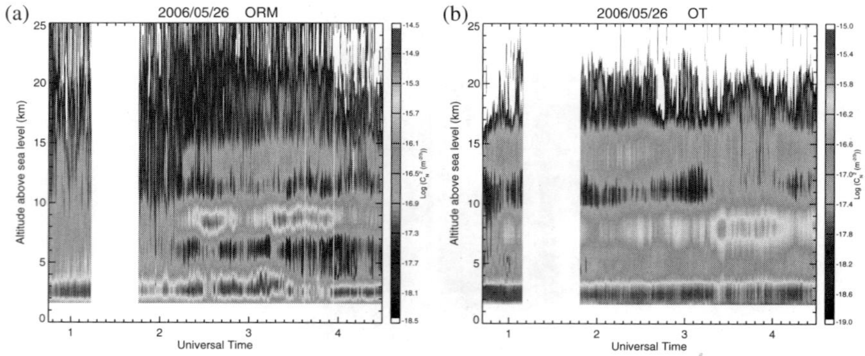

Fig. 2. Evolution of C_N^2(h) as a function of Universal Time and altitude for 2006 May 26-27 night at the (a) Roque de los Muchachos Observatory on the island of La Palma, and (b) Teide Observatory on the islnad of Tenerife. Colours represents C_N^2(h) in logarithmic scale as coded by the scale on the right. The dome contribution has been removed from the profiles following the algorithm in Fuensalida, García-Lorenzo & Hoegemann (2008).[4]

and a lack of human recourses. From the obtained data, we can derive statistical turbulence profiles that characterized La Palma site. Combining all the turbulence profiles observed the same month in different years, an statistical view of the seasonal evolution of turbulence at the Roque the los Muchachos Observatory has been obtained (figure 3(even)).

From Novemeber 2002 to July 2008, 140 nights of useful SCIDAR data have been obatined at the OT. The total number of useful profiles is 87412 above this site. The statistical turbulence profiles that constitute the seasonal evolution of turbulence profiles at the OT site (figure 3(odd)) were obtained by combining all the profiles observed the same month. The statistical significance of these profiles is still very poor because sometimes the number of individual profiles is small, such is the case for January or November.

We find that the turbulence is concentrated close to the ground level for any month at both sites, although the turbulence strength evolves along the months. In general, all the turbulence stratified in layers evolve in time and altitude. Seeasonal trends with smaller statistical significance were already reported (Fuensalida et al. 2004[6]; García-Lorenzo, Fuensalida, & Rodríguez-Hernández 2007[7]). Simulations of C_N^2 above San Pedro Mártir have also revealed a seasonal evolution of the turbulence above this site (Masciadri & Egner 2006[8]). Other turbulence parameters, such as turbulent kinetic energy or eddy dissipation rate, also presents seasonal trends (Nastrom & Eaton 2005[9]).

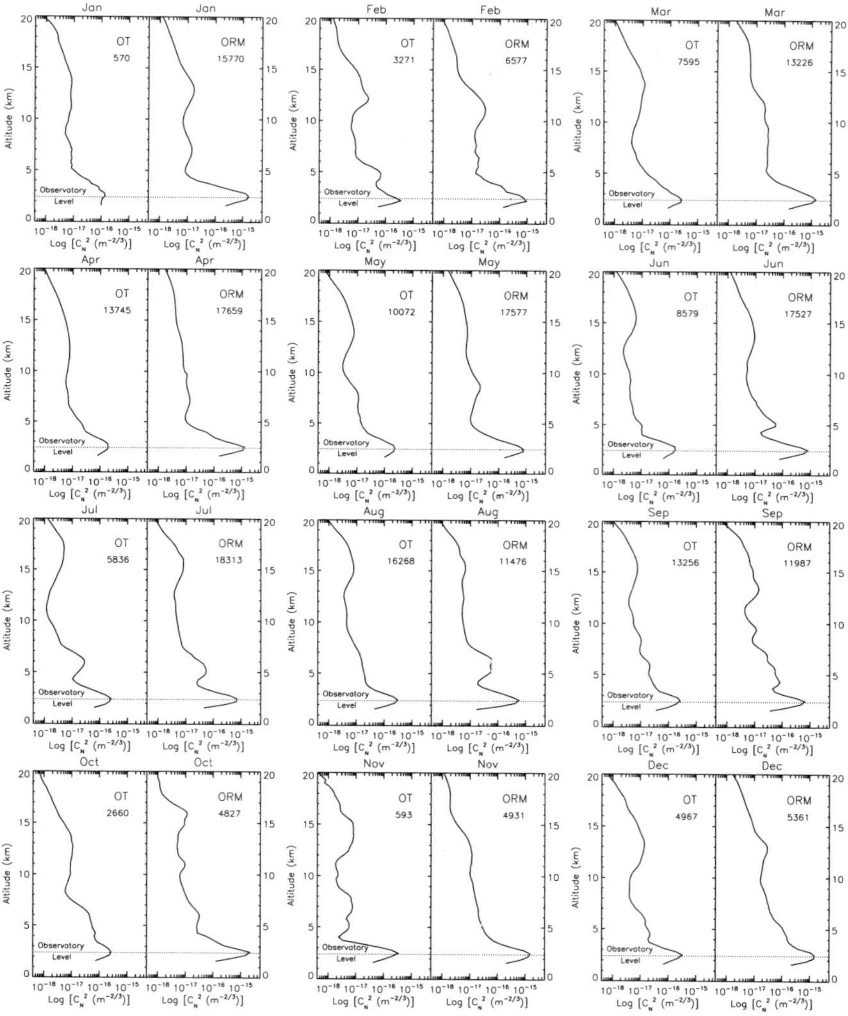

Fig. 3. Monthly average profiles of C_N^2 for Roque de los Muchachos (ORM) and Teide observatories obtained from G-SCIDAR measurements. The horizontal axis represents the C_N^2 in logarithmic scale in units of $m^{(-2/3)}$. The vertical axis corresponds to the altitude above the sea level. The horizontal line indicates the observatories altitude (∼2400 meters). Dome seeing was removed from the individual profiles following the procedure proposed in Fuensalida, García-Lorenzo & Hoegemann 2008.[4] The total number of individual turbulence profiles used to compute these statistical C_N^2 profiles are indicated in each plot.

Statistically, most of the turbulence is concentrated in the first three kilometers over the ORM site in March, and April with not a clear detection of any turbulence concentrated in a layer in highest altitudes. In June and July, a clear turbulence layer is presented at about 15 km (over sea level). The statistical profiles for July, August, and September and even October are caracterized by a relative strong turbulence layer between 5 and 6 km, that seems to be split in two thinner turbulence layers in August and September. November and December show a faint turbulence layer at 13 km, extending to January and February.

In spite of the differences in the statistical significance of the data obtained at the ORM and OT, the derived mean profiles for these two sites present similar features (see Fig. 4). The average profile for OT shows similar structures than the corresponding to ORM, but some times in slightly lower altitudes.

5. Summary and Conclusions

Atmospheric turbulence profiles have been monitored at the two astronomical observatories of the Canary Islands: 150 nights at the Roque de los Muchachos and 140 nights at the Teide Observatory. This systematic monitoring of the atmospheric turbulence structure was carried out using the G-SCIDAR technique at the 1m Jacobus Kaptein Telescope at the ORM and and the 1.5 Carlos Sánchez telescope at the OT. Turbulence inside the dome was removed from the turbulence near the ground outsite the dome following the procedure proposed by Fuensalida, García-Lorenzo & Hoegemann (2008).[4] The monthly statistical turbulence profiles at ORM and OT that constitute the seasonal evolution of turbulence structure at these sites have been derived combining the individual measurements of $C_N^2(h)$. The comparison of the statistical profiles for the different years reveals the presence of stable turbulence layers in the atmospheric structure at the ORM and OT. The statistical analysis of C_N^2 profiles at astronomical sites is of relevant importance for a proper design and operation of efficient adaptive optics systems and crutial for the site selection for the next generation of extremely large telescopes. The statistical study perfomed in this paper reveals the excellent conditions at the ORM and OT observatories, showing a seasonal evolution of the turbulence structure and the stability of such evolution along different years for the atmosphere at the Canary Islands observatories.

Acknowledgments

This paper is based on observations obtained at the Carlos Sánchez Telescope operated by the Instituto de Astrofísica de Canarias at the Teide

Observatory and at the Jabobus Kaptein Telescope operated by the Isaac Newton Group of telescopes at the Roque de los Muchachos Observatory. The authors thank all the staff at the observatories for their kind support. We also thank A. Eff-Darwich for help and useful discussions. This work was partially funded by the Instituto de Astrofísica de Canarias and by the Spanish Ministerio de Educacion y Ciencia (AYA2006-13682 and AYA2003-07728). Begoña García-Lorenzo also thanks the support from the Ramón y Cajal program by the Spanish Ministerio de Educacion y Ciencia.

References

1. J. J. Fuensalida, B. García-Lorenzo, J. M. Delgado, C. Hoegemann, M. Verde, M. Reyes and J. Vernin, *The Newsletter of the Isaac Newton Group of Telescopes* **8**, 15(September 2004).
2. M. A. C. Rodriguez Hernández, J. Jiménez-Fuensalida, B. M. García-Lorenzo, J. M. Delgado, E. Hernandez, C. Hoegemann and H. Vázquez Ramió, The hybrid Shack-Hartmann/G-SCIDAR instrument, in *Society of Photo-Optical Instrumentation Engineers (SPIE) Conference Series*, Society of Photo-Optical Instrumentation Engineers (SPIE) Conference Series Vol. 6747Octtober 2007.
3. H. Vázquez Ramió, M. Reyes, J. M. Delgado, E. Hernández, M. Núñez Cagigal, J. J. Fuensalida, G. Lombardi, F. Derie, J. Navarrete and M. Sarazin, *The Messenger* **132**, 29(June 2008).
4. J. J. Fuensalida, B. García-Lorenzo and C. Hoegemann, *Monthly Notices of the Royal Astronomical Society* **389**, 731(September 2008).
5. B. García-Lorenzo, J. J. Fuensalida, C. Muñoz-Tuñón and E. Mendizabal, *Monthly Notices of the Royal Astronomical Society* **356**, 849(January 2005).
6. J. J. Fuensalida, B. Garcia-Lorenzo, J. Castro, S. Chueca, J. M. Delgado, J. M. Gonzalez-Rodriguez, C. K. Hoegemann, M. Reyes, M. Verde and J. Vernin, Statistics of atmospheric parameters for multiconjugated adaptive optics for the Observatorio del Roque del los Muchachos, in *Society of Photo-Optical Instrumentation Engineers (SPIE) Conference Series*, eds. J. D. Gonglewski and K. Stein, Society of Photo-Optical Instrumentation Engineers (SPIE) Conference Series, Vol. 5572November 2004.
7. B. García-Lorenzo, J. J. Fuensalida and M. A. C. Rodríguez-Hernández, Statistical turbulence vertical profiles at the Roque de los Muchachos Observatory and Teide Observatory, in *Society of Photo-Optical Instrumentation Engineers (SPIE) Conference Series*, Society of Photo-Optical Instrumentation Engineers (SPIE) Conference Series Vol. 6747Octtober 2007.
8. E. Masciadri and S. Egner, *PASP* **118**, 1604(November 2006).
9. G. D. Nastrom and F. D. Eaton, *Journal of Geophysical Research (Atmospheres)* **110**, 19110(Octtober 2005).

C_n^2 PROFILE FROM SHACK-HARTMANN DATA: FIRST STEPS FOR CO-SLIDAR DATA PROCESSING

C. ROBERT*, N. VÉDRENNE, V. MICHAU and J-M. CONAN

Office National d'Études et de Recherches Aérospatiales,
BP 72, 92322 Châtillon, France
**E-mail: clelia.robert@onera.fr*

C_n^2 profile monitoring usually make use of wavefront slope correlations or of scintillation pattern correlations. Scintillation is rather sensitive to high turbulence layers whereas wavefront slope correlations are mainly due to layers close to the receiving plane. Wavefront slope and scintillation correlations are therefore complementary. Slopes and scintillation being recorded simultaneously with a SHWFS, we propose here to exploit their correlations to retrieve the C_n^2 profile. The measurement method named COupled SLodar scIDAR (CO-SLIDAR[1]) uses correlations of SHWFS data from two separated stars. CO-SLIDAR method is presented with results of C_n^2 estimation from simulated SHWFS. Preliminary results with CO-SLIDAR are presented using real data from a binary star.

Keywords: Turbulence; Adaptive optics; Wave-front sensing; Inverse problems.

1. Motivation

Emerging systems of Adaptive Optics (AO) such as MCAO, GLAO, LGS, that are conceived to optimize wavefront correction on different field of view and to optimize their efficiency, require a more precise knowledge of the vertical distribution of optical turbulence.[2–5] The turbulence strength profile C_n^2 (h) may be obtained indirectly from meteorological parameters: for example balloon micro-thermal soundings with a high vertical resolution[6] but spending a few hours. C_n^2 (h) is more often obtained via optical means. These means differ from the number of sources employed and the type of data involved. Using scintillation data like the MASS[7] (Multi-Aperture Scintillation Sensor) C_n^2 (h) is retrieved from correlations of the scintillation pattern produced by a single star in a pupil plane. However vertical resolution is limited to five to six layers. A new method, SCO-SLIDAR[1] operates on a single star using a Shack-Hartmann wavefront sensor (SHWFS).

Although SCO-SLIDAR has been validated only numerically, it conjugates in principle the advantages of MASS and DIMM.[8] The generalized SCIDAR retrieves the C_n^2 profile from correlation of the scintillation pattern produced by a binary star in a pupil plane. The SLODAR retrieves C_n^2 (h) from correlations of wavefront slopes measured on a binary using a Shack-Hartmann wavefront sensor (SHWFS).[9,10]

SHWFS is currently used to measure wavefront slopes only. But it could be sensitive also to scintillation as the average intensity in a given sub-aperture can be obtained by adding pixel intensities in the sub-aperture focal plane up. Therefore, SHWFS delivers instantaneously wavefront slopes and intensities for each frame acquired, and that possibly per star position in the sub-aperture field of view (FOV). We propose to exploit wavefront and scintillation maps in the pupil measured at sub-aperture resolution to retrieve the C_n^2 profile.

We recall here the analytical background of a C_n^2 profile measurement based on the exploitation of the correlations between SHWFS data: Slopes correlations, scintillation correlation and correlations between slopes and Intensities correlations (which are referred further to coupling). Expected performances under weak-turbulence and measurement noise-free conditions are illustrated by computer simulations. Preliminary data processing on a real SHWFS installed on an astronomical telescope are presented.

2. Problem Statement with SHWFS Pseudo-Measurements

Given a star with position $\boldsymbol{\alpha}$ in the FOV, a SHWFS delivers a set of wavefront slopes and intensities per frame. Wavefront slopes are estimated by center of gravity (COG). The COG computed on the mth sub-aperture focal image is a bi-dimensional vector $\boldsymbol{s}_m(\boldsymbol{\alpha})$ with components $s_m^k(\boldsymbol{\alpha})$, for a slope measurement along the k axis ($k \in \{x, y\}$). Star intensities, $i_m(\boldsymbol{\alpha})$, recorded in every sub-aperture leads to scintillation rates $\delta i_m(\boldsymbol{\alpha}) = \frac{i_m(\boldsymbol{\alpha}) - o_m(\boldsymbol{\alpha})}{o_m(\boldsymbol{\alpha})}$ where $o_m(\boldsymbol{\alpha})$ is the time-averaged star intensity.

For two stars separated from $\boldsymbol{\theta}$, correlations of SHWFS data can be empirically estimated from a finite number of frames. Slopes correlations $\overline{s_m^k s_n^l}(\boldsymbol{\theta})$, scintillation rates correlations $\overline{\delta i_m \delta i_n}(\boldsymbol{\theta})$ and their coupling $\overline{s_m^k \delta i_n}(\boldsymbol{\theta})$ can be stacked in a single dimension covariance vector $\boldsymbol{C_{mes}}$, that will be our pseudo measurements in our direct problem statement. $\boldsymbol{C_{mes}}$ covariance vector relates to C_n^2 according to the Eq. (1):

$$\boldsymbol{C_{mes}} = \mathcal{M}\boldsymbol{C_n^2} + \boldsymbol{C_d} + \boldsymbol{u} \tag{1}$$

where $\mathcal{M}$ is the interaction function; $\boldsymbol{C_d}$ is the covariance vector of the noises affecting slope and intensity measurements; $\boldsymbol{u}$ represents uncertainties on $\boldsymbol{C_{mes}}$ due to the limited number of frames. In Rytov regime, $\mathcal{M}$ is a linear operator M and depends on SHWFS geometry, statistical properties of the turbulence, stars separation and distance between sub-apertures.[11] The column vectors of M are formed by the concatenation of weighting functions, for slope correlations and for scintillation and coupling correlations. They can be seen as correlations induced by a turbulent layer at altitude h. Every layer contribution to $\boldsymbol{C_{mes}}$ is illustrated on Figure 1.

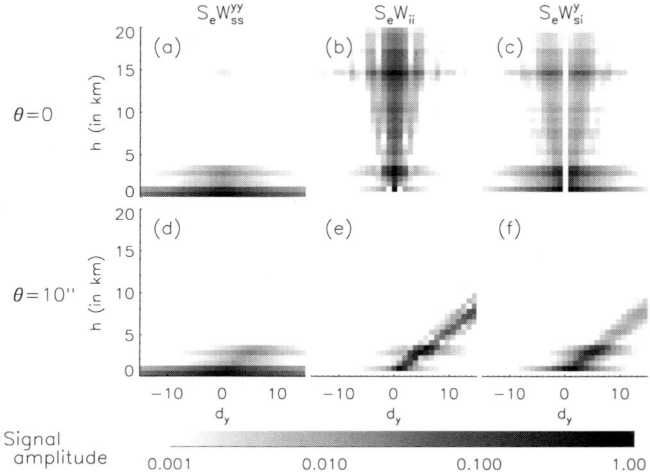

Fig. 1. Contributions of the turbulent layers to C_{mes} terms. Signal amplitude is coded in grey levels with a logarithmic scale. Every plot has been normalized by its maximum value.

Slope correlations ((a) and (d)), scintillation correlations ((b) and (e)) and their coupling ((c) and (f)) are represented for every 32 layers of a typical sampled profile $\boldsymbol{S_e}$, signal amplitude being coded in grey levels with a logarithmic scale. Every plot has been normalized by its maximum value. Slope measurement is along the y axis and so is separation between sub-aperture $(d_x = 0, d_y)$. $\boldsymbol{S_e}$ is a locally averaged version of a 2500 samples balloon sounding profile measured at Haute Provence Observatory (HPO) and is represented on Figure 2. Remote sensing of turbulent altitudes relatively to HPO elevation go from ground to 20 km. Three main layers are identified: a ground layer ranging from ground to 1.3 km (79 % of the turbulence strength), a low altitude layer located around 3 km (16 %) and a high layer at 14 km and above (5 %). The SHWFS counts 16×16 square subapertures (2.5 cm width) and is operated at $\lambda = 0.5\ \mu m$. For plots (d), (e) and (f) the stars separation is $\theta = 10$ $arcsec.$ along the y axis.

Covariances exploited in SLODAR correspond to plots (a) and (d). As can be observed on plot (a), each layer delivers the same signature. Signal amplitude depends on layer strength only. The two first layers, that are the strongest ones, deliver the strongest signals. In SLODAR, data inversion relies exclusively on the shift of the correlation maximum observed on plot (d), which provides a specific signature to each layer. Due to the pupil size (40 cm) and to the angular separation of the stars, this shift can only be exploited up to 7 km. SLODAR requires smaller binary separation or larger telescope diameter to separate layers at altitudes higher than 7 km.

With CO-SLIDAR, data plotted on plots (a) to (f) are exploited. In addition to the ground layer sensitivity of SLODAR type data, CO-SLIDAR benefits from scintillation correlations ((b) and (e)) and from coupling ((c) and (f)). Thanks to scintillation a better sensitivity to distant layers is expected ((b) and (e)). Coupling contributions ((c) and (f)) are strong for low altitude layers, and therefore should also increase detectability.

3. C_n^2 Profile Estimation Method

Pseudo measurements $\boldsymbol{C_{mes}}$ is estimated from measurements, slopes and intensities, affected by detection noise that bias the wavefront slope and scintillation covariances estimate. However these biases may be removed if the system is calibrated. The remaining fluctuation affecting $\boldsymbol{\hat{C}_{mes}} = \boldsymbol{C_{mes}} - \boldsymbol{C_d}$, reduced to statistical convergence, is represented by $\boldsymbol{u}$. $\boldsymbol{u}$ is a centered random vector with Gaussian distribution assuming a great number of frames. The covariance matrix of $\boldsymbol{u}$, $C_{conv} = \langle \boldsymbol{u}\boldsymbol{u}^T \rangle$, is estimated from $\boldsymbol{C_{mes}}$ and $\boldsymbol{C_d}$ as the empirical covariance matrix of a Gaussian random variable vector. The number of frames may be estimated from the exposure time of each frame and the whole duration of the measurement. Exposure time must be small enough to satisfy the frozen turbulence hypothesis, that is between 0.5 ms to 5 ms. Meanwhile, measurement must be short enough to ensure turbulence stationarity: 30 s integrations are exploited for SLODAR. Therefore, a thousand of frames may be used for processing $\boldsymbol{C_{mes}}$.

A sampled estimate of C_n^2, $\boldsymbol{\tilde{S}}$, may be retrieved from the inversion of Eq. 1, providing $\mathcal{M}$ is replaced by M, a matrix, and assuming Gaussian noise. Under positivity constraint, $\boldsymbol{\tilde{S}}$ minimizes the penalized maximum likelihood criterion J:

$$J = (\hat{\boldsymbol{C}}_{\boldsymbol{mes}} - M\boldsymbol{S})^t C_{conv} (\hat{\boldsymbol{C}}_{\boldsymbol{mes}} - M\boldsymbol{S}) + \beta ||H^{-1}\Delta\ \boldsymbol{S}||^2. \qquad (2)$$

Penalization is performed using weighted Laplace's regularization. β is a hyper parameter chosen using the L-plot method.[15] Δ is the Laplace's operator. The weights are given by H, a diagonal matrix considered as an average expected profile. H describes a Night-time Hufnagel-Valley profile[17] whose ground layer strength is adjusted to be conform to slope variances.

4. Numerical Experiment for Validation of C_n^2 Estimation

The ability of both CO-SLIDAR and SCO-SLIDAR to retrieve the C_n^2 profile is illustrated using a numerical experiment. Sources and SHWFS parameters are the same as those used in Figure 1. The source model represents two stars of same magnitude for CO-SLIDAR and a single star for SCO-SLIDAR. Turbulent wavefront slopes and intensities are obtained with a propagation program based on a split-step algorithm[16] . The turbulent volume is decomposed into 32 phase screens whose spatial statistics follow von Karman spectrum. To ensure a good decorrelation of wavefront perturbations on the highest layer and a reasonable processing time we finally chose the outer scale $L_0 = 1.5\ m$. Although constant and small this value should have a minor impact on the consistency of the results since low spatial frequencies of wavefront perturbation have small effect on scintillation but may cause an underestimation of the total wavefront variance. This effect is negligible for a wavefront variance estimated on a sub-aperture area, sub-aperture width being small compared to L_0. The profile used in the simulation is $\boldsymbol{S_e}$ (see Figure 2). Corresponding Fried parameter is $r_0 = 4.9\ cm$ and Rytov log amplitude variance is $\sigma_\chi^2 = 0.04$ at $\lambda = 0.5\ \mu m$. Sub-aperture images are computed without detection noise. COG and intensities for each star are computed separately for the same turbulence occurrence to avoid overlapping in the sub-aperture images. A set of 400 frames is simulated. The inversion process is performed with 32 layers in the sampled turbulence profile $\tilde{\boldsymbol{S}}$.

CO-SLIDAR and SCO-SLIDAR estimation results are reported on Figure 2 together with SLODAR restored profile. SLODAR discriminates the strongest layers from the ground to 7 km. Since this device does not separate layers influence for altitude higher than 7 km, signal from the upper layers is attributed into the 4 km to 7 km region where the signal is weaker.

Both CO-SLIDAR and SCO-SLIDAR recovers the complete profile. Wavefront slope and scintillation correlations enable to separate contribution of the highest layers, responsible for the scintillation signal, from the

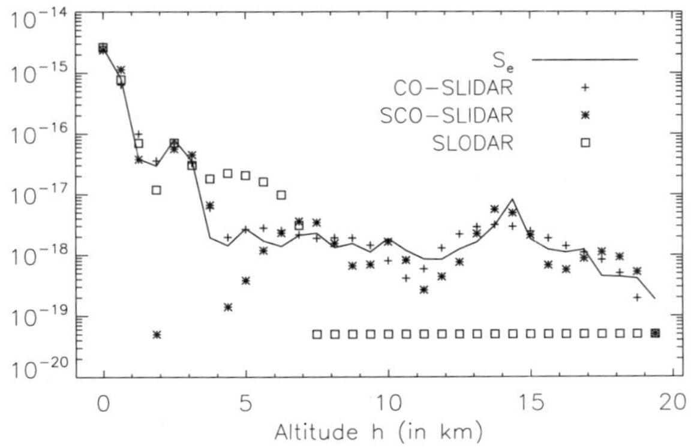

Fig. 2. Inversion results from CO-SLIDAR, SCO-SLIDAR and SLODAR datas obtained in simulation.

lowest ones. Considering mid-altitude layers, in the range 4 km to 13 km, CO-SLIDAR is more precise than SCO-SLIDAR. Unlike SCO-SLIDAR, which has poor discrimination ability from ground to 7 km (where scintillation signal is the weakest), CO-SLIDAR benefits from the stars separation which enables a more precise estimation even in this range where turbulence is weak.

5. Start Up of a SHWFS C_n^2 Profiler in Astronomy

We have initiated the processing on data from a binary star observed on a SHWFS installed on the Carlos Sánchez telescope on El Teide (Canarias). Telescope diameter is $D = 1.5$ m. SHWFS images are made of 10×10 sub-apertures with a central obstruction ($0.4D$). The data sequence contains 5000 frames recorded at 330 Hz. So the sequence duration is 15 seconds. The source is a binary star separated by 5.6 $arcsec$. The sub-aperture image sampling is close to Shannon. Individual slopes are calculated on each star with a local COG, putting them in smaller FOV boxes. As a start up we use only the bi-dimensional slopes from the 2 stars in a restricted data processing to the SLODAR case. In addition we do not use regularization function not to smooth too much the estimated C_n^2 profile. We try to retrieve a set of ten layers across 20 km of atmosphere. To do so four estimators have been used. They are labeled as follows: MC for Least Squares, MCT for truncated least squares, MV for Maximum likelihood and MVP for Maximum likelihood with positivity constraint. Using all of them we retrieve some layers well separated. No symbol plotted means zero or negative values. Using slopes only allows recovering a C_n^2 profile that shows

4 layers well separated. Strong turbulent layers are detected from ground to 2 km and at an altitude of $7-9$ km. Weaker layers are detected at 13 km and 17 km. The altitude distribution seems to be consistent with C_n^2 profiles yet observed at El Teide observatory, even we see here the turbulent layers are sharply discriminated. Moreover the turbulence strength of the second layer is surprisingly comparable to the ground layer leading to small Fried diameter (few cm). We found that the local COG is biased when stars are unresolved. This point must be sort out with other refined position estimators we are going to implement.

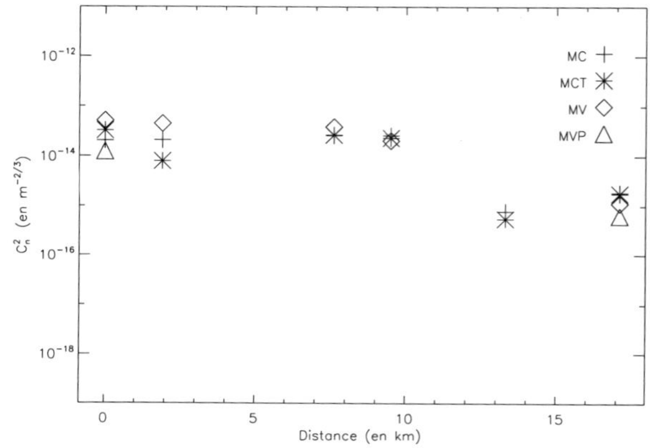

Fig. 3. C_n^2 profile results from inversion of a binary star slopes data recorded on a 10x10 subapertures SHWFS installed at Carlos Sanchez telescope in Tenerife.

6. Conclusion and Perspectives

We have proposed a new approach for C_n^2 profile measurement with a SHWFS reaping the benefit from both slope and intensity data. With CO-SLIDAR, slope correlations recorded on two separated stars deliver low altitude layers sensitivity. Scintillation correlations and coupling with slope deliver high altitude layer sensitivity. We have applied our method to astronomical data from a SHWFS observing a resolved binary star. Further investigations are required in order to improve data slope reduction. The use of star scintillation is required to enable a more precise C_n^2 estimation at high altitudes. We plan to test our concept on experimental SHWFS data recorded on many other SLODAR instrument, in order to quantify its effective performance. Additional numerical simulations could be performed by taking into account detection noise. Outer scale influence on the precision of the methods should be investigated. Determination of wind profile and maybe outer scale should be studied. Adaptation to close binary, moon edge, sun edge should extend the applications.

Acknowledgements

This work has been performed in the framework of a PhD Thesis supported by the French Délégation Générale pour l'Armement. The authors thank M. Mugnier and M. Sauvage for the fruitful exchanges. We thank Begonia García-Lorenzo and Angeles Rodriguez from the Intituto Astrofísico de Canarias all of whom have contributed to record the data on the G-SCIDAR+ SHWFS instrument installed at Telescope Carlos Sánchez.

References

1. N. Védrenne, V. Michau, C. Robert and J.-M. Conan, Optics Letters, **32(18)** 2659–2661 (2007).
2. T. Fusco, J.-M. Conan, L. M. Mugnier, V. Michau, and G. Rousset, Astron. Astrophys. Suppl., **142** 149–156 (2000).
3. T. Fusco, J.-M. Conan, G. Rousset, L. M. Mugnier, and V. Michau, Journal of Optical Society A, **18(10)** 2527–2538 (2001).
4. B. Le Roux, J.-M. Conan, C. Kulcsár, H.-F. Raynaud, L. M. Mugnier, and T. Fusco, Journal of Optical Society A, **21(7)** 1261–1276 (2004).
5. C. Petit, J.-M. Conan, C. Kulcsár, H.-F. Raynaud, T. Fusco, J. Montri, and D. Rabaud, Comptes Rendus de l'Académie des Science, Physique 6, 1059–1069 (2005).
6. J. Vernin and C. Munoz-Tunón., Astron. Astrophys., **284** 311–318 (1994).
7. A. Tokovinin, V. Kornilov, N. Shatsky, and O. Voziakova, Mon. Not. R. Astr. Soc., **343** 891–899 (2003).
8. Kornilov, V. and Tokovinin, A. and Shatsky, N. and Voziakova, O. and Potanin, S. and Safonov, B., MNRAS, **382**, 1268–1278 (2007).
9. R. W. Wilson, J. Bate, J. C. Guerra, N. N. Hubin, M. Sarazin, and C. D. Saunter, Proc. SPIE, **5490** 758–765 (2004).
10. Goodwin, M. and Jenkins, C. and Lambert, A., Optics Express, **15** 14844 (2007).
11. C. Robert, J.-M. Conan, V. Michau, T. Fusco, and N. Védrenne, Journal of Optical Society A, **23(3)** 613–624 (2006).
12. A. Fuchs, M. Tallon, and J. Vernin, Pub. Astron. Soc. Pacific, **110(743)** 86–91 (1998).
13. V. A. Klueckers, N. J. Wooder, T. W. Nicholls, M. J. Adcock, I. Munro, and J. C. Dainty, Astron. Astrophys. Suppl. Ser., **130**, 141–155 (1998).
14. R. W. Wilson, Mon. Not. R. Astr. Soc., **337** 103–108 (2002).
15. P. Hansen, SIAM Review, **34** 561–580 (1992).
16. J. M. Martin and S. M. Flatté, Appl. Opt., **27(11)** 2111–2126 (1988).
17. R. E. Hufnagel, Digest of technical Papers, Topical meeting on Optical Propagation through Turbulence, Boulder, Co, 1974.

HOW WE CAN UNDERSTAND THE ANTARCTIC ATMOSPHERE?

J.W.V. STOREY, M.C.B. ASHLEY, J.S. LAWRENCE
School of Physics, University of New South Wales, Sydney, NSW 2122, Australia

The Antarctic Plateau offers many benefits to astronomers, including dark and transparent infrared skies, long periods of uninterrupted observations, and very low levels of atmospheric turbulence. Efforts to quantify these benefits are ongoing. Characterizing the turbulence is particularly challenging, and requires a different approach to that commonly used at temperate sites. First, the atmosphere has two quite distinct regimes: a free atmosphere that is largely devoid of turbulence, and a thin but highly turbulent stable boundary layer. Second, if heat is used to avoid frost formation on optical surfaces, local turbulence might inadvertently be created by the instrument trying to measure it. In this paper we review the work that has been performed to date, and discuss what is required to advance our understanding of the Antarctic atmosphere.

1. Introduction

The Antarctic plateau, a vast area of several million square kilometers at an elevation of over 3000m, presents astronomers with a remarkable opportunity to build new ground-based telescopes with greatly superior performance to any currently existing [1]. The most promising Antarctic sites are those that are at local maxima in elevation—Dome A, Dome C and Dome F—together with the geographic South Pole which also offers its own set of unique advantages. Characterising these sites is therefore a very high priority. Because of the difficulty of access, and the importance of obtaining good statistics, robotic instruments are greatly preferable to those that require human operators and/or constant maintenance.

For astronomy, the extreme cold, low absolute humidity and highly transparent atmosphere lead directly to improved sensitivity in the infrared and sub-millimetre. Measurement of the sky brightness at unattended sites can be performed with robotic instruments, which can also return estimates of cloud cover throughout the year. At optical wavelengths, CCD cameras are used [2, 3] and these can self-calibrate using known stars in the field of view. In the IR and sub-mm, the "sky-dip" principle is generally used, together with reference hot and cold loads of known temperature [4, 5, 6]. The precipitable water vapour (PWV) is perhaps the most important single parameter to be derived from these longer wavelength measurements, as it can then be fed into model atmospheres to estimate the atmospheric transmission across the entire spectrum [7].

The second major advantage of the Antarctic plateau stems from the very low levels of atmospheric turbulence. The free atmosphere is largely free from the high-altitude jet-stream winds that dominate mid-latitude sites, leading to improved image quality, lower scintillation, and large isoplanatic angles and coherence time. However, as suggested by Gillingham [8], the excellent free-atmosphere conditions will almost always be compromised by an intense but relatively thin boundary layer that results from the strong temperature inversion in the lowest few tens of metres of the atmosphere. Unlike the situation that prevails at all temperate sites, where the boundary layer contributes an inevitable component to the seeing, over much of the Antarctic plateau the boundary layer is sufficiently thin that it is feasible to avoid it by simply placing the telescope above it; for example on a small tower. Characterising the atmospheric turbulence in Antarctica must therefore take into account the two very different regimes—a generally excellent free atmosphere and a generally highly turbulent boundary layer. This ideally requires using one technique that can characterize the free atmosphere uncontaminated by the boundary layer, and a second technique that measures the height of the boundary layer.

The difficulty of making turbulence observations in Antarctica is further complicated by the surprisingly high relative humidity within the boundary layer [9]. Any exposed optical surface will develop a layer of frost on it unless heated a few degrees above ambient. While this gentle heating is of little consequence for instruments measuring sky brightness or transparency, it is clearly important to implement telescope and mirror heating schemes with caution when making accurate measurements of the turbulence.

2. Techniques to Measure the Free Atmosphere

While the first question many observational astronomers ask about a site is *"What is the seeing like?"* in Antarctica this is the wrong question. The seeing from ground level is poor. The true benefits of Antarctica become apparent when the atmosphere is characterized not just in terms of natural seeing, but when all the parameters such as ε_0, τ_0, θ_0, and L_0 of most importance to adaptive optics and interferometry are known, and are measured not just as whole-atmosphere averages but as a function of altitude throughout the atmosphere.

2.1. *DIMM*

Traditionally, the DIMM (Differential Image-Motion Monitor) has been the "gold standard" for characterizing atmospheric turbulence. It is a relatively simple instrument, and returns an apparently unambiguous measurement of the parameter most familiar to astronomers, the "seeing", or FWHM of stellar images. At a traditional temperate site, a DIMM, if carefully calibrated [10, 11] gives a very useful indication of the relative merits of competing sites.

In Antarctica, however, the vast bulk of the atmospheric turbulence is located close to the ground, and this turbulent boundary layer is usually present throughout all of the (interesting) night-time. Thus, any technique that measures the integrated atmospheric turbulence will register poor seeing, most of the time. This was confirmed by the first DIMM measurements at the South Pole [12, 13].

However, during the summer the surface inversion layer is periodically removed by the solar heating, leading to the possibility of exceptionally good seeing even at ground level. Using a DIMM, Agabi *et al.* [14] found that at Dome C the summer-time seeing goes through a daily cycle with minimum levels well below that observed at any other site on earth. This is important not just for solar astronomy, but also for mid-IR observations with telescopes large enough to take advantage of the achievable image quality. The same team has also shown that it is possible to derive an estimate of free-atmosphere seeing statistics and boundary-layer height by using multiple DIMMs at various heights [15], although this technique is very costly in terms of both time and resources.

2.2. *Balloons*

Direct measurement of C_T^2 is possible by using microthermal sensors flown on a meteorological balloon [16]. This technique has been used with great success, both at temperate sites and in Antarctica, but has the disadvantage that a balloon and electronics package must be expended for every measurement. In addition, a minimum amount of wind shear is needed to ensure that the package is free from the balloon wake. Other issues, including the possibility of ice accumulation and a fluctuating radiation load on the sensor, complicate this otherwise very straightforward technique.

In Antarctica, there is a further problem that the balloons have a tendency to burst at a lower altitude than is desirable if the whole atmosphere turbulence profile is to be properly sampled. Nevertheless, balloon-borne microthermal campaigns have already made major contributions to our understanding of the free atmosphere at the South Pole [17] and Dome C [18].

2.3. *Double-Star Scintillation Techniques*

A variety of triangulation techniques (e.g., SCIDAR, SLODAR) have been developed for use at temperate sites to derive an atmospheric turbulence profile. Such techniques can be "generalized" by conjugating the detector to different levels, thereby providing some information about the boundary layer [19]. In general, a larger telescope is required than can conveniently be deployed to a remote site. In this context, the development of a Single-Star SCIDAR [20] that can potentially work with a smaller telescope may prove to be an important development for Antarctica, where at this stage in the development of permanent plateau stations the emphasis must be on small, robotic instruments.

2.4. *MASS*

The Multi-Aperture Scintillation Sensor (MASS) was developed by Tokovinin *et al.* [21]. By correlating the scintillation noise observed in a series of concentric apertures, it is possible to derive a crude turbulence profile. Despite its simplicity, the technique has several powerful advantages that make it ideal for Antarctic deployment: only a single star is required, and only a small (~100 mm) telescope is needed. Most importantly, the technique is tolerant of poor image quality and is largely insensitive to turbulence below about 500 m. The telescope and its housing can therefore be kept warm, and hence frost-free, without compromising the measurements [22]. A robotic MASS operating at Dome C through the first half of 2004 demonstrated that the site could deliver superb image quality, with unprecedented low levels of free-atmosphere seeing [23] and scintillation [24].

2.5. *Modelling*

An alternative to the direct measurement of atmospheric turbulence is to calculate the stability of the atmosphere using, for example, reanalysis of data from the European Centre for Medium-Range Weather Forecasts (ECMWF). This is a particularly valuable way to investigate potential new sites, and to develop an understanding of the reasons for the differences between known sites [25]. Similarly, Swain and Gallée [26] have shown that regional climate models can be used to estimate the boundary layer thickness across the entire plateau.

3. Techniques to Measure the Boundary Layer

As discussed above, Antarctica is unique in possessing a stable boundary layer that persists throughout most of the winter. This boundary layer is strongly turbulent. However, turbulence in stable boundary layers is very poorly understood. The more stable the boundary layer, the greater the probability that the turbulence will be intermittent or sporadic, and that rather than having a fully-developed Kolmogorov character will consist of Kelvin-Helmholtz instabilities, gravity waves, low-level jets and meandering motions. Almost all turbulence-monitoring techniques used by astronomers measure only a single turbulence length scale, and then assume a Kolmogorov distribution; this is likely to be a poor approximation at best in the boundary layer.

3.1. *Tower Microthermals*

By placing microthermal sensors at various heights on a tower it is possible to measure the turbulence at intervals from ground level to the top of the tower. This was the technique used by Marks *et al.* [27] to characterize the boundary layer at the South Pole. Unfortunately, the supersaturated atmosphere results in ice accumulation on the sensors, making this a very labour-intensive technique that does not lend itself readily to automation. To address this problem, Travouillon *et al.* [28] have pioneered the use of sonic anemometers in Antarctica. These sensors are very robust and can be heated between measurements to remove the ice. Although a considerable amount of power is required to keep the sensors defrosted, this method appears to be a very valuable way to probe the boundary layer.

3.2. *SHABAR*

A simple yet elegant way to probe the turbulence profile of the lower atmosphere is to observe the scintillation of an extended object such as the sun or moon with an array of photo-detectors [29]. This is known as a SHABAR, for SHAdow BAnd Radiometer. Recently, Tokovinin *et al.* [30] and Hickson and Lanzetta [31] have developed practical lunar scintillometers, extending the application of the technique to night-time site characterization. Like the MASS, the SHABAR is relatively insensitive to image quality and is insensitive to turbulence that is close to the sensors—simplifying defrosting. Its usefulness is of course limited to those times when the moon is well above the horizon and reasonable full. A prototype lunar scintillometer has been installed at Dome C by Moore *et al.* [32].

3.3. *Acoustic Radar*

Acoustic radars, or sonic radars (SODARs) have been used in Antarctica for many years by atmospheric physicists [33, 34]. Early measurements by

Travouillon *et al.* [35] at the South Pole showed that SODARs are an excellent tool for astronomers in Antarctica, where the low altitude of the boundary layer places it easily within range. In addition, because it is a non-optical technique and does not rely on fine sensors, it is relatively impervious to snow and frosting. Measurements at Dome C with a Remtech PA1 SODAR subsequently demonstrated that, during the night-time, the boundary layer had a typical vertical extent of 30 m or less. However, the limited vertical resolution of this particular model make it unsuitable for a detailed study of boundary-layer height variations.

A new, high-resolution SODAR has therefore been developed by Bonner *et al.* [36] and optimized for Antarctic conditions. This instrument has a vertical resolution of better than 1 metre, and can measure turbulence from within a few metres of the ground to about 100 m. It has been deployed successfully to Dome A, and may be eventually deployed to other sites such as Dome C in future. This instrument can operate over a wide range of frequencies, leading to the possibility of probing a range of turbulence scales in order to explore the various structures within the stable boundary layer [37].

4. The Future

New telescopes on the Antarctic plateau have the potential to change the way we do astronomy. The success of the South Pole Telescope [38] is a particularly good example of what can be achieved. Each of the various Dome sites, plus the South Pole, offers major advantages over existing temperate sites. The costs of building and operating scientific research infrastructure on the plateau are well known, as the US has operated a permanent station (Amundsen-Scott) at the geographic South Pole since 1957. These costs are surprisingly modest.

To fully realize the potential offered by Antarctica, the wider astronomical community and the funding agencies must be presented with clear evidence of the excellence of the sites. It is therefore essential that the site characterisation be carried out as fully and as rigorously as possible. Site testing of any kind requires the acquisition of properly calibrated data over a number of years, in order to understand the effects of year-to-year climatic variations. However, the investment in rigorously carrying out these site-testing studies is repaid many times over when the data result in new telescopes being placed at the best sites, to perform the best science they are capable of. It is important to make measurements in Antarctica that can be directly compared to those from temperate observatories. Ideally, a common suite of instruments would be used.

Measurements of the sky brightness and transparency, PWV and cloud cover have so far revealed few surprises—the Antarctic plateau appears to live up to the high expectations that astronomers are placing on it. Understanding the turbulence structure of the atmosphere is more challenging. Neither the free atmosphere nor the stable boundary layer is well understood at present, with the latter presenting a particularly fascinating theoretical and technical problem. Progress in understanding atmospheric turbulence will thus be most rapid in future if there is close collaboration between astronomers and meteorologists.

Acknowledgments

This research is supported by the Australian Research Council and the Australian Antarctic Division. We thank Eric Fossat and Gilles Durand for sharing their data with us prior to publication. This work has also been assisted by the ARENA Coordination Action (RICA-026150) of the FP6.

References

1. J.W.V. Storey, *Antarctic Science* **17**, 555 (2005).
2. Xiangyan Yuan et al., *Proc. SPIE* **7012**, 70124G (2008).
3. A. Moore et al., *Proc. SPIE* **7012**, 701226 (2008).
4. J.W.V. Storey, M.C.B. Ashley, M. Boccas, M.A. Phillips and A.E.T. Schinckel, *Publ. Astron. Soc. Pac.* **111**, 765 (1999).
5. P.G. Calisse, M.C.B. Ashley, M.G. Burton, M.A. Phillips, J.W.V. Storey, S.J.E. Radford and J.B. Peterson, *Publ. Astron. Soc. Aust.* **21**, 256 (2004).
6. C.A. Kulesa et al., *Proc. SPIE* **7012**, 701249 (2008).
7. J.S. Lawrence, *Publ. Astron. Soc. Pac.* **116**, 482 (2004).
8. P.R. Gillingham, *Proc. Astron. Soc. Aust.* **9**, 55 (1991).
9. G. Durand et al., *EAS Publ. Series* **25**, 77 (2007).
10. M. Sarazin and F. Roddier, *Astron. & Astrop.* **227**, 294 (1990).
11. A. Tokovinin, *Publ. Astron. Soc. Pac.* **114**, 1156 (2002).
12. R.F. Loewenstein, C. Bero, J.P. Lloyd, F. Mrozek, J. Bally and D. Theil, *ASP Conf. Ser.* **141**, 296 (1998).
13. T. Travouillon, M.C.B. Ashley, M.G. Burton, J.W.V. Storey and R.F. Loewenstein, *Astron. & Astrop.* **400**, 1163 (2003).
14. E. Aristidi et al., *Astron. & Astrop.* **444**, 651 (2005).
15. E. Aristidi et al., *Astron. & Astrop.* in press (2008).
16. M. Azouit and J. Vernin, *Publ. Astron. Soc. Pac.* **117**, 536 (2005).
17. R.D. Marks, J. Vernin, M. Azouit, J.F. Manigault and C. Clevelin, *Astron. & Astrop. Suppl.* **134**, 161 (1999).
18. H. Trinquet, A. Agabi, J. Vernin, M. Azouit, E. Aristidi and E. Fossat, *Publ. Astron. Soc. Pac.* **864**, 203 (2008).

19. R. Avila, J. Vernin and E. Masciadri, *Appl. Opt.* **36**, 7898 (1997).
20. M. Chadid, J. Vernin, E. Aristidi, H. Trinquet, M. Azouit and T. Sadibekova, *EAS Publ. Series* **25**, 63 (2007).
21. V. Kornilov, A.A. Tokovinin, O. Vozyakova, A. Zaitsev, N. Shatsky, S.F. Potanin and M.S. Sarazin, *Proc. SPIE* **4839**, 837 (2003).
22. J.S. Lawrence, M.C.B. Ashley, S. Kenyon, J.W.V. Storey, A. Tokovinin, J.P. Lloyd and M.R. Swain, *Proc. SPIE* **5489**, 174 (2004).
23. J.S. Lawrence, M.C.B. Ashley, A. Tokovinin and T. Travouillon, *Nature* **431**, 278 (2004).
24. S.L. Kenyon, J.S. Lawrence, M.C.B. Ashley, J.W.V. Storey, A. Tokovinin and E. Fossat, *Publ. Astron. Soc. Pac.* **118**, 924 (2006).
25. S. Hagelin, E. Masciadri, F. Lascaux and J. Stoesz, *MNRAS* **387**, 1499 (2008).
26. M.R. Swain and H. Gallée, *Publ. Astron. Soc. Pac.* **118**, 1190 (2006).
27. R.D. Marks, J. Vernin, M. Azouit, J.W. Briggs, M.G. Burton, M.C.B. Ashley and J.F. Manigault, *Astron. & Astrop. Suppl.* **118**, 385 (1996).
28. T. Travouillon et al., *Proc. SPIE* **7012**, 70124B (2008).
29. J.M. Beckers, *Solar Physics* **145**, 399 (1993).
30. J. Rajagopal, A. Tokovinin and E. Bustos, *Proc. SPIE* **7013**, 70131P (2008).
31. P. Hickson and K. Lanzetta, *Publ. Astron. Soc. Pac.* **116**, 1143 (2004).
32. A. Moore et al., *Proc. SPIE* **6269**, 62695U-1 (2006).
33. W.D. Neff, *PhD Thesis*, Wave Propagation Laboratory, Boulder Colorado, USA, (1981).
34. S. Argentini, G. Mastrantonio, A. Viola1, P. Pettre and G. Dargaud, *Boundary-Layer Meteorology* **81**, 75 (1996).
35. T. Travouillon, M.C.B. Ashley, M.G. Burton, J.W.V. Storey and R.F. Loewenstein, *Astron. & Astrophys.* **400**, 1163 (2003).
36. C.S. Bonner, M.C.B. Ashley, J.S. Lawrence, J.W.V. Storey, D.M. Luong-Van and S.G. Bradley, *Proc. SPIE* **7014**, 70146I (2008).
37. M.A. Kallistratova, *IOP Conference Series: Earth and Environmental Science* **1**, 012001 (2008).
38. Z. Staniszewski et al., *Astrophys. J.* submitted (2008).

TMT SITE TESTING SURVEY: INSTRUMENTS, METHODS AND OPERATIONS

R. L. RIDDLE*, W. SKIDMORE, T. TRAVOUILLON
TMT Observatory Corporation, Pasadena, CA, USA
**E-mail: riddle@tmt.org*

M. SCHÖCK
TMT Observatory Corporation, Victoria, BC, Canada

S. ELS
Cerro Tololo Inter-American Observatory, La Serena, Chile

The Thirty Meter Telescope (TMT) project has undertaken a large program to select the final site for the telescope. Five candidate sites were selected in Chile, Mexico and Hawaii for testing, and a site testing station was erected on each mountain. The site testing station consists of an extensive suite of instruments that operate robotically on each of the candidate sites. The systems are all monitored and the data archived to a central server system. Testing began in April, 2003, and each site testing station was installed for at least three years. The project is now nearing completion, with the final site selection decision scheduled for mid-2009.

Keywords: Astronomical Instrumentation: robotic telescopes; site testing.

1. The Candidate Sites

Table 1. A list of the five TMT candidate sites selected for testing.

Site Name	Country	Elevation	Latitude [deg N]	Longitude [deg W]
Cerro Tolar	Chile	2290 m	-21.9639	70.0997
Cerro Armazones	Chile	3064 m	-24.5800	70.1833
Cerro Tolonchar	Chile	4480 m	-23.9333	67.9750
San Pedro Mártir	Mexico	2830 m	31.0456	115.4691
Mauna Kea 13N	Hawaii	4050 m	19.8330	155.4810

The TMT project initially considered any potential sites around the world. These sites were then examined through a satellite study,[1] and five sites were selected for further intense study. Table 1 lists the five sites. The three Chilean sites are all located in the Atacama desert region; two are costal sites while one (Cerro Toloncha r) is located near the ALMA site. The site at San Pedro Mártir in Mexico is in the highest region of northern Baja, Mexico. The site on Mauna Kea is not on the summit, but is instead on a plateau to the north of the observatories, about 150 m below the summit.

2. Site Testing Instrument Suite

Fig. 1. This is a typical example of the installation of the testing equipment on a remote mountain site, Cerro Armazones with all the equipment installed. The SFAS SODAR is in the foreground. From left to right, the weather station, the 6.5 m tower with the Halfmann telescope, sonic anemometer and dust sensor, IRMA (to the left of the small block house), ASCA on its 3 m tower, the 30 m tower with sonic anemometers, the required transport in Chile, and the solar panel system.

The scope of the instrumentation suite covers the essential items about a site that are critical to the construction of a large telescope. The major goal was to measure the important turbulence and atmospheric parameters as completely as possible, and with overlapping techniques where applicable as a check on the results from each individual instrument.

Figure 1 shows a typical installation of all of the equipment at one of the TMT candidate sites (in this case Cerro Armazones). As the figure makes clear, a large amount of infrastructure is required to support the site testing program as implemented by TMT; the fact that this equipment was placed in remote locations made the execution of the project even more challenging. The infrastructure is required to support the instrument suite that is necessary to measure all of the physical parameters desired from each site.

The instrument suite used in the site testing program consists of the following instruments:

- **Differential Image Motion Monitors (DIMM)[2]**: The TMT DIMMs are mounted on small (35 cm) but robust custom made telescopes produced by Teleskoptechnik Halfmann in Germany. The telescopes are installed on 6.5 m towers. A DIMM measures the integrated seeing in the air column above the telescope.
- **Multi-Aperture Scintillation Sensors (MASS)[3,4]**: A MASS is a scintillation based instrument which measures 6-layer turbulence profiles, the isoplanatic angle and the turbulence coherence time. They can also be used for atmospheric transparency estimates. The MASS and DIMM are mounted in the same instrument body on the telescope.
- **Sound Detection and Ranging (SODAR) acoustic sounders[5]**: The SODAR is a phased-array acoustic emitter/receiver system which produces low elevation (10 to 800 m) turbulence and wind profiles.
- **Automatic Weather Stations (AWS)[6]**: Commercial weather stations with temperature (air and soil), wind speed and direction, humidity, barometric pressure, precipitation, solar irradiation, heat flux and net radiation sensors were installed at all sites. They are stand alone units, with most of the instruments mounted 2 m above the ground.
- **Sonic Anemometers[7]**: These are mounted at the MASS/DIMM telescope level and/or at several elevations on the 30 m towers. They measure wind speed and direction, a sonic temperature value, and can be used to estimate the in-situ turbulence strength.

- **All-sky cameras (ASCA)**[8]: ASCA provides images of the entire sky in several visible and infrared filters. The data are used for cloud analyses and light pollution studies.
- **30 m Tower**: Sonic anemometers and air temperature sensors are installed on 30 m towers on Cerro Armazones and Cerro Tolonchar. During their site survey, the Large Synoptic Survey Telescope (LSST) project also had a 30 m tower with sonic anemometers installed at San Pedro Mártir.[9] The data from this are available to TMT for site evaluation.
- **Infrared Radiometers for Millimetre Astronomy (IRMA)**[10]: These units measure the flux from the sky at 20 μm. The precipitable water vapor (PWV) content of the atmosphere can be calculated from this using a suitable atmospheric model.
- **Dust Sensors**[11]: Commercial particle counters are mounted at the MASS/DIMM telescope level. They measure the particle density in five different channels for particle sizes of 0.3, 0.5, 1.0, 2.0 and 5.0 μm.

Additional equipment was necessary to support the operation of the site testing instruments.

- **Power system**: The electrical power for all sites aside from SPM is provided by a solar power system. There are two UPS units (one for the telescope and one for the rest of the electronics). Network power switches are used by the robotic system to control the power configuration or power cycle components as necessary.
- **Network**: A continuous internet connection was established at each site so the status of the system can be monitored at all times. All sites have a sufficiently fast connection to allow continuous monitoring of the site and data archiving. The site testing local area network hardware includes a router and fast ethernet switches. The router is configured with a robust firewall; automated access is used for system monitoring and data archiving.
- **Webcams**: Each site has externally accessible web cameras that allow monitoring of the conditions on site. Images are taken during daylight hours at ten minute intervals; these images can be examined individually but are automatically composed into movies. These are used to check the safety of the site equipment, that the equipment is functioning properly, and to observe the weather conditions on site.

- **Firewire disk storage**: The system uses a Firewire drive, shared among all the instruments, to store and archive data. The storage space is large enough to allow the system to run without intervention if the connection to the site is lost for several days.

3. Operations

Each site is a fully autonomous robotic station, and is designed to run without human interference. In several instances, the network connection was lost to a site, and the system still operated properly (including a period at Cerro Tolar where the connection was down for almost a week). The systems were designed that way in order to gather as much data as possible. The systems operated safely, shutting down and keeping the instrumentation safe, through events including blizzards and tropical storms.

The center of the site testing system is the Supervisor Computer (SC). It controls the actions of the telescope, instruments, data management and monitors the status of all systems. The weather station is connected directly to a SC serial port, allowing the system to react quickly to weather conditions. The SC operating system is a stripped down installation of RedHat Linux 9. The SC is a full server system, with a web and mail server communicating information about the status of the system to the outside world; access to the system is limited to web and SSH connections for added security. The web site allows access to the data and displays the status of the system and the data as they are gathered.

The robotic software control system is installed on the SC, and is divided into two separate parts: an extensive set of IDL[12] scripts (with supporting Bash shell scripts and Expect scripts) that control operations of the Halfmann telescope, MASS and DIMM during night operations; and a second set of Bash scripts, C software, plotting software and other bits that control the operation of the other instruments, data management and status updates. The software controls the operation of the other instruments as well, and manages the data flow and backup.

Time is an important factor in any astronomical observation, and the robotic system has to handle time appropriately since a consistent time record is essential for data quality to be assured. A built in ephemeris controls the timing of daily operations, correcting for changes in the rise and set times of the Sun. A network time protocol (NTP) server is installed and all site computers are synchronized to it. This allows the site system to operate properly if the internet connection is disconnected for some reason.

Each day, the data from the site is compressed and archived; the archives are checked for consistency and completeness. These data archives are downloaded to a central server in Pasadena, CA, where the checks are again run to ensure the archive is transferred intact. A second server in La Serena, Chile mirrors the first in case of the destruction of the server; data archives are also periodically written to DVD media as a third backup.

The central server is also used to monitor the sites. The status of all instrumentation and equipment on each site is shared with the central server, and a single page on the central server website is used to monitor all of the site testing equipment for errors or issues. This is critical, for the amount of equipment would be difficult to monitor by examining each item separately.

4. Results

Testing of the candidate sites began in April, 2003 with an initial installation on Cerro Tolar. Table 2 lists the dates of installation of the various instruments on each mountain.

Table 2. The schedule of installation of instruments at the five testing sites. A missing date denotes that the instrument was not installed at the site. Decommission times for the sites are also included; at this writing the station on Cerro Armazones is still in place.

Instrument	Tolar	Armazones	Tolonchar	SPM	MK 13N
DIMM	Oct 03	Nov 04	Nov 05	Oct 04	Jun 05
MASS	Jan 04	Nov 04	Jan 06	Oct 04	Jul 05
Weather station	Apr 03	Jul 03	Nov 05	Oct 04	Jun 05
SODAR	——	Mar 05	Feb 06	Mar 06	Oct 05
Sonic	Feb 06	Feb 06	Mar 06	May 06	Nov 05
30 m Tower	——	Sep 06	Mar 07	Dec 05	——
Dust sensor	Feb 06	Feb 06	Mar 06	May 06	Nov 05
ASCA	Oct 05	Oct 05	Nov 05	Jul 05	Jun 06
IRMA	——	Jan 07	Mar 07	——	Feb 07
Decomissioned	Mar 2007	——	Nov 2008	Aug 2008	Jun 2008

The goal was to install identical equipment onto each of the sites, and to have the sites active for at least two years. This was to avoid having the equipment operate in a season that turned out not to be representative of the average site conditions. As Table 2 shows, the equipment was installed onto all of the sites for three years at least, which was ample time to see some seasonal variations and make clear comparisons between the five sites.

5. Conclusion

The TMT site testing program was a complicated project, with telescopes on two continents, an island, and five different mountain tops, each with unique characteristics and challenges for the placement and operation of robotic instrumentation. The project is now nearing completion, with the final site decision expected in the middle of 2009.

Acknowledgements

The TMT site testing work could not have happened without the contributions of many other people at various institutions. A special thanks to everybody who has supported, and is still supporting, the TMT site testing and selection program, in particular the people at the Cerro Tololo Inter-American Observatory (CTIO), the National Optical Astronomy Observatory (NOAO) and the AURA New Initiatives Office (NIO), the Universidad Católica del Norte in Antofagasta, the Observatorio Astronómico Nacional de San Pedro Mártir, Gemini Observatory, Palomar Observatory, the Submillimeter Array (SMA), the University of Hawaii Institute for Astronomy (IfA), the University of Lethbridge, the University of Moscow, the South African Astronomical Observatory (SAAO), and the entire TMT project including its partner institutions and external reviewers.

The TMT Project gratefully acknowledges the support of the TMT partner institutions. They are the Association of Canadian Universities for Research in Astronomy (ACURA), the California Institute of Technology and the University of California. This work was supported as well by the Gordon and Betty Moore Foundation, the Canada Foundation for Innovation, the Ontario Ministry of Research and Innovation, the National Research Council of Canada, the Natural Sciences and Engineering Research Council of Canada, the British Columbia Knowledge Development Fund, the Association of Universities for Research in Astronomy (AURA) and the U.S. National Science Foundation.

References

1. Erasmus, D., van Staden, C., *A Comparison of Satellite-Observed Cloud Cover and Water Vapor At Mauna Kea and Selected Sites In Northern Chile, the Southwestern U.S.A. and Northern Mexico*, final report, AURA New Initiatives Office (2003).
2. L. Wang, M. Schöck, G. Chanan, W. Skidmore, E. Bustos, J. Seguel and R. Blum, High accuracy DIMM measurements for the TMT site testing program, in *Society of Photo-Optical Instrumentation Engineers (SPIE)*

Conference Series, Society of Photo-Optical Instrumentation Engineers (SPIE) Conference Series Vol. 6267July 2006.

3. S. G. Els, M. Schöck, J. Seguel, A. Tokovinin, V. Kornilov, R. Riddle, W. Skidmore, T. Travouillon, K. Vogiatzis, R. Blum, E. Bustos, B. Gregory, J. Vasquez, D. Walker and P. Gillett, *AO* **47**, 2610(May 2008).
4. S. G. Els, M. Schöck, J. Seguel, W. Skidmore, D. Walker, A. Tokovinin, V. Kornilov, R. Riddle, T. Travouillon, E. Bustos, J. Vasquez, R. Blum, B. Gregory and P. Gillett, The Multi Aperture Scintillation Sensor (MASS) used in the site selection of the Thirty Meter Telescope (TMT), in *Ground-based and Airborne Telescopes. Edited by Stepp, Larry M.. Proceedings of the SPIE, Volume 7012, in press*, Presented at the Society of Photo-Optical Instrumentation Engineers (SPIE) Conference Vol. 7012June 2008.
5. T. Travouillon, SODAR calibration for turbulence profiling in TMT site testing, in *Society of Photo-Optical Instrumentation Engineers (SPIE) Conference Series*, Society of Photo-Optical Instrumentation Engineers (SPIE) Conference Series Vol. 6267July 2006.
6. Monitor Sensors (Aust) Pty Ltd., Website, http://www.monitorsensors.com/ (2008).
7. W. Skidmore, T. Travouillon and R. Riddle, Evaluation of sonic anemometers as highly sensitive optical turbulence measuring devices for the Thirty Meter Telescope site testing campaign, in *Society of Photo-Optical Instrumentation Engineers (SPIE) Conference Series*, Society of Photo-Optical Instrumentation Engineers (SPIE) Conference Series Vol. 6267July 2006.
8. D. E. Walker, H. E. Schwarz and E. Bustos, Monitoring the night sky with the Cerro Tololo All-Sky camera for the TMT and LSST projects, in *Ground-based and Airborne Telescopes. Edited by Stepp, Larry M.. Proceedings of the SPIE, Volume 6267, pp. 62672O (2006).* Presented at the Society of Photo-Optical Instrumentation Engineers (SPIE) Conference Vol. 6267July 2006.
9. J. Sebag, V. Krabbendam, C. Claver, J. Barr, J. Barr, J. Kantor, A. Saha and D. A. Erasmus, LSST site evaluation, in *Society of Photo-Optical Instrumentation Engineers (SPIE) Conference Series*, Society of Photo-Optical Instrumentation Engineers (SPIE) Conference Series Vol. 6267July 2006.
10. R. R. Phillips and D. A. Naylor, Initial operations of an water vapour monitor (IRMA) at Gemini South, Las Campanas Observatories, and in the TMT site testing role, in *Society of Photo-Optical Instrumentation Engineers (SPIE) Conference Series*, Society of Photo-Optical Instrumentation Engineers (SPIE) Conference Series Vol. 6269July 2006.
11. Met One Instruments, Inc., Website, http://www.metone.com/ (2006).
12. ITT Visual Information Solutions, Website, http://www.ittvis.com (2004).

TMT SITE TESTING SURVEY: CALIBRATION AND RESULTS

M. SCHÖCK
TMT Observatory Corporation, Victoria, BC, Canada

S. ELS
Cerro Tololo Inter-American Observatory, La Serena, Chile

R. RIDDLE, W. SKIDMORE and T. TRAVOUILLON
TMT Observatory Corporation, Pasadena, CA, USA

The Thirty Meter Telescope Project has been testing several mountains in Chile, Mexico and Hawaii to determine the eventual placement of the telescope. The site testing project has operated for several years, and employed a variety of methods to measure the suitability of each of the sites. Here, we discuss the calibration of instruments and methods and some top level results from the candidate sites.

Keywords: Site testing; extremely large telescopes.

1. TMT Site Testing

The Thirty Meter Telescope (TMT) recently finished five years of in situ testing on its five candidate sites: Cerros Tolar, Armazones and Tolonchar in northern Chile; San Pedro Mártir on Baja California, Mexico and a site denoted 13 North (13N) approximately 150m below the summit of Mauna Kea, Hawaii. The sites, instrumentation and methods employed were described in the previous paper in these proceedings[1] and are not repeated here. Instead, this paper concentrates on the calibration results obtained for the TMT site testing instrumentation and notes some possible pitfalls that can be encountered when analyzing site data. It also presents some of the top level results from the candidate sites.

Table 1. Summary of the measurement uncertainties as determined from the instrument calibrations. Here, 'n/a' stands for 'not available'. Values given are usually upper limits to the uncertainties of the probability distributions of the respective quantities, although some exceptions apply.

	Relative Error	Absolute Error	Comments
Weather station sensors			
Temperature	$\lesssim 1°C$	$\lesssim 1°C$	
Temperature on 30 m tower	0.1°C	0.1°C	
Wind speed	10%	10%	
Wind direction	$\lesssim 5°$	$\lesssim 5°$	limited by setup accuracy
Humidity	10%	10%	non-linear in mid range (around 30%)
Pressure	1 hPa	1 hPa	vendor quoted accuracy
Solar irradiance	n/a	n/a	not needed
Precipitation	n/a	n/a	used for equipment safety only
Heat flux	5%	5%	vendor quoted accuracy
Net radiation	3%	3%	vendor quoted accuracy
DIMM			
Seeing	$0''.02$	similar	
MASS			
Free-atmosphere seeing	$0''.05$	similar	
Individual layers ($C_n^2 dh$)	10^{-14} m$^{1/3}$	similar	
Isoplanatic angle	$0''.02$	$< 0''.2$	
Coherence time	20%	20%	
SODAR			
Wind profiles	20%	~ 1 m/s	offsets exist between SFAS and XFAS
GL seeing	10%	similar	
Individual layers	$\sim 20\%$	similar	
Sonic anemometer			
Wind speed	$< 5\%$	$< 5\%$	
Wind direction	$\lesssim 5°$	$\lesssim 5°$	limited by setup accuracy
Sonic temperature	$< 3°C$	$< 3°C$	offsets exist
Dust sensor			
Particle count	10%	n/a	
All-sky camera			
Cloud cover	1–4%	1–4%	site dependent
IRMA			
Precipitable water vapor	~ 0.25 mm	n/a	additive offsets exist; site dependent
Satellite data			
Cloud cover	2–6%	2–6%	site dependent
Precipitable water vapor	20%	20%	

2. Instrument Calibrations and Results Verifications

Data taken with our instruments are only considered useful for the TMT site selection process if the reliability and uncertainties of the measurements are known. As a result, we have gone through great efforts to understand all instruments and results, including:

- Side-by-side comparisons of identical instruments
- Side-by-side comparisons of different instruments measuring the same parameters
- Sensitivity analyses of the dependence of the results on input parameters

- Independent verification of all in-house analysis software by at least two people
- Independent verification of all results and statistics by at least two people

Details of the calibration efforts have already been published in some cases[2,3] and exist as TMT internal reports for the other instruments. These will be published shortly. Here, we only present a summary of the uncertainties expected for our instruments (see Table 1). All values given in the table are limits for the statistical properties of the parameters, not for the individual measurements.

A note concerning terminology: We use the term 'absolute error' of a measurement to describe the *accuracy* of the measurement (defined as the closeness of agreement between the average value obtained from a large series of test results and an accepted reference value). It describes how well our measurements agree with the absolute values of the respective parameters (the "truth"). The term 'relative error' is used to describe the expected relative differences between the measurements taken with identical equipment at different sites, the *reproducibility*. Both accuracy and reproducibility are difficult to determine in practice. For our results, they are based on the comparison of different equipment at the same sites, whenever possible. When this is not possible, we estimate them based on investigations of the *repeatability* of the measurements (the expected relative differences between the measurements taken with identical equipment at the same site) and on sensitivity and bias analyses of the instruments.

3. TMT Site Selection Top-Level Results

This section provides a top-level summary of the results from the TMT candidate sites. Many more details are currently being written up in a series of papers to be published shortly. We also describe some of the principles we employ in our data analysis and interpretation and some of the lessons we learned that led us to arrive at these principles.

3.1. *Data Analysis Principles and Lessons Learned*

It is, obviously, impossible to present all aspects of the huge amount of data collected during the TMT site testing work in one paper, or even in a paper series. Beside there simply being so many different properties of that data that might all be interesting for one or another purpose, there are generally also a multitude of different ways of analyzing and interpreting the same subset of data. One example for this is the seemingly simple compilation of the DIMM seeing probability distributions. The most straight-foward method to do this is to weight all individual data points equally. Other methods might involve the calculation of monthly (or seasonal) distributions and averaging those, either with equal weights or weighted by the length of night time and the weather conditions during different parts of the year. All these methods have their advantages and disadvantages depending, for example, on what the results are to be used for, how evenly the data cover each month or season of the year, or whether expected up and down times of the observatory are known as a function of the month/season. The results shown in Section 3.2 show two aspects of the data, the overall statistics including all (night-time) data and monthly time series.

It is also possible to arrive at different numerical results for statistical parameters measured at a site by different data treatment techniques. Obvious examples for this are the reprocessing of MASS turbulence data or the selection of valid DIMM seeing data based on image quality criteria. The former might need to be done if calibration parameters such as detector properties change in time and are measured continuously or periodically during operation (as is the case for the TMT instruments). The latter should be done if telescope and instrument aberrations cannot be ignored (not the case for the TMT instruments), but it is extremely important to take the effect of the seeing itself on the image quality into account also. It is, for example, not correct to use a fixed Strehl ratio as cut-off for valid data. Details on both TMT DIMM and MASS data quality, uncertainties and interpretation have already been published.[2,3]

A second example of possible different interpretations of the same data set is the exclusion of invalid wind measurements, such as times when a mechanical anemometer gives zero wind speed readings because it is frozen over after a winter storm. Obviously, such points should be excluded from the analysis. However, unless an *independent* record exists for when this happened (which, usually, is not the case for a remote site), it is difficult to exclude these and only these events. Anemometers also read zero wind speed in normal operation, when the wind speed is below their detection threshold. This threshold increases in time as dust gets into the gears and lubrication decreases. Thus, excluding all zero wind speed readings might bias the statistics more than leaving frozen-sensor data in the set. Which way of data analysis is preferable depends, again, on the goal of the data analysis (for example, is the high or low wind speed regime of primary interest) and on which conditions exist more frequently at a given site.

A final example is the occurrence of, apparently, negative ground layer (GL) turbulence as calculated from the difference between the MASS and DIMM seeing. One would think that such data should be excluded because they are unphysical. This is, however, an incorrect approach if good data quality for both DIMM and MASS measurements is ensured (that is, if there is no suspicion that many inherently invalid data points exist). Both measurements contain random noise, which can make the MASS seeing larger than the DIMM seeing, in particular for weak GL turbulence conditions. Thus, excluding negative GL turbulence values would bias the statistics toward strong GL conditions. How exactly these cases are treated, depends, as before, on the goal of the analysis and the frequency of weak GL conditions. In the TMT site testing data analysis, we generally keep the data in the statistics as negative values and interpret them as values close to zero. In fact, the spread of these values gives us an estimate of the combined MASS and DIMM measurement noise.

These examples illustrate that there is not one single way of representing the TMT site testing results that is applicable to or useful for all purposes. Over the last five years, we have looked at many aspects of the data sets, using different ways of analyzing, selecting and interpreting the data. Most of the time, we find that the differences are small enough that they will not make a difference for the purpose of TMT site selection (but they might be important for some other application). If significant differences are found, we make sure that we understand their causes and consequences for TMT site selection purposes. It should be understood, however, that these consequences and the level of importance we attribute to them might not apply to other applications of the data.

3.2. *Some Results from the Sites*

Plots comparing some of the main characteristics of the five TMT candidate sites are presented in Figs. 1 and 2. Figure 1 shows the cumulative probability distributions for each site for DIMM and MASS seeing, MASS isoplanatic angle, night-time air temperature at 2 m above the ground and night-time wind speed at 7 m. It can be seen that the total seeing, as measured by the DIMM, is smaller at the Chilean sites than at the North American sites, with Tolar, Armazones and Tolonchar being almost identical. San Pedro Mártir has a slightly larger total seeing at 7 m above the ground than Mauna Kea 13N.

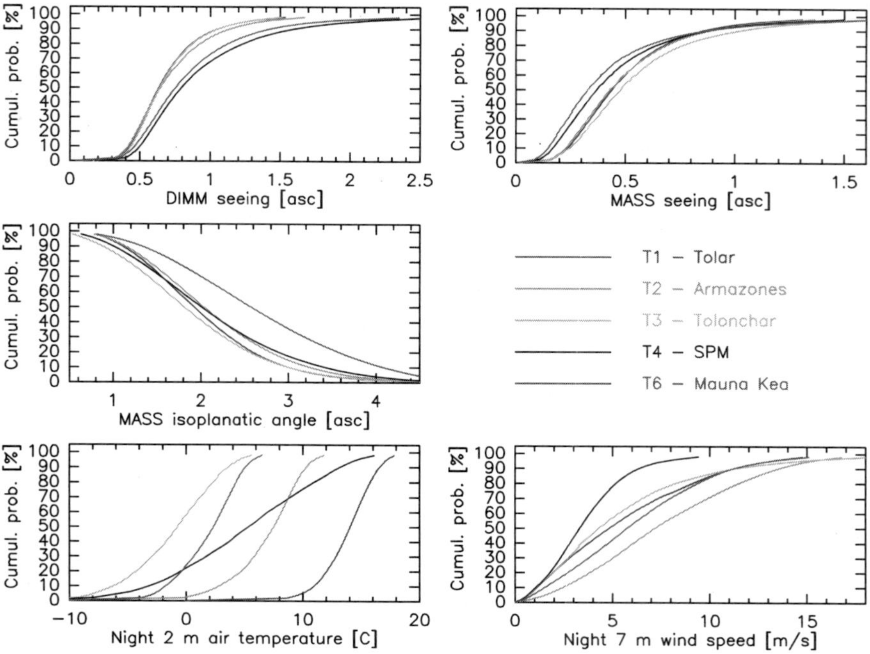

Fig. 1. Cumulative probability distributions for the five candidate sites. DIMM and MASS seeing, isoplanatic angle, temperature and wind speed are shown.

For the MASS seeing (the seeing integrated from ~500 m to the top of the atmosphere), the situation is reversed, with Mauna Kea 13N showing the weakest high-altitude turbulence, followed by San Pedro Mártir, Tolar/Armazones and Tolonchar. This shows that the larger DIMM seeing of the North American sites comes from their having a stronger ground layer than the Chilean sites. This is, at least in part, caused by the North American sites being located in flatter areas with more pronounced surrounding terrain than the Chilean sites, which are all on mountain tops. It also needs to be noted that the DIMM seeing is measured at an elevation of 7 m above the ground. Obviously, the seeing encountered by TMT will be different. At an elevation of 30–60 m, which is more applicable to TMT, the differences in overall seeing between the sites are significantly reduced. More details will be given in the TMT results paper series. The isoplanatic angle behavior is similar to that of the MASS seeing, with the exception that San Pedro Mártir has an isoplanatic angle similar to the Chilean sites.

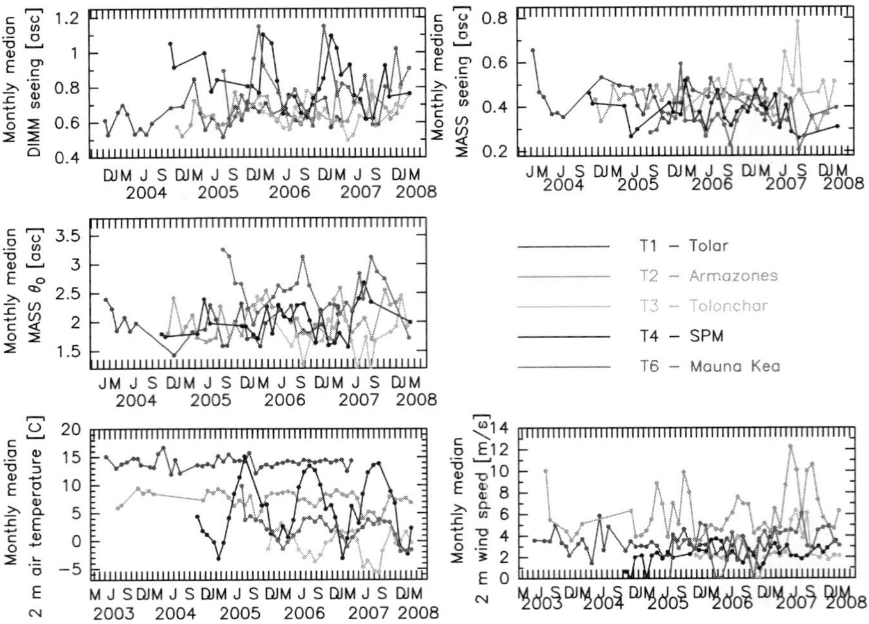

Fig. 2. Monthly median values for the five candidate sites. The DIMM and MASS seeing, isoplanatic angle, temperature and wind speed are shown. Only months for which at least 1000 individual measurements are available (~5 nights of data) are included in the plots. The AWS 2 m wind speeds are shown here instead of the 7 m sonic anemometer measurements because the AWSs have been operating at the sites for much longer than the sonics.

The shapes of the temperature distributions in Fig. 1 are similar for Tolar, Armazones, Tolonchar and Mauna Kea 13N, with the slightly flatter shape of the Tolonchar curve indicating a little more seasonal variability than the other three sites (see also Fig. 2). The median values are a reflection of the site altitudes, Tolar being the lowest and warmest and Tolonchar being the highest and coldest. The median value of the San Pedro Mártir temperature also follows approximately this trend in altitude (it is somewhat colder than Armazones, which is of similar height), while the shape of the curve indicates a larger annual temperature range. This stems from the fact that temperatures at San Pedro Mártir show larger seasonal variations than at the other sites (see Fig. 2).

Finally, the wind speed distributions (sonic anemometer wind speeds at 7 m) show Armazones as the windiest site, followed by Mauna Kea 13N and Tolar. Tolonchar has surprisingly low wind speeds for a high-elevation site. Note that the free-air wind speeds at San Pedro Mártir are higher than indicated by our measurements which are affected by the presence of trees at the site. Also note that, while the 7 m wind speed at Armazones is higher than at the other sites, its wind speed profile is essentially flat above that. This means that the wind speed differences between Armazones and the other sites at the level of the TMT enclosure openings are smaller than at the 7 m levels.

The variations of the monthly median values of the DIMM and MASS seeing, MASS isoplanatic angle, nighttime air temperature and nighttime wind speed at 2 m are given in Fig. 2. The 2 m wind data are given here, as opposed to the 7 m data in Fig. 1, because a longer time series exists for them. Note that the interpretation of outlying points in this figure requires the consideration of the number of data points available for each month. A point could be above or below average simply because it only contains a few days of measurements that are not representative of average conditions. This information cannot be given here without obscuring the plots. It is available in the plots for the individual sites in the final TMT Site Testing Report, which will be made public in the future. Also note that not all the ordinates of the graphs start at zero.

None of the turbulence-related properties of the Chilean sites display a strong seasonal variability, while some pronounced seasonal activity exists for the North American sites. Weather parameters display some obvious seasonal activity, such as the temperature at Tolonchar, Mauna Kea 13N and San Pedro Mártir, and the wind at Armazones. A more detailed analysis of the monthly and seasonal behavior will be given in the TMT results paper series.

4. Summary and Conclusions

TMT has spent the last five years taking in situ data at five candidate sites, accumulating at least two annual cycles of data at each site. The main bottom line result of this effort is that all five candidate sites are excellent sites. It was also determined that not a single site is "perfect" or is the best in every characteristic. A few examples of this are:

- Chilean sites have better DIMM and ground layer seeing, while the North American sites have better high elevation seeing
- Armazones has the largest fraction of clear/usable nights, but is the site with the highest median wind speed

- Mauna Kea 13N has the largest isoplanatic angle and low precipitable water vapor, but a somewhat larger cloud fraction than the other sites

As a result of the TMT site testing work, the TMT Board of Directors down selected to one South American and one North American site in May 2008. These are Cerro Armazones and Mauna Kea 13N. This down select, as well as the final TMT site selection are based on a combination of measured site parameters from the testing program and programmatic criteria. The philosophy underlying this will be described in the TMT results paper series. Site testing reports and data will also be made public in the future.

Acknowledgments

The TMT site testing work could not have happened without the contributions of many other people at various institutions. A special thanks to everybody who has supported, and is still supporting, the TMT site testing and selection program, in particular the people at the Cerro Tololo Inter-American Observatory (CTIO), the National Optical Astronomy Observatory (NOAO) and the AURA New Initiatives Office (NIO), the Universidad Católica del Norte in Antofagasta, the Observatorio Astronómico Nacional de San Pedro Mártir, Gemini Observatory, Palomar Observatory, the Submillimeter Array (SMA), the University of Hawaii Institute for Astronomy (IfA), the University of Lethbridge, the University of Moscow, the South African Astronomical Observatory (SAAO), and the entire TMT project including its partner institutions and external reviewers.

The TMT Project gratefully acknowledges the support of the TMT partner institutions. They are the Association of Canadian Universities for Research in Astronomy (ACURA), the California Institute of Technology and the University of California. This work was supported as well by the Gordon and Betty Moore Foundation, the Canada Foundation for Innovation, the Ontario Ministry of Research and Innovation, the National Research Council of Canada, the Natural Sciences and Engineering Research Council of Canada, the British Columbia Knowledge Development Fund, the Association of Universities for Research in Astronomy (AURA) and the U.S. National Science Foundation.

References

1. R. Riddle et al., "TMT site selection survey: Instruments, methods and operations," these proceedings.

2. L. Wang et al., "High-accuracy differential image motion monitor measurements for the Thirty Meter Telescope site testing program," *App. Opt.* **46**, 6460 (2007).
3. S. Els et al., "Study on the precision of the multiaperture scintillation sensor turbulence profiler (MASS) employed in the site testing campaign for the Thirty Meter Telescope," *App. Opt.* **47**, 2610 (2008).

WAVEFRONT CHARACTERIZATION CAMPAIGN AT PARANAL USING GSM, MOSP, DIMM-MASS, LUSCI AND SCIDAR

WASSILA DALI ALI*, AZIZ ZIAD, AMOKRANE BERDJA, JEROME MAIRE and JULIEN BORGNINO

Laboratoire H. FIZEAU, OCA, CNRS UMR 6525, Université de Nice - Sophia Antipolis, Campus Valrose
F-06108 Nice Cedex 2, France
**E-mail: Wassila.DALI-ALI@unice.fr*

MARC SARAZIN, GIANLUCA LOMBARDI and JULIO NAVARRETE

European Southern Observatory, Karlschwarzschildstr.2, 85748 Garching, Germany
E-mail: msarazin@eso.org

HECTOR VAZQUEZ RAMIO, MARCOS REYES, JOSE MIGUEL DELGADO and JESUS J. FUENSALIDA

Instituto de Astrofisica de Canarias, Via Lactea s/n, 38200 La Laguna, Tenerife, Spain

ANDREI TOKOVININ and EDISON BUSTOS

Cerro Tololo Inter-American Observatory, Casilla 603 La Serena, Chile

In the frame of the EU funded ELT Design Study, a wavefront characterization campaign has been conducted at Paranal VLT Observatory in December 2007 involving concurrent measurements with DIMM, MASS, GSM, MOSP, LuSci and SCIDAR. The campaign goals and instrumentation are described and preliminary results are presented.

Keywords: Seeing; atmospheric turbulence.

1. Introduction

The objective of this campaign is to study the behavior of the wavefront perturbed by atmospheric turbulence at large scales in the perspective of the E-ELT. Thus, different instruments have been used to probe the wavefront at different scales: AMBER (VLTI), NACO (VLT) and seeing monitors (GSM, DIMM). It is also question of the links between the turbulence distribution in the atmosphere and the behavior of the wavefront at large scales. Therefore, different profilers (SCIDAR, MOSP and MASS) have been operating simultaneously with the other instruments during the campaign.

2. Instruments Participating to the 2007 Paranal Campaign

2.1. *Cute-SCIDAR: Scintillation Detection and Ranging*

The principle of generalized SCIDAR is based on the spatial correlation of the binary stars scintillation images. The CCD detector that records these scintillation patterns is placed in the virtual plane of analysis conjugated to some negative altitude making SCIDAR sensitive to turbulence at low altitudes. The upgraded version of the Cute-SCIDAR instrument for Paranal has been developed by IAC.[12] The instrument is remotely controlled and can monitor temporal evolution of turbulence profiles in real time with and without dome seeing contribution. It was mounted at the VLTI Auxiliary Telescope 4 (AT4).

2.2. *ESO-DIMM (Differential Image Motion Monitor)*

A DIMM is based on the statistical analysis of angle-of-arrival (AA) fluctuations of the wavefront. It was the main element of the instrumentation developed for the VLT site evaluation campaign (see Sarazin[10]). The principle of a DIMM is well known (e.g. Sarazin and Roddier[9]). The ESO DIMM uses a 35-cm telescope placed on a tower at 6 m above ground. It works in robotic mode at Paranal since 1998.

2.3. *GSM (Generalized Seeing Monitor)*

The Generalized Seeing Monitor measures atmospheric parameters that characterize the optical properties of wavefront: seeing, outer scale of turbulence, isoplanatic angle and coherence time. These parameters are deduced from the AA spatio-temporal statistics and from scintillation. The GSM instrument consists of several independent synchronized modules measuring AA fluctuations at different points of the wavefront with 10-cm telescopes. Usually, GSM consists of 4 separate modules, but exclusively for this campaign, 6 modules were used.[4] Two modules are placed on the same mount to measure the seeing like with the DIMM configuration. More details on the GSM instrument are provided by Ziad *et al.*[14]

2.4. *LuSci (Lunar Scintillometer)*

To separate the ground layer contribution in the seeing degradation, a new instrument, LuSci, was used in this campaign. It is based on the same principle as SHABAR (SHadow Band Ranging) which is used for the solar scintillation measurement. For LuSci, the Moon is used as extended source. Normalized scintillation covariance is measured over a set of detectors at fixed separations (baselines). Each baseline has a sensitivity to the turbulence in a given altitude range which depends on detector configuration (size, baseline separation and orientation) as well as on the lunar phase. The signal from all baseline is fitted to a continuous $C_n^2(z)$ profile model with 4 parameters, wich is then integrated to derive the ground-layer seeing ϵ_{GL} In this campaign, LuSci had four detectors spaced linearly from 3.5 cm to 38 cm. The instrument was almost totally automated.[7,11]

2.5. *MASS (Multi Aperture Scintillation Sensor)*

The MASS instrument measures the scintillation. The light from a single bright star is received by a system of 4 concentric apertures (the inner aperture A of 2 cm diameter, and the annular apertures B, C, and D surrounding it, with the outer diameter of D about 8 cm). The beam splitting between annular apertures is done by internal optics (Kornilov et al.[2]). The MASS data products are the turbulence profile model consisting of 6 discrete layers logarithmically spaced from 0.5 km to 16 km and quantities derived from it (free-atmosphere seeing, isoplanatic angle) and the estimate of the AO time constant.

2.6. *MOSP (Monitor of Outer Scale Profile)*

MOSP is the only instrument able to extract the outer scale profiles. The principle of the instrument is based on the measurement of AA fluctuations at different angular separations by observation of the lunar limb. Spatio-angular correlations of these fluctuations enables to measure the conditions of turbulence in different atmospheric layers. The AA fluctuations are measured perpendicularly to the lunar limb, leading to transverse AA correlation along the moon edge. The instrument is composed of a small telescope (29 cm diameter) having a large equivalent focal length (10 m), equipped with a PixelFly camera CCD and a Barlow lens. The exposure time was set to 1 ms or 2 ms (Maire *et al.*[5]).

2.7. *NAOS/AO (Nasmyth Adaptive Optics System)*

NAOS is an Adaptive Optics (AO) system installed in the Nasmyth B focus of the 4rd VLT Unit Telescope at the Paranal Observatory in Chile and has been operational since October 2002(Fusco *et al.*[1]). Developed by the ONERA group, this AO system allows the calculation of atmoshreric parameters (seeing, outer scale and coherence time) without opening the loop. This method is based on the reconstruction of the open-loop data from the DM and tip-tilt mirror voltages and the residual wavefront slopes.

2.8. *VLT Interferometer using AMBER (Astronomical Multi-BEam combineR)*

Interferometric measurements with large baselines are appropriate for the wavefront study at large scales. Indeed, measurements of the optical path length difference (OPD) between the two arms of an interferometer lead to the phase structure function which is characteristic of the wavefront. AMBER is a three-beam combiner working in the near-infrared (J, H and K bands), it features a set of monomode fiber into which the incoming light is injected in order to have a clean plane wavefront before combination. A set of collimated and mutually parallel beams are focused by a common optical element in a common Airy pattern that contains fringes with same principle than the GI2T interferometer see (Maire *et al.* and Petrov *et al.*[3,6]). The OPD fluctuations that appear as spatial fluctuations of the dispersed fringes, are related to the value of the outer scale (Roddier[8]). Simultaneous measurements of outer scale by AMBER, GSM and MOSP in this campaign allow the separation of the instrumental contribution from atmospheric one.

2.9. *Meteorological Station*

We collected data from the meteorological station, the tower located at the north side of the Paranal telescope platform where most of instruments were placed. The sampling intervals are 2 seconds for digital sensors (wind speed and direction) and one minute for analogic sensors (T, Rh, P). One minute averages are ingested into the database. The resolution of the various sensors is the following:

- Wind direction: 5.63° (threshold: 0.3 m/s).
- Wind speed: accuracy ~2m/s
- Temperature: accuracy ~0.1°.
- Humidity: linearity about 1%

3. Conditions of Campaign Observations

The conditions were adequate for this campaign, all nights were clear. The observations started at the expected date, i.e. 17th December 2007 and continued for 10 nights until 26th December with most instruments observing except for some short interrupts. A common list of targets was provided for the instruments observing single stars. All the instruments have been simultaneously in operation from the night of Dec. 19th to the 24th. We focus on this period to make cross-comparisons of data produced by the instruments. We summarize the operational status of all instruments in Figure 1.

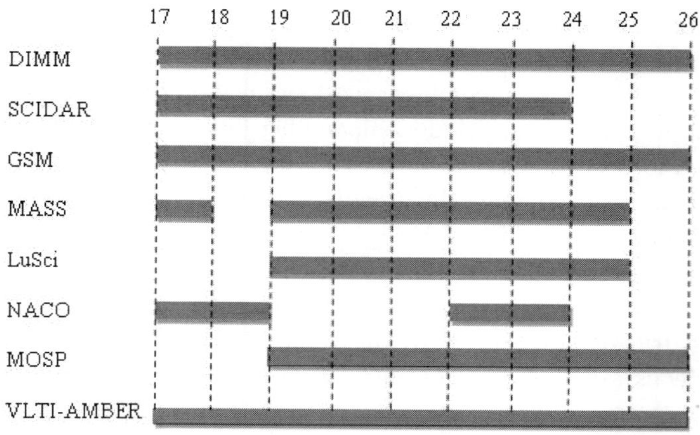

Fig. 1. December 2007 running period of the Paranal Campaign instruments.

The instruments GSM, MOSP, MASS and LuSci were located at around 1.5 m high from the ground level and were installed as shown in Figure 2 near the 6-m ESO-DIMM tower.

In Table 1 we present the parameters monitored by each instrument. In the next section, we make a cross-comparison of the same parameters measured by different instruments.

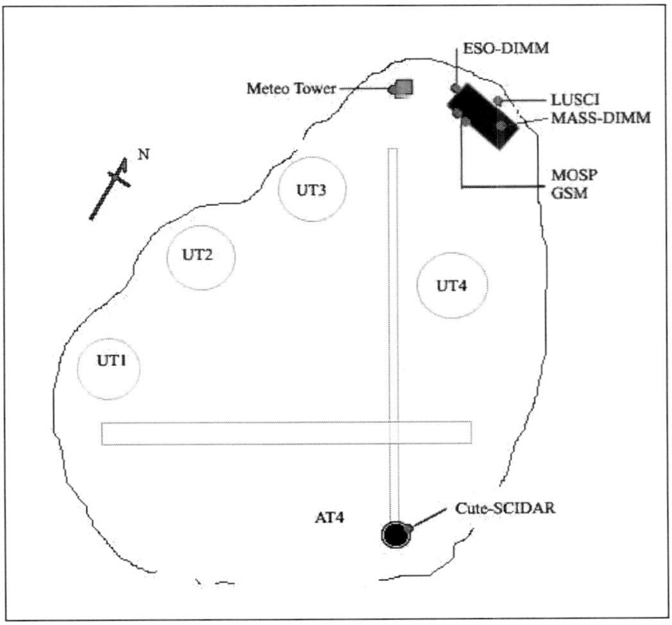

Fig. 2. Paranal astronomical site monitors were located at the Northern side of the VLT platform. GSM, MOSP, LuSci and MASS were placed near the ESO-DIMM tower. The Cute-SCIDAR was located in the south-eastern side of the platform at VLTI J2 station.

Table 1. Parameters measured by the instruments used during the Paranal campaign.

Instrument	Parameters				
Cute-SCIDAR	ϵ_0	$C_n^2(h)$	θ_0	–	–
ESO-DIMM	ϵ_0	–	θ_0	τ_0	–
GSM	ϵ_0	–	θ_0	τ_0	L_0
LuSci	ϵ_{GL}	$C_n^2(h)_{GL}$	–	–	–
MASS	ϵ_{FA}	$C_n^2(h)$	θ_0	τ_0	–
MOSP	ϵ_0	$C_n^2(h)$	–	–	$L_0(h)$
NAOS/AO	ϵ_0	–	–	τ_0	L_0
VLTI/AMBER	ϵ_0	–	–	–	L_0

where $C_n^2(h)$ is the refractive index structure constant at altitude h, L_0 the outer scale, ϵ_0 the seeing, θ_0 the isoplanatic angle and τ_0 is the AO coherence time.

4. Presentation of the Campaign Results

In next sections we present the result of instruments measurements except for AMBER and NACO for which the data are not yet available.

4.1. *Outer Scale Measurements*

To estimate the turbulence outer scale profile $L_0(h)$ during this campaign, we analyzed the wavefront AA fluctuations from image motion of the Moon limb obtained by means of the MOSP instrument. For the extraction of the height dependence of outer scale, we used a simulated annealing algorithm using $C_n^2(h)$ profiles measured simultaneously with Cute-SCIDAR. The profiles of $L_0(h)$ are shown in Figure 3, as we can see that the outer scale values in the surface layer are typically few meters.

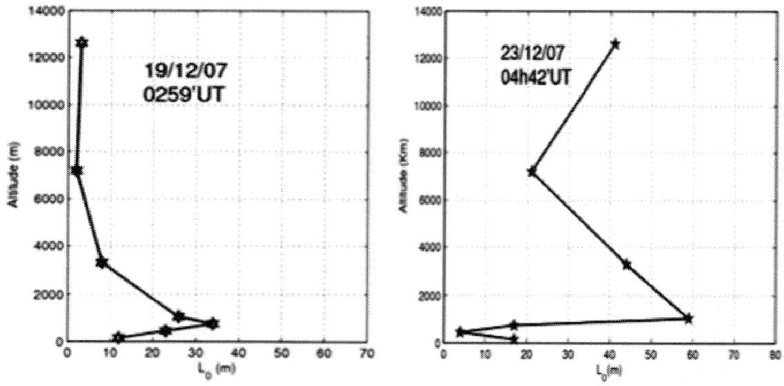

Fig. 3. Outer scale profiles extracted from MOSP data for two nights 19th and 23rd of December during the Paranal 2007 campaign

The calculation of the AA longitudinal covariance with GSM at different baselines allows the estimation of the outer scale. Note that during this campaign a new configuration of 15 baselines was used. We present the outer scale versus the Fried parameter values delivered by GSM in Figure 4. We do not observe any correlation between these parameters, as was found in the previous GSM campaigns at other sites as well.

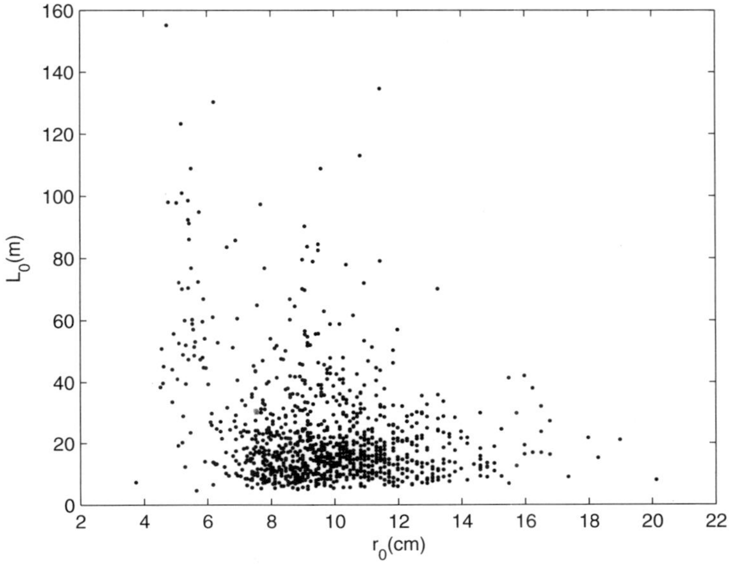

Fig. 4. Variation of outer scale versus the Fried parameter at $\lambda = 0.5\mu m$ measured by GSM during the Paranal campaign of Dec. 2007.

4.2. *Seeing Measurements*

Figure 5 shows the total seeing (including both free atmosphere and surface layer contributions) measured simultaneously by Cute-SCIDAR, GSM, and ESO-DIMM. We show also the surface layer contribution measured by LuSci that we compare with the seeing integrated on the first 300-m thick layer with Cute-SCIDAR.

4.3. *Coherence Time Measurements*

The AO coherence τ_0 can be estimated from the DIMM seeing using the wind at 200mb extracted from the 6-hourly ECMWF forecasts products,

$$\tau_{0,DIMM}^{-5/3} = \frac{0.314 \times r_{0,DIMM}}{MAX(wspeed_{30m}, 0.4 \times wspeed_{200mb})}. \tag{1}$$

For MASS, the $\tau_{0,MASS}$ measurement is based on the temporal analysis of the scintillation signal in the smallest aperture. It does not include the ground layer (GL) turbulence. A correction factor of 1.27 is applied to the

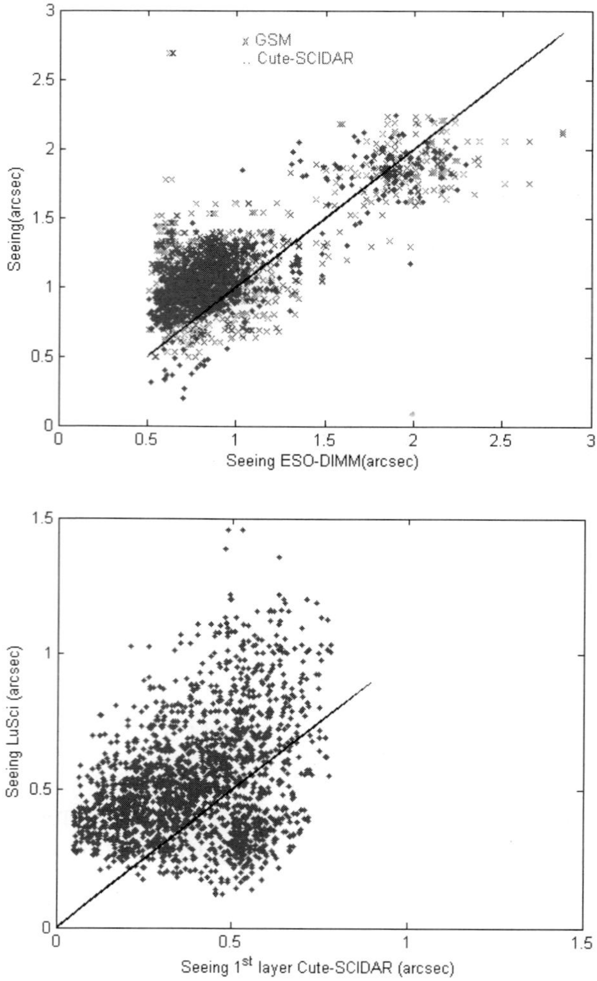

Fig. 5. Seeing cross-comparison between GSM, Cute-SCIDAR and ESO-DIMM (left) and comparison of seeing integrated on the first 300m thick layer of Cute-SCIDAR and the global seeing delivered by LuSci.

$\tau_{0,MASS}$ values delivered by the software[a]. The time constant is corrected for the GL contribution by using the DIMM data as $\tau_0^{-5/3} = \tau_{0,GL}^{-5/3} + \tau_{0,MASS}^{-5/3}$ with:

[a]http://www.ctio.noao.edu/~atokovin/profiler/timeconst.pdf.

$$\tau_{0,GL}^{-5/3} = \frac{0.314 \times r_{0,GL}}{wspeed_{30m}}. \tag{2}$$

and

$$r_{0,GL}^{-5/3} = r_{0,DIMM}^{-5/3} - r_{0,MASS}^{-5/3}. \tag{3}$$

For GSM, the method used for estimating the coherence time τ_{AA} consists of processing the AA temporal structure function $D_\alpha(r,\tau)$ where α indicates the AA measured with GSM with time difference τ. This temporal structure function saturates for large values of τ and the point at which the structure function reaches the $\frac{1}{e}$ of this maximum corresponds to the GSM τ_{AA}.[13]

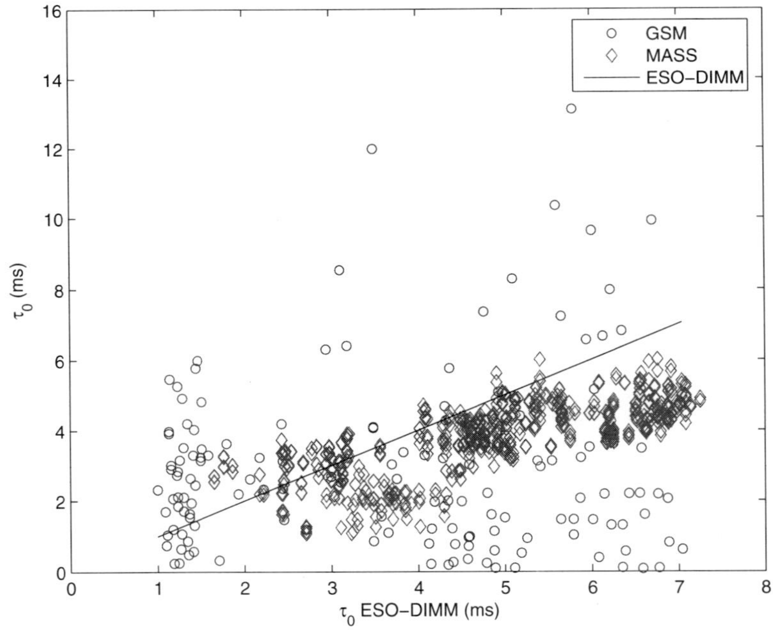

Fig. 6. Coherence time cross-comparison between values delivered by GSM and MASS with those estimated from the ESO-DIMM seeing and wind speed.

4.4. *Isoplanatic Angle Measurements*

The isoplanatic angle θ_0 is calculated from the turbulence profiles measured by Cute-SCIDAR and MASS. However, θ_0 can also be derived from the scintillation index in a 10-cm aperture which is measured with the GSM and the ESO-DIMM.[12] Then for the Paranal 2007 Campaign we have simultaneous estimations of θ_0 with 4 different instruments. We note that as coherence time, the isoplanatic angle from ESO-DIMM are an indirect estimates. The results presented in Figure 7 show that the median value given by Cute-SCIDAR is $\theta_0^{med} = 2.05''$. The regression line for each data set is plotted and dispersion about these lines were calculated then for ESO-DIMM we find $0.87''$, $0.70''$ for GSM and $0.51''$ for MASS.

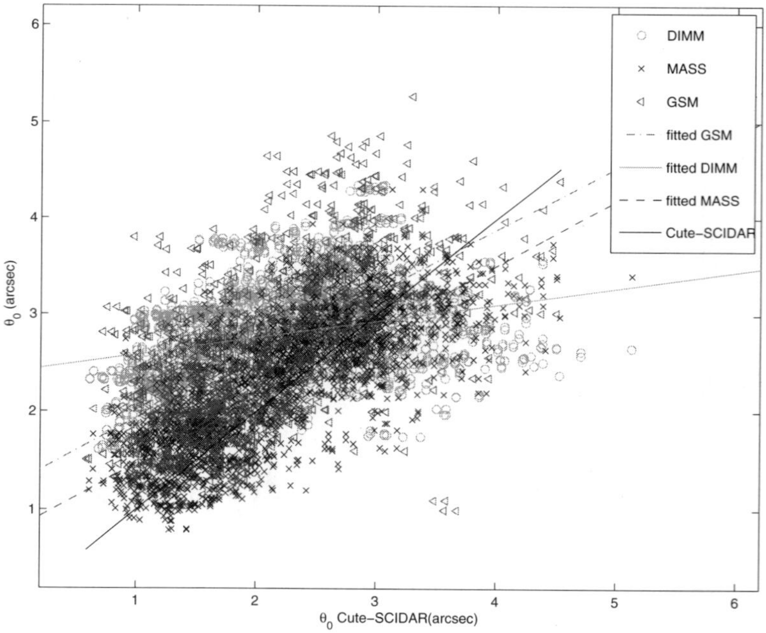

Fig. 7. Isoplanatic angle measured by 4 instruments. Here we plot the GSM, MASS and ESO-DIMM measurements compared with isoplanatic angle values delivered by Cute-SCIDAR.

5. Summary of the Paranal Campaign Results

We summarize in the Table 2 the measurements of the parameters obtained for all instruments during the campaign.

Table 2. Summary of the campaign results: for each instrument, we give min-median-max value of the measured parameter over all the nights of observation.

Instrument	$\epsilon_0(arcsec)$	$\theta_0(arcsec)$	$\tau_0(ms)$	$L_0(m)$
Cute-SCIDAR	0.20-1.01-2.24	0.59-2.01-5.13	–	–
ESO-DIMM	0.37-0.81-2.84	1.37-2.77-4.35	0.88-4.11-7.36	–
GSM	0.50-1.05-2.69	0.83-2.45-5.27	0.03-2.24-19.30	4.6-17.1-
-				-174.3
LuSci	0.12-0.43-2.23	–	–	–
MASS	0.13-0.33-1.27	0.79-2.24-4.31	0.86-6.61-49.74	–
meteo tower	$Temp.30m(^oC)$	$Temp.02m(^oC)$	$windspeed30m(m/s)$	
-	8.96- 11.64- 13.24	8.09 - 11.49- 13.28	0.33- 7.46- 14.03	

6. Conclusion

The preliminary results presented in this paper demonstrate that various instruments based on different physical principles can be used to characterize the wavefront properties. The collected data will allow to perform a detailed analysis of the turbulence in the surface layer as well as to obtain the first statistics of the vertical profiles of the outer scale of the turbulence at Paranal.

Acknowledgments

The GSM-MOSP campaign and the Cute-SCIDAR were funded by the European Union in the frame of the 6th Framework Programm for Research and Technology Development under contract number 011863.

References

1. T. Fusco et al., *SPIE.* **5490**, 118 (2004).
2. V. G. Kornilov et al., *Proc. SPIE*, **4839**, 837 (2003).
3. J. Maire, A. Ziad, J. Borgnino and F. Martin, *MNRAS.* **377**, 1236 (2007).
4. J. Maire, A. Ziad, J. Borgnino and F. Martin, *MNRAS.* **386**, 1064 (2008).
5. J. Maire et al., *A.A.* **448**, 1225 (2006).
6. R. Petrov et al., *A.A.* **464**, 29 (2007).
7. J. Rajagopal et al., *SPIE.* **7013**, 60 (2008).
8. F. Roddier, *Progress in optics.* **19**, 281 (1981).
9. M. Sarazin and F. Roddier, *A.A.* **227**, 294 (1990).
10. M. Sarazin, *SPIE.* **628**, 138 (1986).
11. A. Tokovinin, *RMXAA.* **31**, 61 (2007).
12. H. V. Ramio et al. *SPIE.* **7012**, 701245 (2008).
13. A. Ziad et al., *Appl. Opt.* **39**, 5415 (2000).
14. A. Ziad et al., *Appl. Opt.* **43**, 2316 (2004).

TURBULENCE OUTER SCALE FOR HIGH ANGULAR RESOLUTION TECHNIQUES

A. ZIAD*, J. BORGNINO, J. MAIRE, A. BERDJA and F. MARTIN

UMR 6525 H. Fizeau, Université de Nice Sophia-Antipolis,
CNRS, Observatoire de la Côte d'Azur,
Parc Valrose 06108, Nice, France
**E-mail: ziad@unice.fr*

Atmospheric turbulence outer scale is a relevant parameter for High Angular Resolution techniques as Adaptive Optics and Interferometry. This paper reviews in a non-exhaustif way different techniques and instruments for the measurement of the wavefront coherence outer scale which should be distinguished from the in-situ geophysical parameter.

Keywords: Outer scale; Atmospheric turbulence; Adaptive Optics; Interferometry.

1. Introduction

Wavefront outer scale is a relevant parameter for the experimental performance evaluation of large aperture telescopes. The actual size of the outer scale has long been controversial, with measured values ranging from less than 10 m to more than 2 km.[1] What is not controversial is the conclusion that when the diameter of the telescope approaches or exceeds the size of the outer scale, the optical consequences of atmospheric turbulence are changed dramatically from their traditional Kolmogorov behavior. In particular, power in the lowest Zernike aberration modes, e.g., tip and tilt and the overall stroke required for an adaptive-optics system can be much reduced. A finite outer scale has implications for interferometry as well. With the current interest in the design of extremely large ground-based optical and infrared telescopes, reliable estimates of the outer scale profile have assumed considerable importance.

Different techniques have been used for the outer scale estimation. In situ measurements have given very small local outer scales (see Table 1 in Avila et al.[1]). The outer scale has also been estimated directly at the ground from the wave-front analysis from HAR techniques using interferometric[2,3] or Shack-Hartmann or more generally AO systems data.[1,4,5] Dedicated instruments have been also developed for the outer scale monitoring which are the Generalized Seeing Monitor (GSM)[6] and the Monitor of Outer Scale Profile (MOSP).[7,8] The measured values of outer scale from HAR techniques,[2,3] GSM[6] and MOSP[7,8] are somewhat coherent and are larger than the in-situ results. The main explanation of this difference comes from the definition of outer scale itself. In section 2, a discussion on this aspect is presented. Section 3 is dedicated to the description of GSM and MOSP, the dedicated instruments of outer scale measurement. In the same section, some results of observations and comparisons between the different techniques are presented and discussed.

2. Which Outer Scale for HAR Techniques?

For a long time the size of the outer scale has been controversial due to a huge dispersion ranging from few meters to 2km (see Table 1 in Avila et al.[1]). This dispersion of the outer scale values is mainly due to the definition of this parameter. Indeed, the in-situ measurements of the outer scale are based on the Tatarski definition relative to temperature measurements while the wavefront coherence outer scale is deduced from the structure function of the refractive index. Indeed, Tatarski[9] introduced the mixing scale Δz_0 as the height at which the equality between the temperature systematic difference and the random one is reached:

$$[\langle T(z)\rangle - \langle T(z+\Delta z_0)\rangle]^2 = \langle [T(z) - T(z+\Delta z_0)]^2\rangle \tag{1}$$

The height Δz_0 is related to the atmospheric turbulence geophysical outer scale as:

$$L_0 \simeq \frac{\Delta z_0}{\alpha^{3/4}} \tag{2}$$

where α is the ratio between the viscosity coefficient and the thermal diffusivity coefficient.

Using the temperature structure function, one may deduce the expression relating the temperature structure constante C_T^2 and the temperature gradient:

$$C_T^2 \simeq \alpha L_0^{4/3} (\frac{dT}{dz})^2 \tag{3}$$

In the same way the refractive index structure constant C_N^2 and the potential temperature θ gradient are related as:

$$C_N^2 \simeq \alpha' L_0^{4/3} (\frac{d\theta}{dz})^2 \tag{4}$$

where $\alpha' = 0.36$. This last equation is used to estimate the outer scale from in-situ soundings.

In the other hand the refractive index structure function obtained in the case of Kolmogorov model has a 2/3 power law of the baseline.[9] This expression is only valid in the inertial range for baselines small than the outer scale. But for the von Kàrmàn model, the refractive index structure function is given by:[9]

$$D_N(r) = \frac{2\Gamma(7/6)}{\sqrt{\pi}} C_N^2 \mathcal{L}_0{}^{2/3} [1 - \frac{2^{2/3}}{\Gamma(1/3)} (\frac{r}{\mathcal{L}_0})^{1/3} K_{1/3}(\frac{r}{\mathcal{L}_0})] \tag{5}$$

Where $K_{1/3}$ is the modified Bessel function. This expression is separated into 2 terms: an energy term related to the C_N^2 and the second term is a distortion function which depends on the outer scale $\mathcal{L}_0$ for large baselines. More exactly the departure of the saturation correspond to that we call the wavefront coherence outer scale $\mathcal{L}_0$.[10,11] Now, an obvious question to ask: Is this parameter the relevant outer scale for HAR techniques?

The best way to answer this question, is to carry out direct measurements from HAR techniques to highlight the saturation of the structure function which is characteristic of $\mathcal{L}_0$. This is possible by using an interferometer to measure the phase φ structure function D_φ at different baselines B. Indeed, the fringe excursion of an interferometer (OPD) is directly related to phase structure function as:

$$\sigma_{OPD} = \frac{\lambda}{2\pi} \sqrt{\langle [\varphi(\vec{r}) - \varphi(\vec{r} + \vec{B})]^2 \rangle} = \frac{\lambda}{2\pi} \sqrt{D_\varphi(B)} \tag{6}$$

where the phase structure function is given by:

$$D_\varphi(B) = 4\pi \int df\, f W_\varphi(f) [1 - J_0(2\pi B f)] [\frac{2J_1(\pi D f)}{\pi D f}]^2 \tag{7}$$

where f is the modulus of the spatial frequency, D the pupil of diameter D and W_φ is the phase spectrum which in the case of the von Kàrmàn model shows explicitly the outer scale as:

$$W_\varphi(f) = 0.0229 r_0^{-5/3}(f^2 + \frac{1}{\mathcal{L}_0^2})^{-11/6} \quad (8)$$

When $\mathcal{L}_0$ tends to infinity we retrieve the Kolmogorov model.

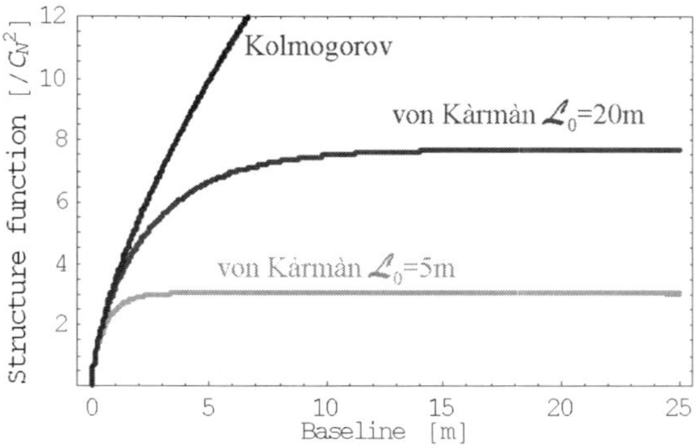

Fig. 1. (color on-line) Refractive index structure function for infinite outer scale (blue) and for $\mathcal{L}_0 = 5m$ and $\mathcal{L}_0 = 20m$ (respectively in green and red) in the case of von Kàrmàn model.

The pertinence of outer scale for HAR techniques will be discussed further in Section 3.

3. Instruments for Wavefront Outer Scale $\mathcal{L}_0$ Estimation

The GSM is the only instrument dedicated to this outer scale measurement. Developped by our team TOMI [a] at Fizeau Laboratory, since 1997 this instrument visited the major sites over the world (Paranal, La Silla, Pachon, Mauna Kea, La Palma, Dome C, San Pedro, Palomar, Maydanak...)[6]

The GSM consists of evaluating the optical parameters of the perturbed wavefront by measuring AA fluctuations. Indeed, the GSM uses the same principle than a Shack-Hartmann, i.e., measuring AA at different points of the wavefront. Computing spatio-temporal correlations of these measured AA leads to estimates of the seeing ϵ_0, outer scale $\mathcal{L}_0$, isoplanatic angle θ_0 and coherence time τ_0. Outer scale values provided by the GSM are model-dependent deduced in the case of the von Kàrmàn model.

The instrument consists of different identical units equipped with 10cm telescopes installed on equatorial mounts and pointing at the same star.

[a]Turbulence Optique Modelisation & Instrumentation: http://fizeau-tomi.unice.fr/

Each telescope measures the AA fluctuations by mean of flux modulation which is produced by the displacement of the observed star image over a Ronchi grating. A detailed description of the GSM instrument is given by Ziad et al.[6]

Fig. 2 shows the global histograms and the cumulative distributions of the outer scale measured at Paranal Observatory in Dec. 1998[6] and at Dome C site from March to May 2006.[12] One can see that the outer scale has a log-normal distribution. With $\sim 7.5m$ median outer scale, Dome C presents an outer scale of $\sim$ 3 times lower than Paranal. The small value of $\mathcal{L}_0$ at Dome C is due to a prevailing surface layer of about $\sim 30m$ height.

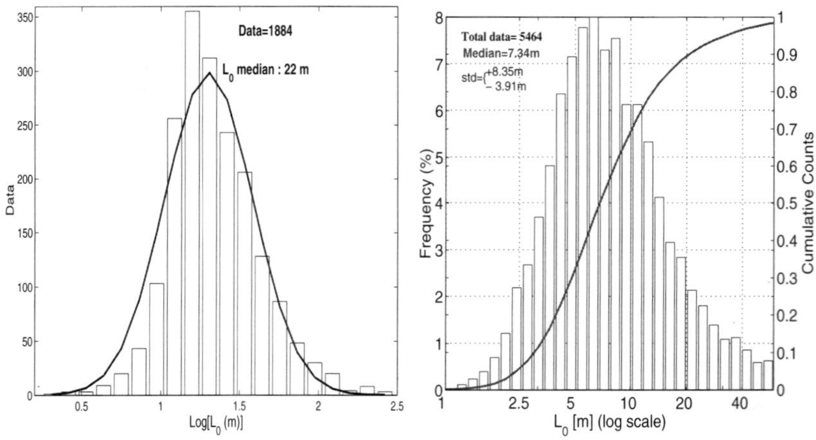

Fig. 2. Histograms of outer scale measurements with GSM at Paranal (Dec. 1998) and at Dome C (Mars to May 2006).

The GSM Outer scale measurements have been confirmed by processing data obtained with existent High Angular Resolution instruments like optical interferometers and adaptive optics systems.[2,4] Indeed, in 2003 and 2004 we organized at Calern Observatory in south of France simultaneous observations between the GI2T interferometer and the GSM.[3] An example of the results obtained during these observations are shown in Fig. 3. In blue we have the mean value of the OPD measured with the GI2T interferometer for 3 different baselines: 12, 14 and 32m. The error bars correspond to the standard deviation of measurements. As can be remarked, the departure from the Kolmogorov model is clear and significantly pronounced for large baselines. Using extreme values of Fried parameter and median outer scale measured by GSM simultaneously in Eq. 6, we plotted in Fig. 3 the OPD versus the baseline in the case of Kolmogorov and von Kàrmàn models. An excellent agreement is obtained for von Kàrmàn model which justifies the pertinence of this model and therefore the $\mathcal{L}_0$ measured by GSM.

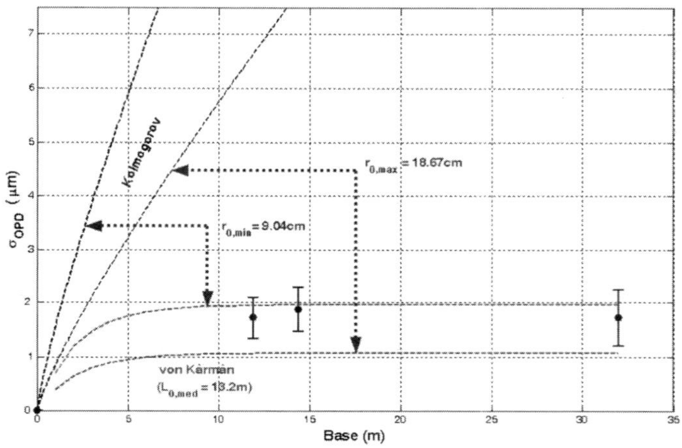

Fig. 3. Comparison of the phase structure function or σ_{OPD} (Eq. 6) measured by the GI2T interferometer and modeled from simultaneous GSM measurements using Eqs 6, 7 and 8 in the case of Kolmogorov and von Kàrmàn models.

In the same way, other results have been obtained with the Palomar Testbed Interferometer (PTI) always for simultaneous measurements with the GSM at the Palomar Observatory on September 2001.[2] This time the results concern the fringe jitter. The results of these observations are shown in Fig. 4. In red we have the jitter measured directly with the PTI at 2.2micron for 110m baseline. And in green we have a modeling of this jitter using GSM data (seeing and outer scale). Once again an excellent agreement is obtained for von Kàrmàn model which justifies the validity of this model and therefore the $\mathcal{L}_0$ measured by GSM.

The multi-conjugated AO systems specifications require the knowledge of atmospheric turbulence profiles including outer scale one. For this reason, a new instrument MOSP has been developed by our team for the measurement of the outer scale profile.[7,8] We retrieve the vertical distribution of wavefront outer scale by analyzing angular correlation of wavefront AA fluctuations deduced from Moon's limb image motion. We use simulated annealing algorithm to deduce the height dependence of the wavefront outer scale with given C_N^2 profiles simultaneously measured with a profiler as

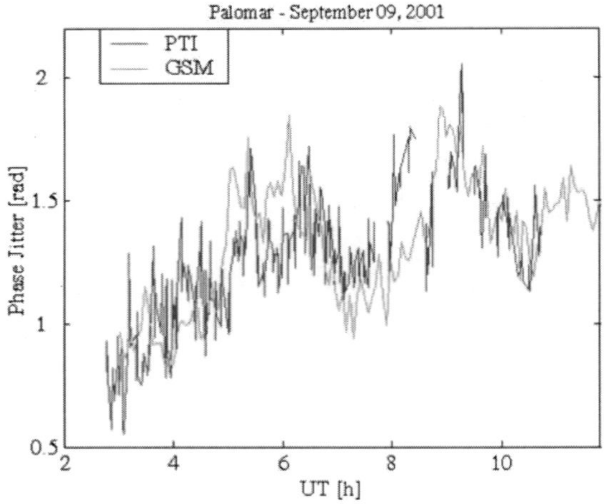

Fig. 4. (color on-line) Comparison of fringe jitter measured with the PTI interferometer (red line) and modeled using the GSM data (green line). These data have been obtained during the night of September 9, 2001 at the Palomar Observatory.

the SCIDAR instrument.[7] If C_N^2 profiles are not available, it is possible to extract simultaneously both of outer scale and C_N^2 profiles. We present in Fig. 5, results obtained during observations at the Mauna Kea Observatory in July 2005. A good agreement of C_N^2 profile is obtained between MOSP and G-Scidar. One can remark that the resolution is similar in the ground layer between the two instruments. Estimated outer scale profiles from the MOSP exhibit smaller values in the boundary layer than in the free atmosphere. Comparisons with GSM outer scale measurements are possible and give good agreement.[7]

4. Conclusions

The wavefront outer scale $\mathcal{L}_0$ relevant for HAR techniques as Adaptive Optics and Interferometry has to be distinguished from in situ geophysical outer scale. As described in section 2 this difference has its origin in the definition of the outer scale. Two dedicated instruments are now available for wavefront outer scale $\mathcal{L}_0$ measurements which are GSM and MOSP leading respectively to the $\mathcal{L}_0$ estimation at ground level and to the $\mathcal{L}_0$ profile. We have measured the outer scale $\mathcal{L}_0$ simultaneously using the GSM and HAR techniques. The results lie mainly in the same range and are consistent among the various methods, despite the fact that the spatial scales of the respective instruments themselves span two orders of magnitude. These

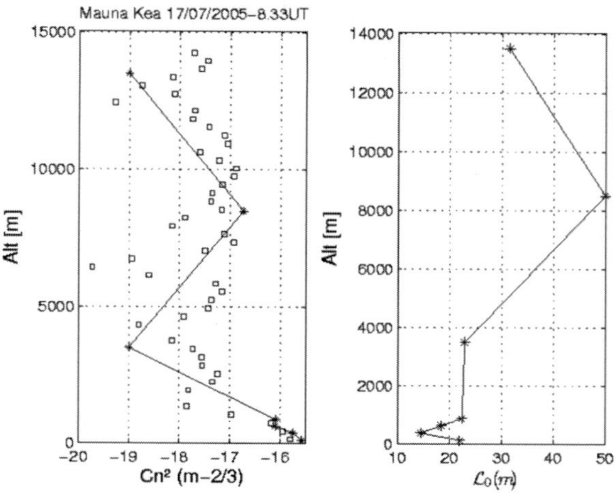

Fig. 5. Example of C_N^2 and outer scale profiles (solid line) that can be retrieved simultaneously using the simulated annealing method. Comparisons with simultaneous SCIDAR profile (square markers) are in close agreement despite different vertical resolutions of these instruments.

results are a conclusive proof for the validity of the outer scale measured by GSM and MOSP instruments. The $\mathcal{L}_0$ measurements at all major sites are generally small[6,12] and then it will be essential to take outer scale effects into account when predicting the optical performance of extremely large optical and infrared telescopes.

References

1. R. Avila et al., *J. Opt. Soc. Am. A* **14**, 3070 (1997).
2. A. Ziad et al., *Appl. Opt.* **43**, 2316 (2004).
3. J. Maire et al., *A&A* **448**, 1225 (2006).
4. M. Schöck et al., *Appl. Opt.* **42**, 3705 (2003).
5. T. Fusco et al., *J. Opt. A: Pure Appl. Opt.* **6**, 585 (2004).
6. A. Ziad et al., *Appl. Opt.* **39**, 5415 (2000).
7. J. Maire et al., *Mon. Not. R. Astron. Soc.* **377**, 1236 (2007).
8. A. Ziad, J. Maire and J. Borgnino, *Proc. of "Symposium on Seeing"* held in Kona, Hawaii, 20-22 March 2007.
9. V. I. Tatarskii, *Israel Program for Scientific Translations*, 73 (1971).
10. J. Borgnino, *Appl. Opt.* **29**, 1863 (1990).
11. J. Borgnino, F. Martin, A. Ziad, *Optics Communications* **91**, 267 (1992).
12. A. Ziad et al., *A&A* **491**, 917 (2008).

RETRIEVING HIGH LAYER ATMOSPHERIC TURBULENCE STATISTICS ON E-ELT SCALES

C. ARCIDIACONO*, R. RAGAZZONI, J. FARINATO, G. GENTILE, A. BARUFFOLO, M. DIMA, C. METTI and V. VIOTTO

Osservatorio Astronomico di Padova, INAF, Padova, Italy
**E-mail: carmelo.arcidiacono@oapd.inaf.it*

E. DIOLAITI

Osservatorio Astronomico di Bologna, INAF, Bologna, Italy

In the framework of the Extremely Large Telescope design study, the Work Package (WP) 12000 is studying the Site Characterization for an European Extremely Large Telescope. In particular, INAF is in the WP 12300 group for the Large scale atmospheric properties study. Previous studies done in many astronomical sites have been optimized on spatial scales comparable with 3m-4m to 10m class telescopes. The strong interest of the Astronomical Community in giant telescopes imposes a different site characterization opportune for 30-40m class telescopes. One of the central point in the Adaptive Optics for Extremely Large Telescopes is given from the achievable sky coverage. Generally speaking, sky coverage is dominated by the high altitude layers correction. In other words ground layer adaptive optics has a sky coverage much larger than other kind of corrections. That means that ways to improve the sky coverage in the sensing of high altitude layers can be very effective in terms of overall performances. Moreover, there are good reasons to translate high coherence time of flowing layers, in a generalized Taylor assumption, into larger sky coverages. This paper presents the optical design of TOE, The Onduline Experiment, a WaveFront Sensor for sensing a Very Large Field of View on-board the VLT and possibly other telescopes as GranTeCan in Canary islands. Such a WFS is to be intended as a tool to probe the atmospheric parameters in the free atmosphere (i.e. far from the ground layer) on a linear scale of the same order of magnitude of the diameter of the ELTs currently in the design phase.

Keywords: Site Characterization: Wavefront sensing; Adaptive Optics; Extremely Large Telescope.

1. Introduction

In this paper we outline the opto-mechanical design of TOE, acronym for The Onduline Experiment, a WaveFront Sensor (WFS) for sensing optical turbulence on Very Large Field of View (FoV) on-board the VLT and possibly other 8m-class telescopes. Such a WFS is to be intended, in the framework of the WorkPackage 12300 in the ELT-Design Study, as a tool to probe the atmospheric parameters in the free atmosphere (i.e. far from the ground layer) on a spatial scale of the same order of the diameter of the ELTs[1,2]. In the following such a baseline length is assumed to be D=42m. The experiment is designed to probe the atmospheric turbulence at altitudes within 5 and 15 km from ground. A new concept wavefront sensor will measure the first derivative of the phase delay on linear scale of about 40m thanks to a large FoV optical system. Such a system is intended to measure the signal generated by a large number of reference stars on field of 7arcmin Field of View (for example on a bright open cluster) on a single observation. Frame rates larger than 10Hz offer the possibility to build first derivative temporal series on ELT diameter scale using a single VLT telescope. Such a frequency has to be larger than the typical crossing frequency of r_0 patch size on a 40 m scale. The instantaneous first derivative map will be used a posteriori to compute an average power spectrum (PS): the goal is to verify if the average PS slope follows Kolmogorov $k^{-11/3}$ power law (k is the frequency) and to identify possible deviation from this law, especially on very low spatial frequencies (corresponding to the ELT ones): this may have a large impact on the ELT adaptive optics (AO) design defining a reference value for the maximum atmospheric stroke expected. But most interesting results will come from the temporal series analysis, which are intended to test Taylor hypothesis: measuring coherence time of wavefront aberration structures, of different size, crossing the sensor Field of View it is possible to build a coherence time versus spatial frequency function. This function will be used to define the improvement on the limiting magnitude for adaptive optics reference stars generated by the possibility to average for different frames the optical aberration, which is crossing over the telescope aperture.

2. Wavefront Sensor Concept

The basic idea is to place on the focal plane an optical window with a transmissivity that varies sinusoidally along one axis between 0% (Intensity $= I = I_0\,(1 - \sin\phi)\,/2$, with $\phi = \pi/2$) and 100%($\phi = 3/2\pi$, ϕ being the phase value) and a detector on a subsequent re-imaged pupil plane, in order to reconstruct the global and local tilt of the WF in only one direction. Let's consider the light from a single star, in absence of atmosphere: the effect on the image of the pupil (which is uniformly illuminated, because the light from the star is collimated) of the presence of the sinusoidal pattern is shown in Fig. 1. If the light of the star focuses where the transmissivity of the pattern is minimum (0%), the pupil will be dark, but moving the pattern along the axis orthogonal to the sinusoidal modulation the luminosity of the pupil varies uniformly as a function of the point of the sinusoid where the image of the star lies. The sinusoidal pattern intensity modulation allows us to detect a signal which is proportional to the movement of the object on the focal plane (tilt). Let's now introduce a atmospheric originated tilt

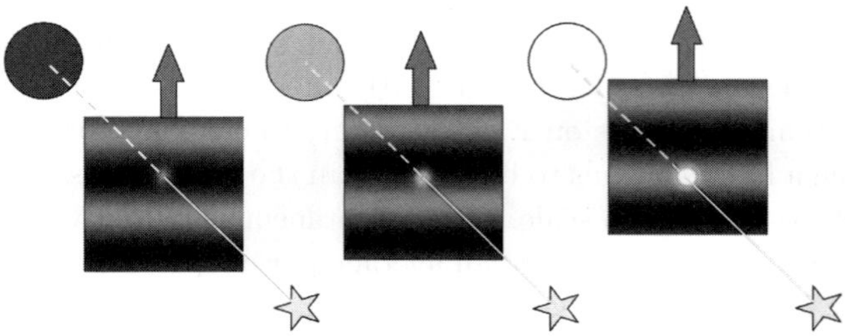

Fig. 1. Focal plane image is refracted by the sinusoidal pattern. Moving it along arrow direction, the light incidence point will change and the pupil illumination, represented by the circles, will change accordingly. In absence of aberration, one can measure the position of the barycenter of the star on the focal plane as a function of the intensity of the re-imaged pupil.

on the WF. The effect is proportional to the movement of the star image on the focal plane. So, if the sinusoidal pattern is fixed, the result is the same we've just described for a moving pattern in absence of atmosphere. For a given tilt, if the image of the star forms where the spatial derivative of the transmissivity of the pattern is maximum (where the phase of the sine is $\pm\pi$ and the transmissivity is then 50%), then the detected signal for tilt will be maximum too. In order to obtain the tilt in both axis, one can divide the light from the star in two parts (e.g. using a beam splitter) and analyze them by the introduction of two orthogonal sinusoidal patterns. As the detector is placed in the pupil plane, it is possible to divide it many sub-apertures where, for the purpose of the WF sensing, we are interested solely in tip-tilt. Each sub-pupil behaves exactly like the pupil in the previous example, so a time-variable tilt of the WF in the sub-pupil will produce the same modulation in intensity predicted for the entire pupil. So, in order to detect the WF high-orders due to the atmosphere, one can analyze the pupil illumination after the sinusoidal pattern. The signal one can obtain from each sub-pupil will be proportional to the first derivative of the WF, which is the local tilt, on that sub-aperture. In other terms, the WFS gives a map of the local tip-tilt in each sub-aperture. If the star is on the point of maximum variation of the transparency with the tip-tilt movement, or close to it, the gain will be maximized to detect the local tip-tilt. So, in the case of a single reference star, one will try to place the sinusoidal pattern in order to have the star in the central position between the minimum and the maximum of transmissivity, to maximize the output signal. An intuitive way to describe the response can be expressed using vectors: on a two dimensional plane the signal intensity is given by the sum of two vectors both with modules equal to the residual intensity after the transmissivity pattern, but one fixed along the y axis, the second with xy-projections respectively proportional to $\cos\phi$ and $\sin\phi$. The step of the sinusoidal pattern should be enough large to avoid wrapping around of this information and not so large that the signal becomes too weak. As a rule of thumb the rms of the tip-tilt fluctuation should be of the same order of one radian phase: the step of the pattern will be, at least, of the order of 2π times the expected seeing. What happens if a large number of stars is now in place? Each of these stars will contribute to the formation of the pupil image and each of it will be modulated by local tip-tilt in the same fashion. Because of the sinusoidal transparency of the pattern, the amount of light from a star that reaches the detector is the sum of two above described vectors. While the half of the light of the whole set of stars will permanently reach the detector the other half is piled up, generally speaking, in a random-walk summation way. The sensitivity of the WFS is equivalent to the average of the star in the FoV (however with a depth of focus given by the field of view where all the stars are collected, provided a large enough number of collected stars is involved). The signals originated by different reference stars are optically superimposed in a Layer Oriented fashion[3], in fact if the CCD is exactly conjugated to the pupil plane then all the stars pupils overlap making the WFS sensitive to ground layer turbulence, while slightly defocusing the CCD the pupil images corresponding to different stars move, replicating the pupils super-imposition geometry such as it is actually realized at different altitude conjugating the sensor at the desired plane.

3. Laboratory Experiment

We realized an optical setup to test the wavefront sensing ability of such as WFS. The first part of the optical setup foresees a system with a plate holding some optical fibers to simulate the stars over a 2' FoV. The divergent beam coming from the fibers is then made parallel by a lens, to simulate the light coming from infinity. After the lens, a stop is simulating the telescope entrance pupil and, just before it, turbulent screens can be positioned (using a custom made holder) simulating the atmospheric layers ranging from a few hundred meters to about 15Km. A lens positioned after the stop is making the beam converging into a focal plane, where the sinusoidal pattern plate can be inserted on a linear stage. A commercial objective is finally re-imaging the telescope entrance pupil on a CCD camera, see Fig. 2. Using one single fiber, in order to obtain the largest signal

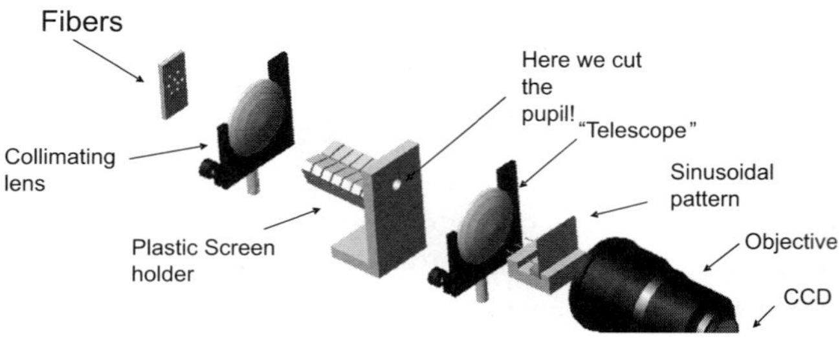

Fig. 2. The laboratory experiment arrangement.

and the most linear response we placed the sinusoidal pattern in order to project the star focal image on the sine flexure point ($\phi = \pi/2$), where the transmission spatial derivative is maximized. The task of the experiment is to show that we can measure a turbulent wavefront (created by positioning one turbulent screen in the dedicated holder). To accomplish such a task, we repeated several measurements of the same turbulent screen shifting it along an axis orthogonal to the optical one, moving it of known step, simulating somehow an evolution of the atmosphere. At the end through cross correlation we were able to identify the actual step applied and with very high correlation value (of the order of 80%).

4. Simulations

Numerical simulations analysis had two different aims: to verify the wavefront sensing concept and to check the ability of the system, such as it has been configured, to sense atmospheric layers spectral behavior. Atmospheric turbulence has been simulated by a phase screen generated using $k^{-11/3}$ power spectrum, the telescope and cameras have been simulated in a purely diffractive mode such as Fourier transformation of the phase screen masked by the obstructed pupil. For each reference the complex array corresponding to focal plane image has been multiplied by an array representing the sinusoidal transmissive pattern (transmissivity from 0 to 1) and Fourier transformed back in order to obtain intensity modulation on pupil image. Finally the different reference pupil images are added together. A strong relation between sensitivity and the position of the reference stars focal plane image with respect to sine phase of the pattern came out from simulation, see Fig. 3. According to Layer Oriented scheme the super-imposition of the pupils depends on stars field positions and conjugation altitude: first derivative measurements of the conjugated layer are correctly super-imposed on the CCD, but with different amplification (positive or negative depending on the corresponding transmissivity spatial derivative) and weighted for the reference stars intensity. This kind of simulation take into account the diffraction effect "grating-like" due to the transmissivity pattern: such as effect being evident on small sine wavelength (again see Fig. 3) leading to a minimum $\approx 10"$ value. The need for ground layer subtraction from the high layer signal came out from a detailed numerical analysis of the system using a real multi reference approach with real stars positions and magnitudes (NGC2215, on a FoV of 7'). In fact since ground is the strongest layer it will introduce on the wavefront measurement large phase errors which are super-imposed to the high layer corresponding signal. This test repeated for 500 atmosphere realizations evidenced the capability to measure correctly the injected turbulence power spectrum with average correlation between simulated measurements and simulated phase screens of the order of 70%.

5. System Overview

The instrument has been designed to be mounted at Nasmyth adapter rotator of VLT. Anyway, it preserves an intrinsic versatility to interface it with other 8-meter class telescopes: from the mechanical point of view, changing the connecting flange; from the optical point of view, it depends of course from the F/number of the considered focal plane; if the F/number is simi-

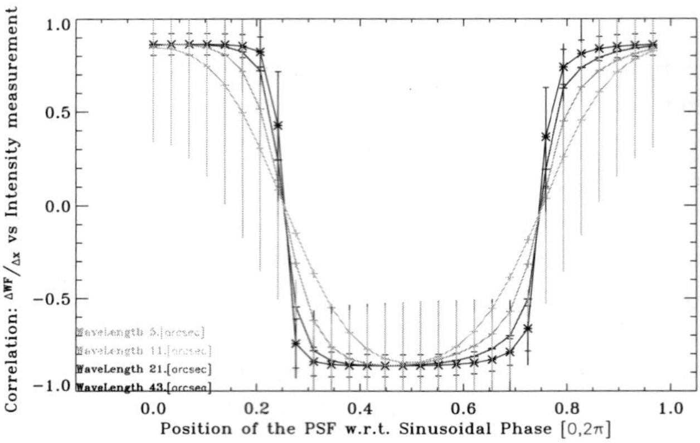

Fig. 3. This plot presents the correlation value between first derivative of the original phase screen and simulated signal as function of the position on the sinusoidal pattern, ϕ. The function has been computed for 4 different sinusoidal wavelengths (5", 11", 21" and 43"). We used 1" seeing phase screen, one reference star, averaging over 20 atmospheric layer realizations. This plot shows the relation signal and transmissivity spatial derivative: in fact $\phi = 0$ and $\phi = \pi$ the signal and the first derivative of the WF are respectively very well correlated (+1) and anti-correlated (-1) according to the sign of the derivative.

lar to the F/15 VLT one, the retrofit is almost straight forward, otherwise additional optical components shall be considered. We decided to split the FoV in four parts at the Nasmyth focal plane level, using a system based on smaller optics. Thus, there are four commercial photographic objectives just after the VLT Nasmyth Focal plane, dividing a 7'FoV into four 3' subfields. The instrument is sensitive to the WF derivative along one axis only, placing four parallel sinusoidal pattern plates exactly on the Nasmyth focal plane and just before the four objectives. These patterns can be moved using linear stages along the direction orthogonal to the lines of the grating, in order to maximize the instrument sensitivity. Each objective produces a pupil image on a CCD camera (in total four), which can be moved along a direction parallel to the optical axis of the instrument by motorized linear stages, to conjugate the sensor to altitudes ranging from ground to about 20Km. Three objectives will be conjugated to the same high altitude layer, while the fourth will be conjugated to the ground, sensing the contribution of the ground layer to be removed off-line. In order to sense turbulence on the 40 meter scale on a layer at 4.5km TOE should be optically conjugated at 9 km, pointing the VLT at 60 degrees from the Zenith.

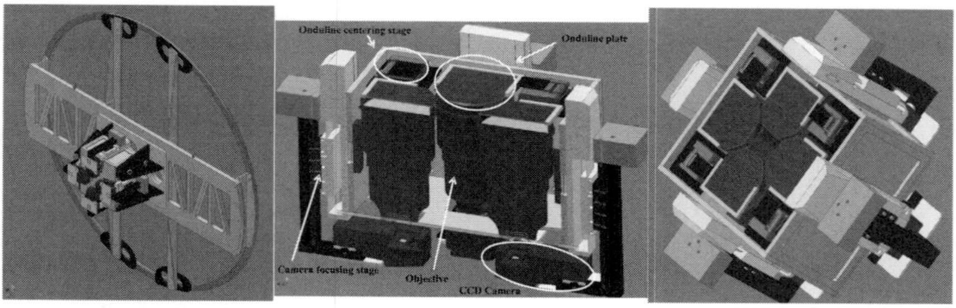

Fig. 4. A mechanical overview, from the left: the sensor mounted on the adapter rotator flange, an overall cross-section view of the instrument and a front view.

6. Conclusions

We presented the TOE experiment design, which aims to measure the free atmosphere turbulence characteristics on a 40 meter scale by the means of a new concept WFS. Laboratory tests verified the WF sensing concept and numerical simulations of the multi reference approach assessed the instrument capability to derive turbulence power spectrum. Although the sensor has been designed in order to be mounted on VLT Nasmyth adapter rotator, can be easily modified to be mounted on other telescopes of similar characteristics. TOE experiment will pose the limits for possible limiting magnitude gain of adaptive optics system aboard ELTs by measuring the coherence time of the frozen turbulence (Taylor) hypothesis.

References

1. R. Ragazzoni, A. Baruffolo, C. Arcidiacono, E. Diolaiti, J. Farinato and R. Soci, Wavefront sensing on 100-m scale, in *Advancements in Adaptive Optics*, eds. D. Bonaccini Calia, B. L. Ellerbroek and R. Ragazzoni, Proc. of SPIE, Vol. 5490Octtober 2004.
2. C. Metti, G. Gentile, M. Dima, J. Farinato, C. Arcidiacono, A. Baruffolo, V. Viotto, E. Diolaiti and R. Ragazzoni, TOE-The Onduline Experiment: a new kind of wavefront sensor to characterize astronomical sites for Extremely Large Telescopes, in *Ground-based and Airborne Telescopes II*, eds. L. M. Stepp and R. Gilmozzi, Proc. of SPIE, Vol. 7012August 2008.
3. R. Ragazzoni, J. Farinato and E. Marchetti, Adaptive optics for 100-m-class telescopes: new challenges require new solutions, in *Adaptive Optical Systems Technology*, ed. P. L. Wizinowich, Proc. SPIE, Vol. 4007July 2000.

THE PARANAL SURFACE LAYER

J. MELNICK*, M. SARAZIN, G.-L. LOMBARDI and J. NAVARRETE

European Southern Observatory,
Garching, D85748, Germany
** E-mail: jmelnick_@eso.org*

The wind pattern over Paranal changed suddenly in 1998 when winds from the NNE-NE became much more frequent than before. At the same time the seeing measured by the Paranal DIMM became significantly worse than hitherto. The seeing change, however, was not seen in 1999 when the first VLT Unit Telescope (UT) was commissioned, and while the DIMM seeing continued to deteriorate in subsequent years, the image quality delivered by the UT's remained constant, or even improved with time. In this contribution we describe our attempts to understand the origin of the discrepancy between DIMM and UT's, and the reasons for the dependency of wind direction on DIMM seeing.

Keywords: Seeing; DIMM:Differential Image Motion Monitor; MASS:Multi-Aperture Seeing Scintilometer; Atmosphere: Surface Layer; Atmosphere: Richardson Number.

1. Basic Facts about the Paranal Surface Layer

The seeing measured by the Paranal Differential Image Motion Monitor (DIMM), has degraded systematically since the beginning of the VLT site testing campaign in the late 80's until now. However, the image quality delivered by the VLT 8m unit telescopes (UT's) has remained constant since first light in 1999, and in fact if anything the image quality delivered by the VLT UT's has even improved since first light. It has been shown by Sarazin et al. (2008; see also Martin et al., 2000) that this rather inconvenient discrepancy arises from a very thin layer of turbulence on the surface that is seen by the DIMM (mounted on 6m platform), but not by the UT's (whose primary mirror is at 12m height, secondary at 20m, and dome roof at 30m). This turbulent surface layer was only rarely present before 1997, but became very frequent after 1997 when the dominant surface winds over Paranal rotated about 45 degrees to the East. The aim of this contribution is to described our efforts to unveil the nature of the Paranal surface layer, and ultimately to understand the causes for the change in the predominant winds in 1997.

2. The Wind-Rose of Paranal

The predominant winds over Paranal changed suddenly in 1997. This is shown in Figure 1 that compares the wind-roses of Paranal in 1997 and in 1998. The main difference is that in 1998 (and as we will see below ever thereafter) the NNE winds became much more frequent.

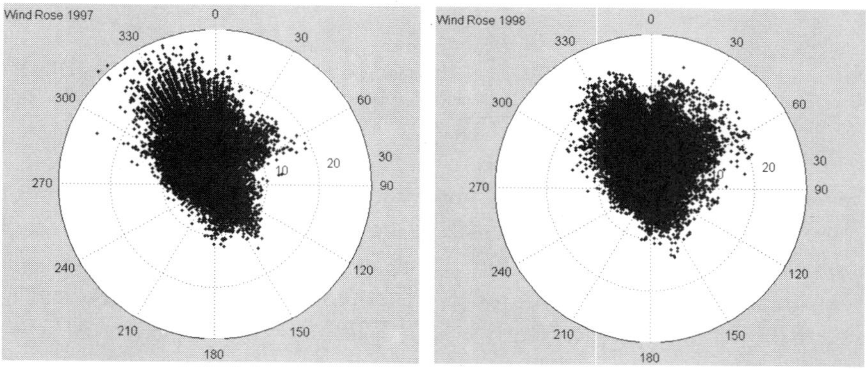

Fig. 1. The wind-rose of Paranal in 1997 (left) and in 1998 (right). The Paranal winds 'jumped' to the East in 1998 and the distribution of wind directions has remained essentially constant ever since.

3. The Inconvenient Discrepancy

The inconvenient discrepancy between DIMM and UT's is defined as $ds = \left(\epsilon_{\mathrm{DIMM}}^{5/3} - \epsilon_{\mathrm{UT}}^{5/3}\right)^{3/5}$ where ϵ_{UT} is the image quality of the VLT UT telescope(s) measured using the size of the Shack-Hartman sub-aperture spots of the telescope active optics sensor. Figure 2 shows the wind-rose of Paranal color-coded according to the value of ds. The right-panel of the figure shows the same wind-rose, but this time color coded by the temperature gradient in the first 28m above the surface, ΔT = T30m – T2m in degrees Celsius.

The thing to notice in these figures is that while the inconvenient discrepancy is maximal when the wind blows from the NNE, the temperature gradient near the surface is largest when the wind blows from the NE. Both

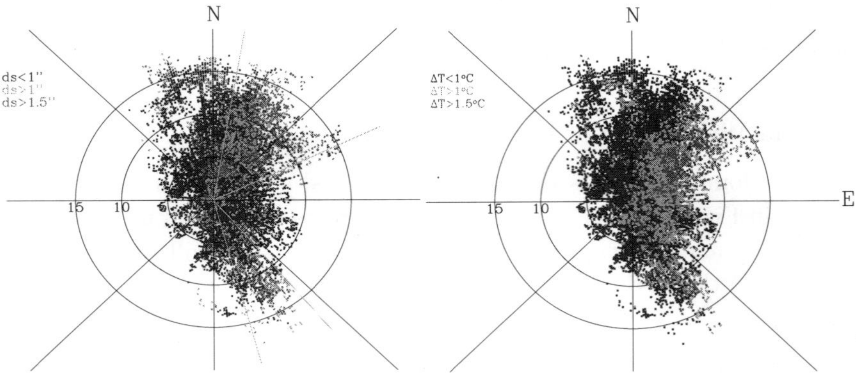

Fig. 2. (color on-line) The wind-rose of Paranal color coded according to the (DIMM-UT) discrepancy ds defined in the text (left); and according to the gradient in air temperature between 2m and 30m, $\Delta T = T30m - T2m$ (right).

vectors are not aligned, at least not well aligned. Whatever the origin of these features, they explain why the seeing degraded after 1997 as we shall see below. Of course the other important feature to notice is that there is a temperature inversion over Paranal: the potential temperature gradient is positive, with an average (yearly) value of $d\theta/dh = +0.017^o$K/m. At least the first 30m over Paranal are convectively stable, so the relevant stability criterion should be provided by the (bulk) Richardson number (e.g. Stull, 1988),

$$\mathcal{R}_i = \frac{g}{\theta} \frac{\frac{d\theta}{dh}}{(\frac{dU}{dh})^2} \tag{1}$$

where g is the acceleration of gravity, θ the mean potential temperature, and U the mean wind speed at an altitude h. Experimentally it is verified that flows are unstable when $\mathcal{R}_i \leq 0.25$, and that once a flow becomes unstable, it does not stabilize again until $\mathcal{R}_i > 1$. The mean (bulk) Richardson number of the Paranal surface layer is $\mathcal{R}_i = 0.19 \pm 0.33(\sigma)$, so most of the time the layer should be turbulent, which, if the layer is indeed very thin ($\Delta h < 20$m), would explain the discrepancy between DIMM and the UT's. If that is the case, we would expect the Richardson number to be significantly *lower* when the wind blows from the NNE where the (DIMM-UT) discrepancy is largest. However, this is not the case. The cyan lines in Figure 2 delineate the wind-cone for which ds is largest: roughly between 20 and 65 degrees (East). The corresponding Richardson number is

$\mathcal{R}_{i,NNE} = 0.66 \pm 0.81$ substantially larger than the mean for all winds, albeit also with a much larger dispersion. So mean winds and temperatures do not tell the whole story of the Paranal surface layer. While the median relative humidity on Paranal is known to be very low ($RH \sim 10\%$), we also checked the humidity gradient to be sure that water vapor was not playing a role, but we found very low values. So the effect of moisture in the air can be neglected with very high confidence.

3.1. *Turbulence*

Figure 3 shows the wind-rose of Paranal this time color-coded according to the total kinetic energy of the wind (TKE) measured by the rms dispersion of the horizontal wind velocity (i.e. the U & V components). Besides standard thermometers at 2m and 30m, and spin-wheel anemometers at 10m and 30m, the Paranal weather tower is equipped with a sonic anemometer at 20m. This instrument is configured to read the wind velocity every two seconds, while the wind-rose presented in Figures 1 and 2 depict one-minute averages. So we used the rms dispersion of the 2s measurements about the 1min averages to estimate the TKE.

The result is unequivocal: the wind turbulence is highest when the wind blows from the NNE, and is remarkably well aligned with the direction of maximum inconvenient discrepancy. Thus, the correct criterion for turbulent stability is not the *bulk* Richardson number, but the *flux* Richardson number because wind speeds stronger than $\sim$ 5m/s turbulence is the dominant transport mechanism for kinetic and thermal energy. The study of the flux Richardson number is beyond the scope of this presentation, in particular because the set of data used here has high (temporal) resolution wind velocities, but not temperature. We can only say that after seeing these results we mounted a set of 4 sonic wind and temperature sensors on the Paranal weather mast but the data are still in the process of being analyzed.

4. Going with the Flow

The surface-layer on Paranal is more turbulent when the wind blows from the NNE and it also has the temperature gradient required for turbulence to create significant fluctuations in the index of refraction (C_n^2: optical turbulence) that we associate with seeing. (For very strong winds pressure fluctuation might contribute to C_n^2, but we lack the instrumentation to look into this aspect experimentally.) We have seen above that the temper-

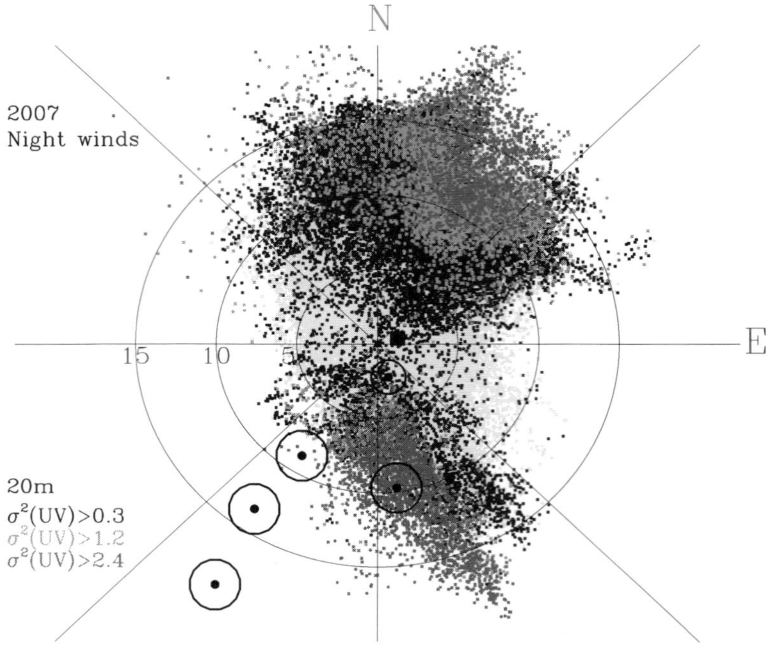

Fig. 3. (color on-line) The Purple-Rose of Paranal: The wind-rose color-coded according to the kinetic energy of the wind calculated from the rms dispersion of the horizontal wind velocity measured every 2 seconds about 1 minute averages.

ature gradient is highest when the wind blows not from the NNE, but from the NE. This must surely be telling us something.

4.1. *Seeing is Believing*

We can use the seeing data to gain more insight into the nature of the optical turbulence of the surface layer on Paranal. Figure 4 shows a 'synoptic map' of the atmospheric conditions over Paranal on a given day/night, in this case February 27, 2006. This rather busy figure presents a minute-by-minute comparison between temperature, humidity, wind-speed and direction with the seeing measured by the DIMM, the MASS, and the Shack-Hatmann sensors of UT1 (see figure caption for a full description of these plots). The horizontal axis shows the time since sunset in hours with a vertical line indicating sunrise. The *inconvenient discrepancy* in this figure is the separation between the blue points (DIMM) and the red points (UT1) in the lower panel. About 5 hours after sunset we see a huge outburst of discrepancy with a DIMM seeing of about 3″ and a much lower UT1 seeing of about 1″. At the same time, the wind speed rises above 10m/s and the temperature gradient reaches almost $\Delta T = 2^o$C. A couple of hours later (7h after sunset) the discrepancy drops to close to $ds = 0$, but the wind is still very strong (about 15m/s) and very turbulent as indicated by the large fluctuations on one-minute intervals. But the temperature gradient has dropped to $\Delta T \sim 0$ so in spite of the large wind turbulence there is very little optical turbulence.

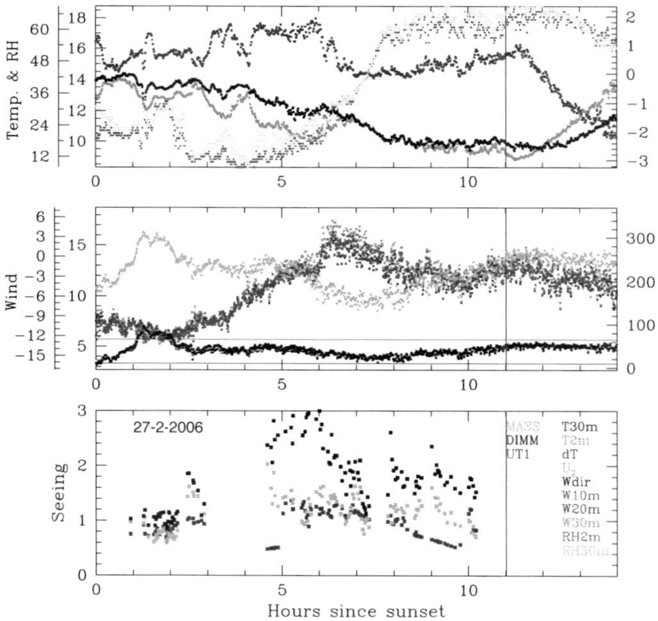

Fig. 4. Synoptic plot of the atmospheric conditions over Paranal on February 27, 2006. The horizontal scale shows time in hours since sunset and the vertical line at 11hrs indicates the time of sunrise. The color-key is given in the bottom panel. The top panel shows the minute-by-minute evolution of temperature and humidity at 2m and 30m altitude. The middle panel shows the horizontal wind speed at 10m, 20m, and 30m, and the vertical wind speed at 20m. The two wind-speed scales on the left of the plot correspond to the vertical wind velocity (outer) and the horizontal wind speed (inner). The wind direction is also given with the scale in degrees in the right-hand side of the figure (0 corresponds to winds from the North and 90 to winds from the East). The bottom panel shows the DIMM seeing, the MASS seeing, and the UT1 image quality from the Shack-Hartmann spots.

Figure 5 shows the synoptic map of two nights earlier when the discrepancy was very small. On this night both the temperature gradient and the wind speeds are low, and the wind blows mostly from the NNW-N (turning to the N-NNE towards the end of the night).

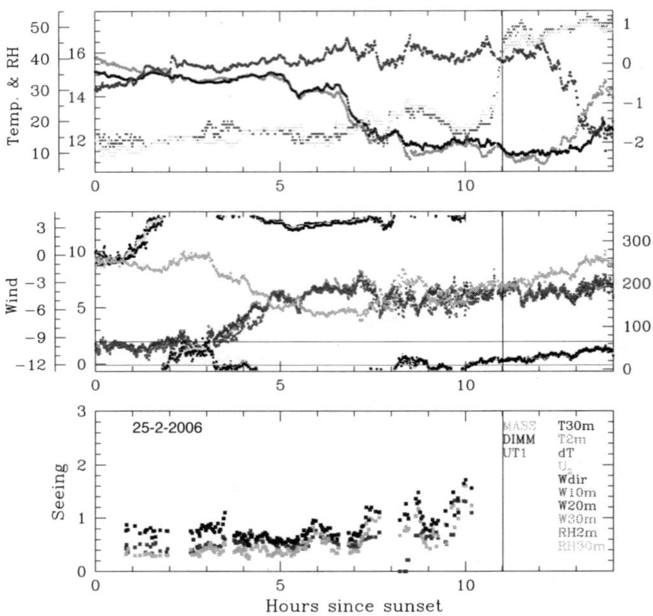

Fig. 5. Similar to Figure 4 for the 25th of February. In this night the discrepancy is small consistent with the small temperature gradient and the relatively low wind speed. The wind blows from the N-NNW during most of the night and turns to the NNE shortly before sunrise.

5. Conclusions

The synoptic plots shown in the previous section illustrate the origin of the inconvenient discrepancy between DIMM and UT's: winds from the NNE-NE are significantly more turbulent than the good seeing winds (N-NNW), and bring a steep temperature stratification over the summit. The latter probably results from the advection of cold katabatic air trapped at night time in the deep valleys adjacent to the observatory, while the turbulence is caused by the orographic obstacles windward of Paranal in the NNE direction. Fortunately, the temperature inversion is lower than the UT domes and therefore does not affect, or affects only marginally, the overall image quality delivered by the UT unit telescopes. Thus, the conventional wisdom for observatories is today truer than yesterday: select sites with clear horizons at least in the direction of the predominant winds, and erect your telescope as high as possible above the ground!

References

1. Martin, F., et al. (2000), *A&A Sup. Ser.*, **144**, 39.
2. Stull, R.B. (1988), *An Introduction to Boundary Layer Meteorology*; Kluwer.
3. Sarazin, M., Melnick, J., Navarrete, J., Lombardi, G. (2008), *The Messenger*, **132**, p. 11.

THE MESOSCALE METEOROLOGICAL MODELS MESO-NH AND AROME

C. LAC[1], T. MARIC[1], J.P. PINTY[2], J. CUXART[3], V. MASSON[1], P. TULET[1]
[1]*CNRM/GAME, Toulouse, France.* [2]*Laboratoire d'Aérologie, Toulouse, France.*
[3]*Univ. de les Illes Balears, Spain*

1. Introduction

The meteorological modelling is now a common way to explore the optical atmospheric turbulence since the pioneering article [1]. [2,3] among others have largely contributed to progress in this way, by using progressively more sophisticated and finer scale meteorological models. Indeed, astronomical observatories are generally installed on the top of high mountains and local effects are expected as lee waves and gravity waves as well as catabatic flows. Furthermore, it is known that optical turbulence is triggered by phenomena such as lee waves, gravity waves and wind shear, and is mainly concentrated in the boundary layer, that requires high resolution and sophisticated parameterization in the numerical meteorological models.

The paper will focus on the characteristics of two mesoscale meteorological models Meso-NH and AROME, currently available for optical turbulence studies and prediction, by highlighting their strengths and limitations. It will be organized as follows : section 2 provides general consideration on meteorological models and gives an overview of some current mesoscale models. Section 3 details the Meso-NH model and its externalized surface for the characteristics concerning optical turbulence. Section 4 is devoted to the new operational mesoscale model AROME for which the data assimilation system is detailed in [4].

2. General Consideration on Meteorological Models

Meteorological phenomena concern a broad range of space and time scales, from tens of thousands of kilometers to millimetres, and from tens of days to seconds. A comprehensive understanding of the atmosphere includes a treatment of all these scales of motion, including their interaction.

Consequently, meteorological models extend from the Global Climate Models (GCM) to Large Eddy models with the general characteristics of resolution given in Table 1. Direct Numerical Simulations (DNS) that carry gridspacings at the Kolmogorov microscale (millimeters to centimetres) are commonly not used as they are unable to deal with meteorological problems of interest.

The range of space and time scales of a model are a key issue as they determine the needs in physical parameterization. Indeed, physical parameterizations are intended to represent the physical processes that occur at scales smaller than the scales resolved by the model. The word "parameterization" means that we do not attempt to reproduce these processes in detail, but only their statistical feedback onto the larger scales. The physical parameterizations include radiation, surface schemes, turbulence, cloud microphysics, convection and gravity waves. If the orographic gravity wave drag only concerns global models, the convection parameterization is required until horizontal grid size of about 1km. All the other parameterizations remain necessary, whatever the resolution. Even with Large Eddy Simulations (LES) corresponding to resolution finer than 100m, for which the resolved scales are assumed to contain most of the energy of turbulent motion, turbulence scheme remains necessary near the surface or where temperature inversions exist. Only DNS would not need parameterization. The optical turbulence can be treated by Limited Area Prediction models up to Large eddy models.

Table 1 : Space and time scales for the different kinds of meteorological models.

	Horizontal scales	**Vertical scales**	**Time range**
1.1.1. *Climate models*	200km	1000m	100 years
Global weather prediction	20km	500m	10 days
Limited area weather prediction	10km	500m	2 days
Cloud Resolving Models	500m	500m	1 day
Large Eddy models	50m	50m	5 hours

For scales finer than 10km resolution, that we call *Meso-scale models*, vertical accelerations cannot be ignored compared to gravitational and pressure gradient forces, and the equation set is non-hydrostatic. This characteristic is very important for optical turbulence as it allows to better represent the gravity waves induced by convection or orography, for example the slantwise propagation of gravity waves downstream of a mountain, impacting the turbulence activity. A non exhaustive list of meso-scale models is given in Table 2, included some of the most commonly used MM5 [5], WRF [6], RAMS [7], ARPS[8], LM[9], UM[10], Meso-NH[11] and AROME[12]. All the models are non-hydrostatic and most of them are fully compressible. Only Meso-NH is anelastic, which means that acoustic waves are filtered. There are two sets of models, the first one corresponding to operational NWP based on spectral space, semi-lagrangian transport and semi-implicit temporal scheme (SL-SI), implying longer time steps. The second one are in grid point space, eulerian transport and explicit temporal scheme, with nested-grid capability. Most of all these models have flexible resolution and a variety of physical parameterization options. But only a few of them have the capability to run LES, owing to a sophisticated turbulence scheme and the possibility of cyclic lateral boundary conditions (ARPS, WRF and Meso-NH).

Table 2 : Main dynamical characteristics of some common meso-scale models.

	MM5	WRF	RAMS	LM	Meso-NH	ARPS	UM	AROME
Spectral (SP) /Grid point (GP)	GP	GP	GP	GP	GP	GP	SP	SP
ANelastic/Full Compressible	FC	FC	FC	FC	AN	FC	FC	FC
Transport : Eulerian or Semi-Lagrangian	EU	EU	EU	EU	EU	EU	SL	SL
Temporal scheme	Split Explicit	Split Explicit	Split Explicit	Split Explicit	Leap-Frog	Split Explicit	Semi-Implicit	Semi-Implicit
Nesting (1-way or 2-way)	2	2	2	1	2	1	1	1
Data assimilation	3D/4D-Var	3D Var				3D Var	3D/4D-Var	3D Var

3. The Meso-NH Model and the Externalized Surface

3.1. *Dynamics*

As cited above, Meso-NH is a non-hydrostatic numerical model based on the anelastic approximation, that can simulate the time evolution of atmospheric 3D motions ranging from the synoptic scale down to LES. Eulerian advection schemes have been recently improved in Meso-NH in terms of stability and accuracy by implementing a 4th order centred advection scheme for momentum components and the PPM (Piecewise Parabolic Method) advection scheme [13] for meteorological scalar variables. Therefore, the temporal scheme is an hybrid scheme, consisting of forward time differencing for the thermodynamic variables and leap-frog differencing for the velocity components. A stringent test case to examine the robustness of these schemes is a low-level trapped lee-waves case which has been observed in nature [14] and studied [15] (Fig. 1). Gravity waves excited by an isolated mountain can lead to trapped lee waves propagating horizontally downstream, creating quasi-periodic phase oscillations. The wave trains are associated to turbulence, and clouds if a microphysical scheme is considered, making the test interesting for optical turbulence. Indeed, the 4th order centred advection scheme for momentum components and the PPM for meteorological scalar variables lead to an accurate and stable solution (Fig. 1). A flux-corrected 2nd order scheme for thermodynamic variables instead of PPM induces residual instabilities in the wave trains (Fig. 2), while the 2nd order centred scheme in addition for the velocity components instead of the 4th order one leads to a numerical explosion sooner in the simulation (not shown).

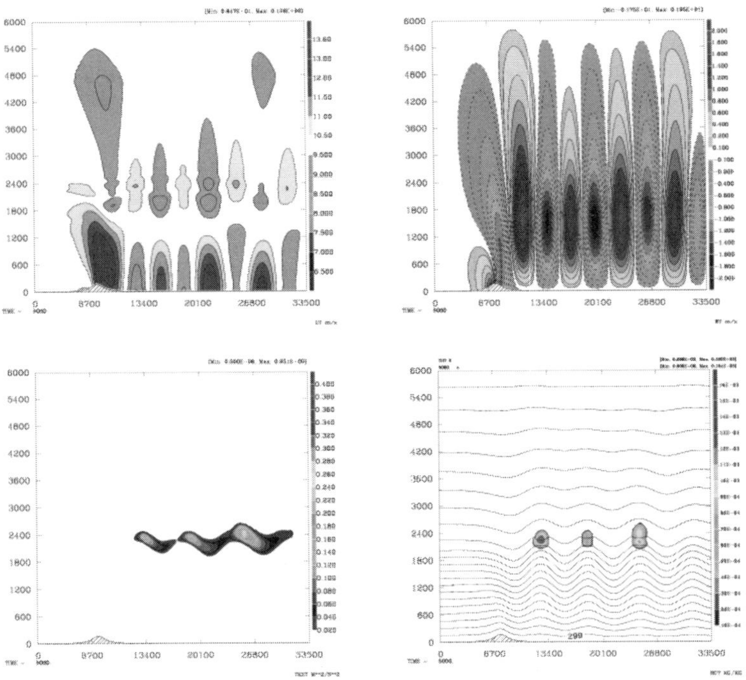

Figure 1 : Vertical cross-sections of zonal wind component (a), vertical velocity (b), Turbulent Kinetic Energy (c) and cloud mixing ratio (d) after 5000s of simulation with a 4th order centred scheme for momentum and PPM for meteorological scalar variables.

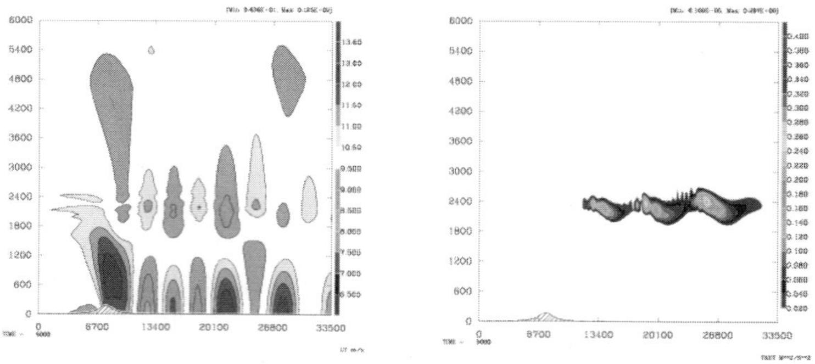

Figure 2 : Same figure as Fig. 1 for zonal wind component (a) and Turbulent Kinetic Energy (b) with a 4th order centred scheme for momentum and a flux-corrected 2nd order scheme for meteorological scalar variables.

3.2. *Turbulence*

3.2.1. *The turbulence scheme*

The turbulence scheme, based on a 1.5 order closure, is accurately presented in [16]. It relies on the turbulent kinetic energy equation:

$$\frac{de}{dt}=-\overline{u'_i u'_j}\frac{\partial \overline{u_i}}{\partial x_j}+\frac{g}{\theta_{vref}}\overline{u'_3\theta'_v}+\frac{1}{\rho_{dref}}\frac{\partial}{\partial x_j}(C_e\rho_{dref}L\sqrt{e}\frac{\partial e}{\partial x_j})-C_\varepsilon\frac{e^{3/2}}{L}$$

where u_i is the velocity in the i direction, primes denote turbulent components, e is the Turbulent Kinetic Energy $e=\frac{1}{2}(\sum_i \overline{u'^2_i})$, θ_{vref} and ρ_{dref} are the reference temperature and density profiles respectively and are specified once and for all. L is the mixing length corresponding to the characteristic length scale of the sub-grid eddies, and C_e and C_ε are closure constants. The source terms are respectively the shear production, the buoyancy production, the diffusion and the dissipation. This 3D turbulent scheme can be reduced to a 1D mode, for mesoscale simulations (*Δx>1km*), by not considering the horizontal gradients.

Optical turbulence diagnostics have been added to the turbulence scheme to simulate all the principal atmospheric parameters that are necessary for the exploitation of a telescope and the selection of different observing modes [3,17,18].

3.2.2. *Diurnal cycle of the ABL*

It is well-known that flow in the boundary layer over land is primarily controlled by the diurnal cycle of the earth's surface energy budget. During the day, under weak synoptic scale conditions, the Atmospheric Boundary Layer (ABL) is mainly governed by thermal effects, and buoyancy is the dominant mechanism driving turbulence. During the night, turbulent mixing is greatly reduced or even completely removed, or else becomes intermittent.

The structure of the ABL is usually classified into four characteristic types : convective, neutral, stable and cloud-topped. The LES technique has been especially helpful to study the different kinds of ABL. Examples of such simulations with Meso-NH are presented below.

3.2.3. *Simulations of stable ABL*

Optical turbulence for astronomy is mainly concerned by nocturnal ABL, associated to stable conditions. These conditions are the most hardly to simulate for the turbulence scheme. When the stability of stratification is weak (presence of moderate wind and/or clouds) the turbulence is more or less continuous. But for moderate to strong stability (usually in weak pressure gradient conditions and clear skies), the turbulence is weak, sporadic or intermittent, and highly dependent on the physiographical characteristics of every particular location. The stratification induces also a strong anisotropy of the turbulent eddies, with a relatively large horizontal size compared to the vertical one. This feature compels to use very high vertical resolution.

Concerning LES, [19] have recently studied a steady-state low-level jet (LLJ) observed in the SABLES-98 experiment. The model distinguishes between two separated layers, one below the maximum of the wind with strong stratification and shear, and the other above, with weaker stable stratification and shear (Fig. 3 and b). The upper part extends as high as 3 times the height of the maximum of the wind. The lower layer has a maximum of turbulence, twofold the upper one, that is mainly subgrid, while the turbulence in the upper part is mainly resolved (Fig. 3). This underlines the fact that at high horizontal resolution (Δx=6m and Δy=4m), LES allow to resolve much of the eddies, except in the presence of strong stable stratification or near the ground , where most of the eddies remain subgrid.

At mesoscale (Δx=1km), several simulations of clear-sky and weak synoptic pressure gradients have been run for different locations : the Majorca island, the Duero Basin in the Iberian Peninsula and Iceland [20,21]. All these simulations illustrate complex flows due to the cooling near the ground added to the topographic effects, conditions that are quite frequent in the telescope surroundings. Over the Duero Basin, MM5 and Meso-NH were evaluated and inter-compared for two nights : one night with a well-defined synoptic wind and

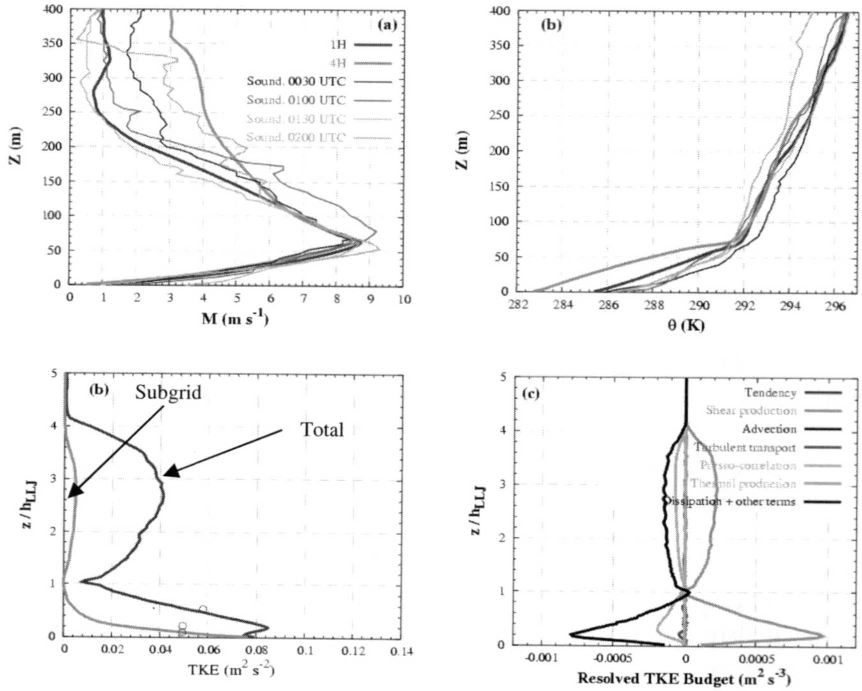

Figure 3 : From [19]. Vertical simulated profiles during the SABLES-98 Experiment of wind intensity (a), potential temperature (b) at 2 simulated instants (1H and 4H) compared to soundings. In (c), vertical profile of the Turbulent Kinetic Energy at 1 instant partitioned between total (resolved+subgrid) and subgrid parts. The vertical scale is normalized by the LLJ height. In (d), vertical profile of the budget of TKE contributions.

a second one with practically no horizontal pressure gradient [22]. If the results are quite similar in the first case, with a good agreement with the observations, differences are larger for the more stable night, where the characteristics of the turbulence parameterizations may be more evident. The fact that MM5 imposes a minimum value of $0.2m^2.s\text{-}2$ for the TKE reinforces the flow on the mountain slopes and remove the LLJ in the valley (Fig. 4). Meso-NH allows the LLJ to persist, but tends to generate a two-layer structure that is not evident in the tower measurements.

3.2.4. *Simulation of dry convective ABL*

Water vapour is a crucial parameter for several major areas in atmospheric sciences including astronomical observations, even if most of these observations

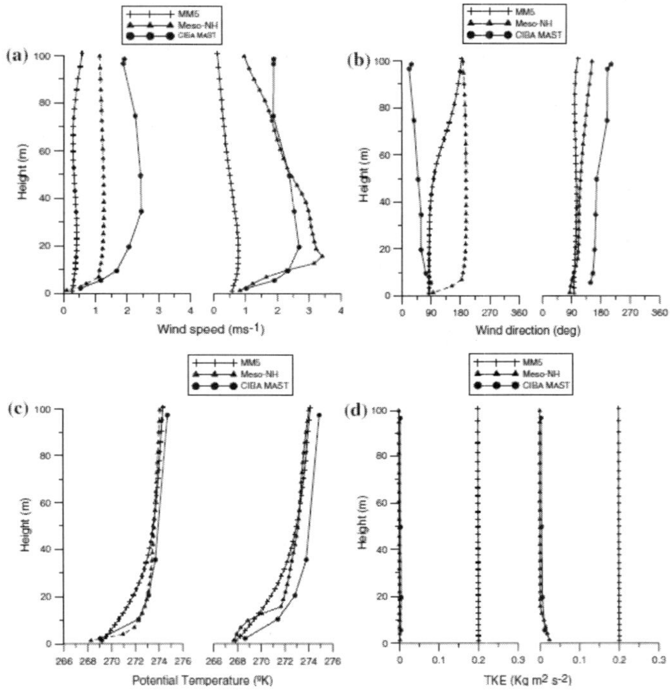

Figure 4 : From [22]. Simulated and measured vertical profiles of (a) wind speed $(m.s^{-1})$, (b) wind direction, (c) potential temperature and (d) turbulent kinetic energy at 2300 (on the left) and 0400 (on the right) UTC respectively, during the night of 14-15 January 2003.

operate during the night. [23] have analyzed the sources of variability of water vapour in a daytime growing convective boundary layer over land, from LES simulations ($\Delta x=\Delta y=100m$). The probability density functions (PDF) of vertical velocity and water vapour mixing ratio derived from the LES show good agreement with aircraft in-situ measurements (Fig. 5). The water vapour variability is found to be important, and a negative skewness of the water vapour distribution is observed and simulated within the growing convective ABL.

3.3. *Surface Coupling*

Surface atmosphere exchanges, mainly momentum, water and heat surface fluxes, drive the boundary layer evolution. The modelling of these fluxes is performed by an externalized surface model, named SURFEX, including ISBA for the vegetation [24], TEB for the urban effects [25], sea and lake parameterizations. This surface model is independent to the atmospheric model

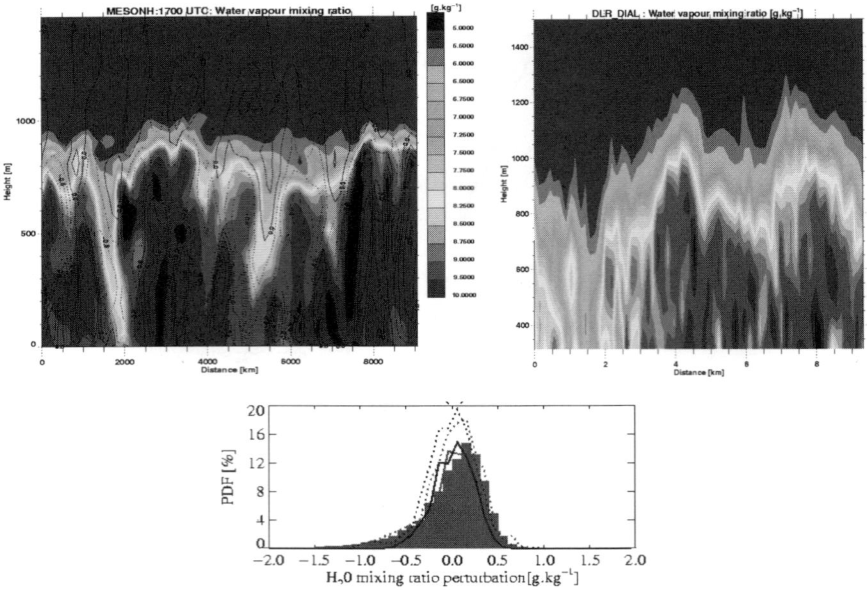

Figure 5 : From [23]. Vertical cross-sections of water vapour mixing ratio (g.kg^{-1}) on June 14th 2002 12UHT during IHOP campaign simulated with Meso-NH (a) and observed (b). In (c), PDF of water vapour mixing ratio perturbation at mid-altitude of the ABL simulated (in grey) and observed (in lines).

coupled above. The coupling method is applied on a single layer : the surface scheme is forced by the lowest atmospheric layer and respond to atmospheric variables at this level (temperature, wind, humidity, incoming radiation) and produce average upwards turbulent fluxes and radiative quantities (albedo, emissivity, surface temperature).

The surface strongly impacts the development of the ABL, mainly through the turbulent fluxes. For instance, [26] focus on the impact of the vegetation and pointed out the ABL height differences between forest and cropland during the CarboEurope experiment.

Another possibility of SURFEX and Meso-NH that could be interesting for astronomical studies concerns the transport of atmospheric dusts and their radiative impact [27]. Dusts are emitted at the surface and follow saltation fluxes in the SBL, and are then transported in the atmosphere, undergoing wet and dry deposition and influencing both short wave and long wave radiation. Figure 6 illustrates the dust plume with large optical depths emitted during the 24th and 25th September 2000, ready to leave the coast of Mauritania and Senegal at noon the 25th. By the 26th, the plume has been transported over the ocean between

Mauritania and the Cap Verde Islands with optical depths between 1 and 2, in agreement with the satellite images.

The prediction of the spatial concentration and distribution of the dust is a fundamental issue for the ground-based astronomy in which mesoscale models might play a crucial role. Many observatories are indeed located on the summit of mountains located in dry and/or desert regions and the quality of the transparency of the sky is essential for the efficiency of the astronomical facilities.

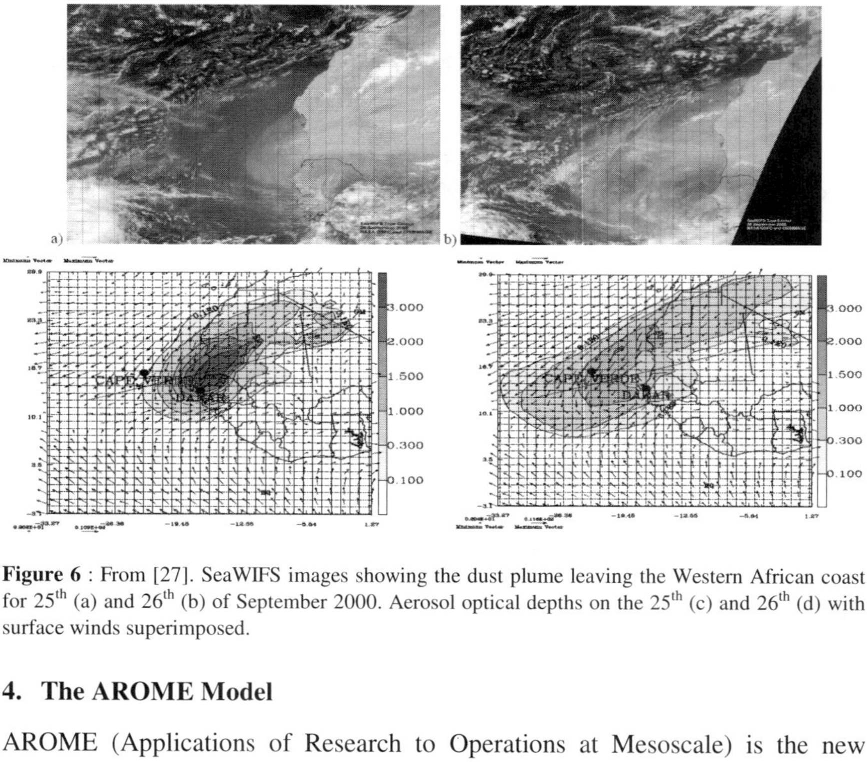

Figure 6 : From [27]. SeaWIFS images showing the dust plume leaving the Western African coast for 25th (a) and 26th (b) of September 2000. Aerosol optical depths on the 25th (c) and 26th (d) with surface winds superimposed.

4. The AROME Model

AROME (Applications of Research to Operations at Mesoscale) is the new operational weather forecast model of Meteo-France [12], running with an horizontal resolution of 2.5km on France, including a 3D-Var meso-scale data assimilation system [3]. The model is built upon the ALADIN-NH dynamical core, spectral with a semi-implicit, semi-lagrangian timestep solver, and compressible dynamical equation. The physical package is exported from Meso-NH in its 1D version, including microphysics, turbulence, convection and statistical cloud scheme. This limited-area model is nested to ALADIN model that runs at 10km horizontal resolution on Western Europe (Fig. 7).

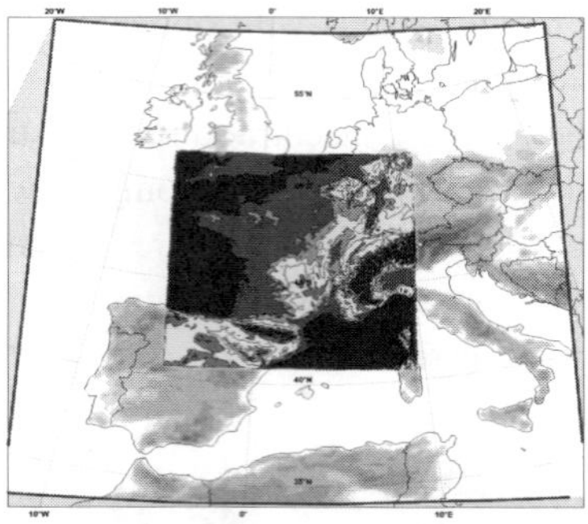

Figure 7 : (color on-line) AROME domain (orography in color, 2.5km resolution) embedded in the ALADIN domain (with blue contour, 10km resolution).

One of the main objectives of AROME is to improve heavy precipitation forecasts with strong emphasis on Mediterranean flash-floods, but also the prediction of local events (fog, breeze, urban effects, orography). This new model can be an interesting platform for optical turbulence predictions or climatological studies.

5. Conclusion

Among the large community of meteorological meso-scale models, the Meso-NH model has been presented in the context of optical turbulence simulations. We focus on the accuracy of the new advection schemes recently implemented and on the numerous LES in different boundary layer conditions. The externalized surface model, coupled to Meso-NH, has known a lot of progress last years, and integrates sophisticated surface schemes to take into account the impact on the ABL. Moreover, Masciadri et al. largely demonstrated Meso-NH performances related to the astronomical application of seeing forecasting. Therefore, Meso-NH could be more extensively used for site testing studies.

AROME, owing to the Meso-NH physics, a performing time computation and a complex data assimilation system, has the feasibility of the seeing forecasting, in a complementary way to Meso-NH studies, supposing reliable radiosoundings in the vicinity of the site. Climatology studies for different sites comparison could be also an interesting issue.

References

1. Coulman, C., P.R.Gillingham PASP, **98**, 376-387 (1986).
2. Bougeault, P., C. De Hui, B. Fleury and J. Laurent *Applied Optics*, **34**, 3481-3488 (1995).

3. Masciadri, E., Vernin, J., Bougeault, P., A&ASS, **137**, 185 (1999).
4. Brousseau P., and L. Auger, ATMO08, Proceedings (2008).
5. Grell, G.A., J. Dudhia and D.R. Stauffer, NCAR (1995).
6. Skamarock, W.C., et al., NCAR (2005).
7. Pielke, R.E., Cotton, W.R., Walko, R.L., Tremback, C.J., Lyons, W.A., Grasso, L.D., Nicholls, M.E., Moran, M.D., Wesley, D.A., Lee, T.J., and Copeland, J.H. *Metero. Atmos. Phys.*, **49**, 69-91 (1992).
8. Xue, M., K.K. Droegemeier, and V. Wong : *Meteor. Atmos. Physics.*, **75**, 161-193 (2000).
9. Steppeler, J., Doms, G., Schättler, U., Bitzer, H.W., Gassmann, A., Damrath, U., Gregoric, G. *Meteorology and Atmospheric Physics*, **82** (1-4), 75-96 (2003).
10. Davies, T.; Cullen, M.J.P.; Malcolm, A.J.; Mawson, M.H.; Staniforth, A.; White, A.A.; Wood, N., *Q.J.M.R.S.*, **608**, 1759-1782 (2005).
11. Lafore, J.P., J. Stein, N. Asencio, P. Bougeault, V. Ducrocq, J. Duron, C. Fischer, P. Hereil, P. Mascart, J.P. Pinty, J.L. Redelsperger, E. Richard, and J. Vila-Guerau de Arellano *Annales Geophysicae*, **16**, 90-109 (1998).
12. Bouttier, F., G. Hello, S. Malardel, Y. Seity, Meteo-France (2006).
13. Colella, P. and P.R.Woodward, *J. Comput. Phys.*, **54**, 174-201.
14. Shutts, G. and A. Broad, *Q.J.M.R.S.*, **119**, 377-408 (1993).
15. Nance, L. and D. Durran, *J. Atm. Sci.*, **55**, 1429-1445 (1998).
16. Cuxart, J., Bougeault, Ph. and Redelsperger, J.L., *Q. J. R. Meteorol. Soc.*, **126**, 1-30 (2000).
17. Masciadri E. & Jabouille P. A&A, **376**, 727 (2001).
18. Masciadri E., Avila R., Sanchez L.J.: RMxAA, **40**, 3 (2004).
19. Cuxart, J., and M.A. Jiménez, *Mon. Weather Rev.*, **64**, 1666-1679 (2007).
20. Cuxart, J., M.A. Jiménez, and D. Martínez, *Mon. Weather Rev.*, **135**, 918-932 (2007).
21. Ágústsson, H., J. Cuxart, A. Mira, and H. Ólafsson, *Meteorol. Z.*, **16**, 99-110 (2007).
22. Bravo, M., T. Mira, M.R. Soler, and J. Cuxart, *Bound.-Layer. Meteor.*, **128**, 77-101 (2008).
23. Couvreux, F., F. Guichard, J.-L. Redelsperger, C. Kiemle, V. Masson, J.-P. Lafore, and C. Flamant, *Quart. J. Roy. Meteor. Soc.*, **131**, 2665-2693 (2005).
24. Noilhan, J., and S. Planton. *Mon. Wea. Rev.*, **117**, 536-549 (1989).
25. Masson V. *Bound. Layer Meteor*, **1994**, 357-397 (2000).
26. Sarrat, C., J. Noilhan, P. Lacarrère, S. Donier, C. Lac, J.-C. Calvet, H. Dolman, C. Gerbig, B. Neininger, P. Ciais, J.D. Paris, F. Boumard, M. Ramonet, and A. Butet, *J. Geophys. Res.*, **112**, D12104, doi:10.1029/2006JD008107, (2006).
27. Grini, A., P. Tulet, and L. Gomes, *J. Geophys. Res.*, **111**, D19205, doi:10.1029/2005JD007007, (2006).

INTRODUCTION TO DATA ASSIMILATION IN METEOROLOGY

P. BROUSSEAU and L. AUGER
Meteo-France, CNRM/GAME, Toulouse, France

Data assimilation is the process by which observations are inserted into a forecast model to prevent it from drifting away from the true state of the atmosphere. It is not a trivial process and involves specific observations treatment as well as errors statistics definition for both model and observations. In the context of fine scale forecast, useful for predicting parameters such as optical turbulence, appropriate data assimilation system can be designed, with high resolution observations such as radar data and synoptic observations. That kind of system is used for a few years in Meteo-France with an operational status, and it proved us that the maximum improvement is mostly due to the high density of small scale observations.

1. Introduction

Local meteorological meso-scale model provides precise simulation of atmospheric variables inside troposphere useful in the context of optical turbulence simulations [6]. Such limited-area simulations require initial conditions to be as good as possible. So far, initial conditions were derived from a global atmospheric model through a process that adapts fields to finer grid resolution, so called «dynamical adaptation mode». For global model, initial conditions for a given time, also called analysis, is a subtle blend between the last short-term forecast and every observation available inside a specific time-window, that process being named « data assimilation ». Nowadays initial conditions for local meso-scale models can also be obtained through meso-scale data assimilation, showing some improvement compared to the dynamical adaptation mode.

The first part of this paper will be devoted to a general and comprehensive introduction of data assimilation for meteorological models. Then we will review the different kinds of data focusing on specificities of local area models. The last part will show the improvements expected for data assimilation mode versus dynamical adaptation mode with the example of Meteo-France's AROME project.

2. General Ideas on Data Assimilation

2.1. *Principle*

In meteorology, the basic objective information that can be used to produce the analysis is a collection of observed values provided by observations of the true state. If the model state is over-determined by the observations, then the analysis reduces to an interpolation problem. In most cases the analysis problem is under-determined because data is sparse. Moreover data is somehow indirectly related to the model variables, making the resolution even more complex. In order to make it a well-posed problem it is necessary to rely on some background information in the form of an a priori estimate of the model state. The background information can be a climatology or a trivial state ; it can also be generated from the output of a previous analysis followed by a forecast. As shown on Figure 1, each observation available inside a specific time-window, named assimilation window, are combined with the previous model forecast valid at the assimilation time, to produce the new analysis. This analysis is used to initialize an new model forecast used at the next assimilation time to assimilate new observations, the whole process is called an assimilation cycle.

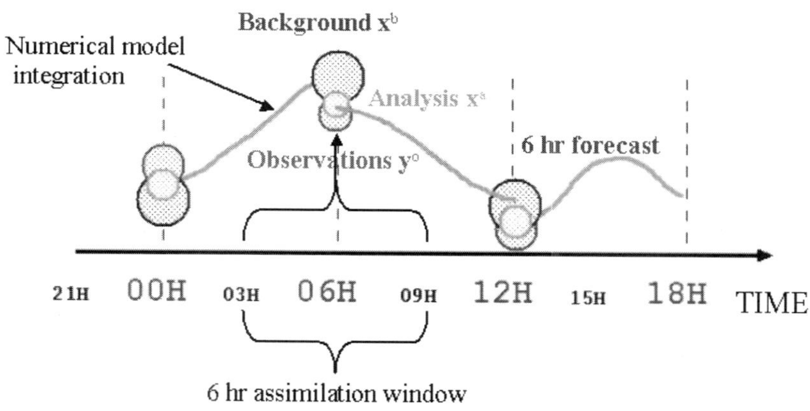

Figure 1 : Schematic representation of a 6 hr assimilation cycle.

2.2. *The Linear Statistic Estimation*

All data are important sources of information, but at the same time we do not trust any of them completely, so we must make compromises when necessary. There are errors both in the model and in the observations, so we can never be sure which one to trust. This uncertainty; (represented by the circles on

Figure 1), is modeled using probabilistic concepts in the frame of the linear statistic estimation.

notations :

$\mathbf{x}^b$ background model state
$\mathbf{x}^a$ analysis model state
$\mathbf{y}^o$ vector of observations
$\boldsymbol{H}$ observation operator
$\mathbf{H}$ linear observation operator
$\mathbf{B}$ covariance matrix of the background errors
$\mathbf{R}$ covariance matrix of observation errors

Under the assumption that the background and observation errors are unbiased and mutually uncorrelated, and supposing probability density functions, the Best Linear Unbiased Estimate of the analysis is obtained adding the background and a correction $\delta\mathbf{x}$ depending on the difference between observations and background :

$$\begin{aligned}\mathbf{x}^a &= \mathbf{x}^b + \delta\mathbf{x} \\ &= \mathbf{x}^b + \mathbf{BH}^T (\mathbf{HBH}^T + \mathbf{R})^{-1} (\mathbf{y}^o - H(\mathbf{x}^b)) \qquad (1) \\ &= \mathbf{x}^b + \mathbf{Kd}\end{aligned}$$

Where
$\mathbf{K} = \mathbf{BH}^T (\mathbf{HBH}^T + \mathbf{R})^{-1}$ is called the gain matrix of the analysis
$\mathbf{d} = (\mathbf{y}^o - H(\mathbf{x}^b))$ is called the innovation.

As observed in equation (1), the background-error statistics **B** determine how observations modify the background to produce the analysis, filtering and propagating innovation. B statistics should contain some information about the uncertainty of the guess, which depends on, among other things, the model, the domain and the meteorological situation of the day (flow and initial conditions). Determining this uncertainty is a major problem in data assimilation [2].

2.3. *The Variational Formulation*

In meteorology, the size of the problem is so big that is not possible to resolve it numerically without simplifying assumptions. The principle of a variational formulation is to avoid the computation of the gain matrix **K** completely by looking for the correction $\delta\mathbf{x}$ as an approximate solution to the equivalent minimization problem defined by the cost function :

$$J(\delta\mathbf{x}) = J_b(\delta\mathbf{x}) + J_o(\delta\mathbf{x})$$
$$= \delta\mathbf{x}^T\, \mathbf{B}^{-1}\, \delta\mathbf{x} + (\mathbf{d} - \mathbf{H}\delta\mathbf{x})^T\, \mathbf{R}^{-1}\, (\mathbf{d} - \mathbf{H}\,\delta\mathbf{x}) \qquad (2)$$

In equation (2), the first and the second terms represent respectively the distance to the background and to the observations. The solution to this problem is the same as in equation (1). More details concerning data assimilation could be found in Bouttier and Courtier [3].

3. Different Kinds of Observation

Observations available are very numerous and diverse. They differ in measured parameters, spatial and temporal coverage, resolution. Some of them are very useful to describe large scale phenomena and are very informative in a global data assimilation system. This is the case, for example, of radio-sounding observations, providing information on geopotential, temperature, wind and humidity fields. They are very accurate but available only twice a day and have an irregular spatial coverage (Figure 2). Observations made by instruments providing radiance measurements on polar satellites have a good spatial coverage but their availability over a limited area domain depends on satellite orbits (Figure 3). Other observation networks, with a high temporal and spatial frequency, like surface observations (Figure 4) or radar measurements (Figure 5), are able to provide information on meso-scale events. Only a meso-scale data assimilation system can benefit from such data.

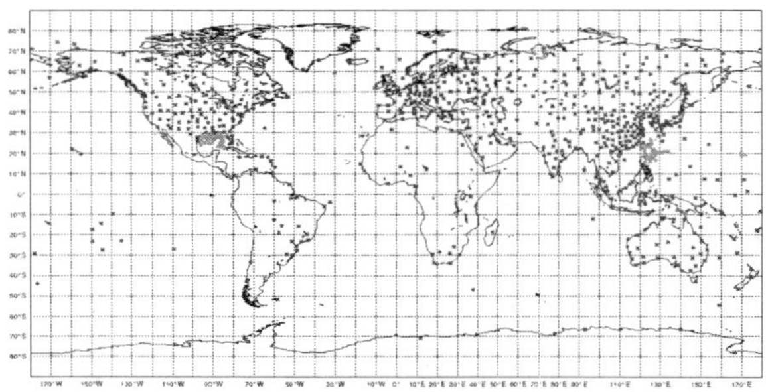

Figure 2 : (color on-line) Data coverage for radio-sonde observations entering the ARPEGE data assimilation system on the 2008 september 7, 00 UTC. Observations from ground stations (red) and dropsondes (green).

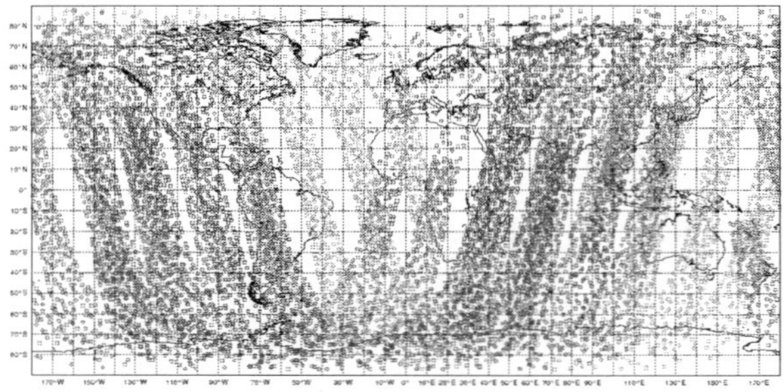

Figure 3 : (color on-line) Data coverage for ATOVS AMSU-A instrument entering the ARPEGE data assimilation system on the 2008 September 11, 00UTC. Observations from NOAA 15 (red), NOAA 16 (brown), NOAA 18 (pink), AQUA (orange) and METOP (blue).

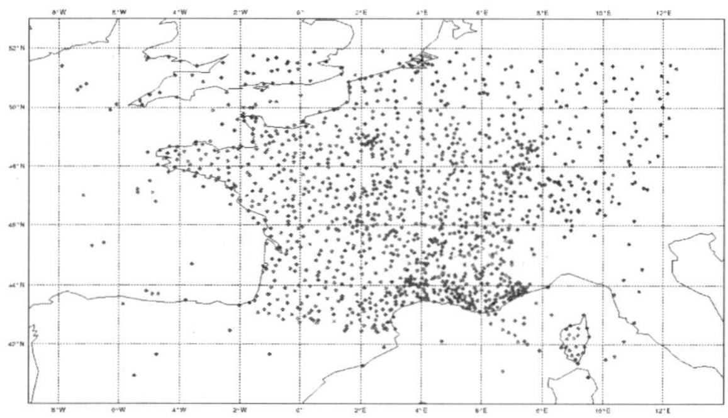

Figure 4 : (color on-line) Data coverage for surface measurements entering the AROME data assimilation system on the 2008 September 11, 00UTC. Observations from the Global Transmitting System ground stations (red), ships (brown), French network (pink) and mountain stations (green).

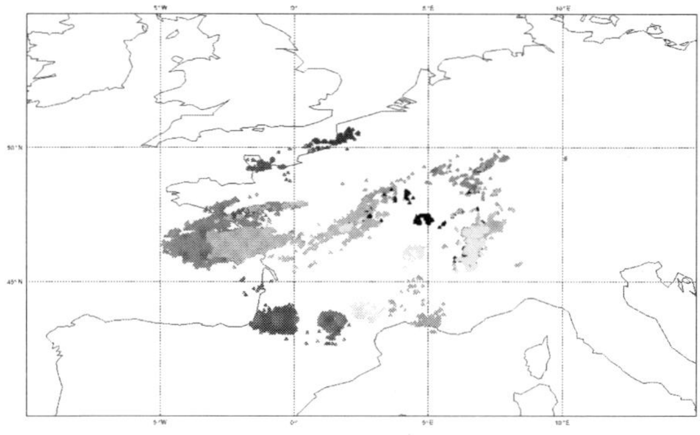

Figure 5 : (color on-line) Data coverage for doppler wind measurements entering the AROME data assimilation system on the 2008 September 11, 00UTC from different RADAR (corresponding to the different colors).

4. A New Meso-Scale Assimilation System

Presently, The French National Weather Prediction System contains a global model (ARPEGE) and a regional one (ALADIN) that have a resolution respectively of 15 km and 10 km over Europe. The AROME project aims at completing this system with a 2.5 km resolution meso-scale model covering the French territory in order to improve local meteorological forecast of potentially dangerous events. During the forecast, such local model needs lateral boundary conditions from a larger model giving meteorological information. At the initial time of the forecast, the state of the atmosphere inside the domain could also be provided by a larger model, downscaling large scale analysis to the thinner grid model. But in order to take advantage of observation network informative at meso-scale, local data assimilation systems have been developed.

4.1. *AROME Data Assimilation System*

The AROME data assimilation system [4] is based on the ALADIN-FRANCE 3D-VAR scheme which is operationally running at Meteo-France at regional scale since the end of 2005 [5]. It is based on the idea that forecasts initialized with more recent observations will be more accurate and on the fact that we want to use high temporal and spatial frequency observations (RADAR measurements for example) so we will take the best possible advantage of mesoscale data.

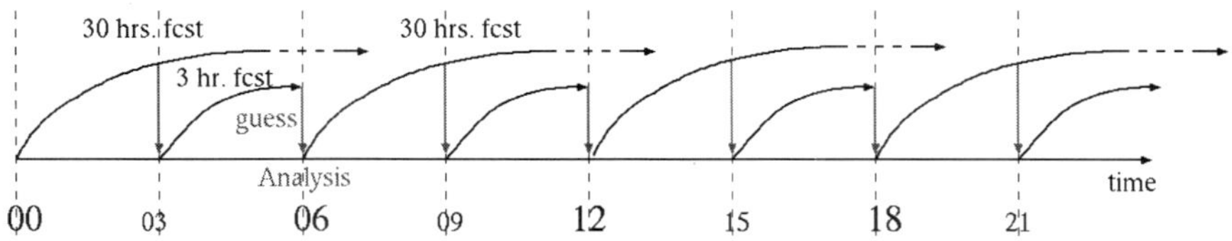

Figure 6 : Schematic representation of the AROME 3h Rapid Update Cycle.

Recent observations are assimilated using the previous model forecast as a background to produce a new estimate of 3-D atmospheric fields in a 3h Rapid Update Cycling (Figure 6). This system analyses the two wind components, temperature, specific humidity and surface pressure fields at the actual model resolution (2.5 km). Others model fields (such as rain) are cycled from the previous AROME forecast without any modification.

4.2. *General Benefit of the AROME Analysis*

In order to evaluate the interest of this meso-scale assimilation system, objective scores compared to radio-sonde measurements have been calculated. These scores, plotted on Figure 7, show an improvement of the bias and the root mean square error for the AROME analysis compared to the ALADIN one for temperature, wind and specific humidity fields all over the troposphere. The general benefit of the AROME analysis appears during the first 12-h forecast ranges, then lateral conditions largely take over the model solution. Objective scores of two forecasts (not shown), initialized either with the AROME analysis or ALADIN one, and using the same lateral conditions become very similar. Subjective evaluation (mainly localization of rain events) confirms forecast improvement during the first 12-h forecast ranges. In some particular cases, this benefit can also be observed after this time-range.

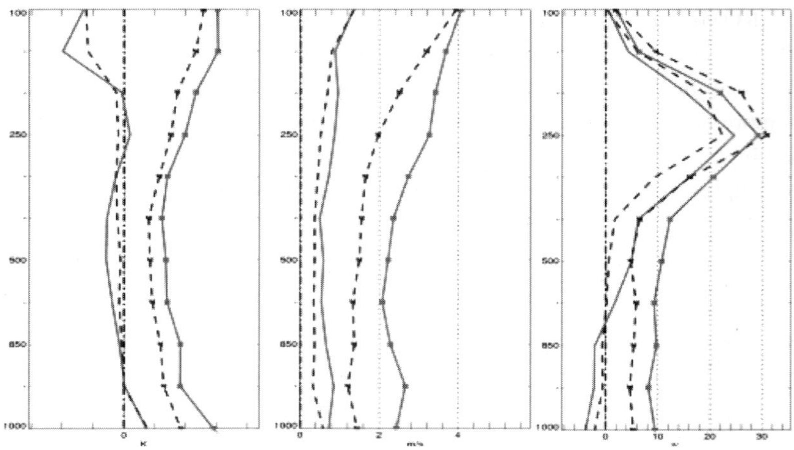

Figure 7: (color on-line) Vertical profile of Bias (simple line) and Root Mean Square Error (cross line) of ALADIN analysis (plain pink line) and AROME analysis (dashed blue line) compared to radio-sonde for temperature (left panel, °K), wind (middle panel, ms^{-1}) and relative humidity (right panel, %) fields. y-axis : pressure (hPa).

4.3. *An Assimilation Experiment*

The effect of the AROME analysis is now illustrated. Two AROME simulations starting on the 2008 February 7, 00 UTC are performed, initialized either with the AROME analysis or ALADIN analysis. Low cloud cover field (Figure 8a and 8b) at 9-hr forecast range compared to the Meteosat 9 image (Figure 8c) show that only the forecast using the AROME analysis is able to simulate the fog event observed in the south-west and north-east of France. In this case, the assimilation of surface parameters, temperature and humidity at 2-m high, allows a better simulation of this boundary layer phenomenon.

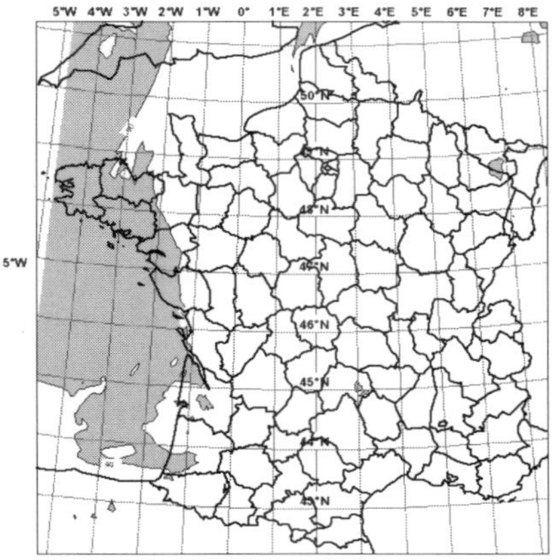

Figure 8a : Low cloud cover field simulated on the 2008 February 7, 09 UTC by an AROME forecast initialized with an ALADIN analysis.

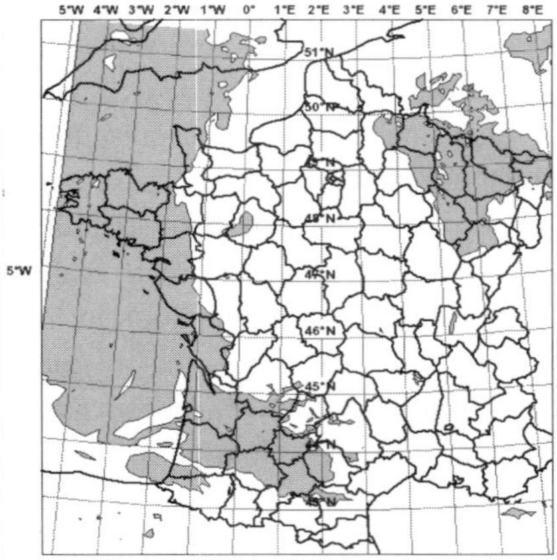

Figure 8b : Low cloud cover field simulated on the 2008 February 7, 09 UTC by an AROME forecast initialized with an AROME analysis.

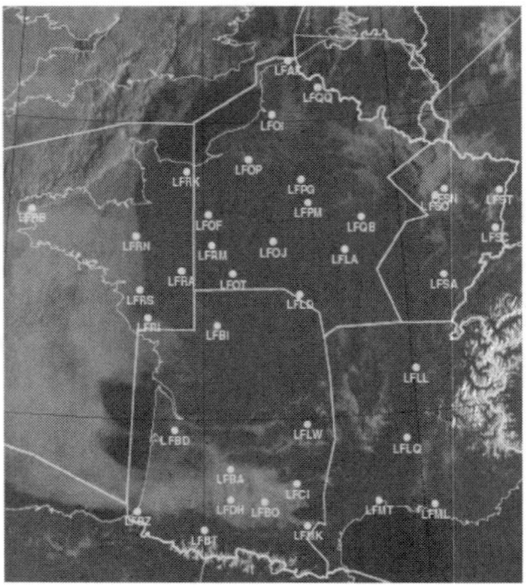

Figure 8c : Visible channel image observed with Meteosat 9 on the 2008 February 7, 09 UTC.

5. Conclusion

Data assimilation is a way to draw regularly a forecast system towards observations in order to stick as much as possible to the true state of the atmosphere (otherwise meteorological models would quickly drift away and give an erroneous information). That operation is done every 6 hours for a global model (more frequently for local area models) and consists in a clever averaging of available observations and a previous 6 hours model forecast.

Only local models that present a finer resolution can be helpful for giving optical turbulence forecast. These models, usually initialized from global models can also perform their own assimilation with specificities : a more frequent cycle (3 hours or less), small scale observations (such as radars or surface observations). Data assimilation in local models shows improvements up to 12-18 hours forecast. This is mostly due to the specific use of proper high density observations over the whole domain, even if the local aspect of the assimilation system provides less spin-up for the forecast (meaning that the model forecast is faster in a balanced state). As a consequence, a specific local model on a remote area would not give a better forecast compared to its initialization from a global model without a dense observations network.

References

1. Benjamin S.G., et al. : An hourly assimilation-forecast cycle : The RUC, Mon. Wea. Rev., 132, 4959-518 (2004).
2. Berre L. : Estimation of synoptic and mesoscale forecast error cavariances in a limited area model, Mon. Wea. Rev., 128, 644-667 (2000).
3. Bouttier F. and Courtier P. : Data assimilation concepts and methods, Meteorological Training Course Lecture Series, ECMWF (2002).
4. Brousseau, P., F. Bouttier, G. Hello, Y. Seity, C. Fischer, L. Berre, T. Montmerle, L. Auger, S. Malardel : A prototype convective-scale data assimilation system for operation: the AROME-RUC. Hirlam technical report N.68 on the SRNWP workshop on data assimilation, 21-23 March 2007, 23-30 (2008b).
5. Fischer et al. : An overview of the variational assimilation in the ALADIN/France numerical weather-prediction system, Quart. J. Roy. Meteo. Soc., 131, 3477-3492 (2005).
6. Lac, C., T. Maric, J.P. Pinty, J. Cuxart, V. Masson, P. Tulet, C. Sarrat, F. Couvreux, F. Bouttier : The mesoscale meteorological models Meso-NH and AROME, ATMO08, Proceedings.

ADAPTATION OF FORCE RESTORE[1] ISBA MODEL TO POLAR CONDITIONS

P. LE MOIGNE* and J. NOILHAN
CNRM, Centre National de Recherches Meteorologiques,
Toulouse, France
** E-mail: Patrick.LeMoigne@meteo.fr*

E. MASCIADRI* and F. LASCAUX
INAF, Osservatorio Astrofisico di Arcetri,
Firenze, Italy
** E-mail: masciadr@arcetri.astro.it*

I. PIETRONI*
ISAC, Istituto di Scienze dell'Atmosfera e del Clima,
Consiglio Nazionale delle Ricerche (CNR),
Roma, Italy
** E-mail:i.pietroni@isac.cnr.it*

This article describes how the simple ISBA[2,3] soil vegetation atmosphere transfer scheme has been adapted to polar conditions of Antarctica to retrieve surface and deep temperatures of polar cap. The main goal of the study is to improve the initialization of surface temperature of the French mesoscale model meso-NH.[4,5]

Keywords: Surface scheme; surface temperature; energy budget; calibration.

1. Introduction

In the frame of Arome[a],[6] a new project of Meteo-France in terms of numerical weather prediction, joint efforts of CNRM (Centre National de Recherches Meteorologiques, Toulouse, France) research departments, Aladin[7] community and Laboratoire d'Aerologie (Universite Paul Sabatier, Toulouse, France) will contribute to the development of a new meso-scale weather prediction model for research activities, operational data assimi-

[a]Applications de la Recherche a l'Operationnel a MesoEchelle

lation and forecast (2008). In the development of Arome, the modeling of surface processes has been externalised in order to have a full compatibility between the surface representation in research and operational models. One of the objectives of the externalisation is to better take into account the heterogeneity of the surface with specific physical parameterisations over town, vegetation, lake and open water. The part of the numerical code concerning the interaction between soil, biosphere and atmosphere (ISBA), that characterizes the exchanges of energy between the atmosphere and the vegetation, has been adapted to Polar conditions in order to be used at DomeC (Antarctica). This specific study has been conceived in the context of a collaboration set-up between the mesoscale modeling group of CNRM and the ForOT Team of INAF. The latter are, at present, involved in researches aiming to characterize and quantify the optical turbulence developed above potential astronomical sites in the Internal Antarctic Plateau.

2. Observations

A meteorological dataset (starting on the 28th of February) is available at DomeC for the year 2005 (Osservatorio Meteo-Climatologico at DomeC, PNRA, Italy). The automatic weather station (AWS) measures wind speed and direction at 3 meters high, surface pressure and finally temperature at 1.2 meters above ground. The time series of these parameters are shown on figure 1.

Total precipitation and humidity near the ground are unknown but they're required to realize the numerical simulations. On the Antarctic Plateau, accumulated precipitation (snow in this region) are weak and smaller than 50mm a year ([8]). For that reason, precipitation have been neglected. Specific humidity near the ground can be retrieved from the knowledge of air temperature and surface pressure, assuming that in antarctic regions, the relative humidity is nearly constant. It has been set to 50%. Formulae of Goff-Gratch (1961) are used to retrieve saturation vapour pressure over ice.

$$\log_{10}(e_i) = a - b * T - c/T + d \log_{10}(T) \tag{1}$$

Equation 1 gives the expression of $\log_{10}(e_i)$ as a function of temperature T with coefficients $a = 2.07023$, $b = 0.00320991$, $c = 2484.896$ and $d = 3.56654$.

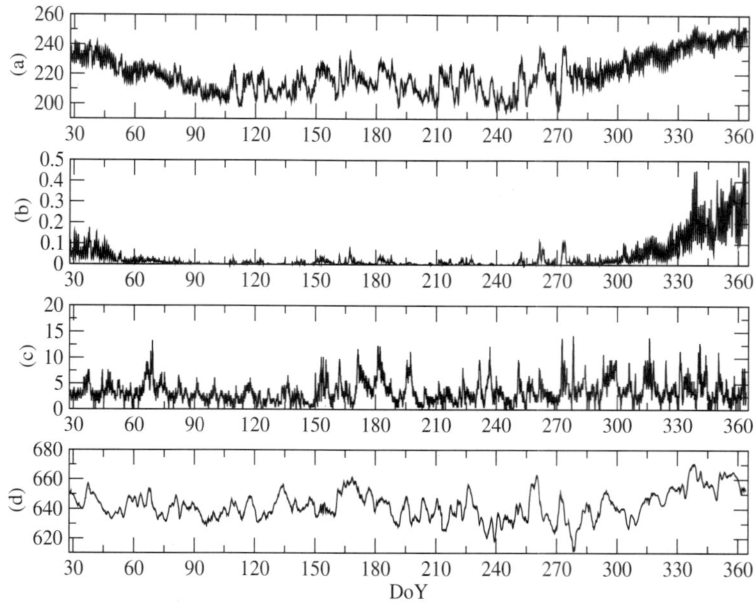

Fig. 1. Observed parameters as a function of Day of Year (DoY) used to force surface scheme: (a) air temperature (K), (b) air specific humidity (g/kg), (c) wind speed (m/s), (d) surface pressure (Pa)

$$e = e_i * RHU \tag{2}$$

In equation 2, RHU represents relative humidity (50% here) and e is the vapour pressure:

$$q = \frac{0.622e}{p - 0.378e} \tag{3}$$

The order of magnitude of specific humidity is very low compared to our latitudes and reflects the dryness of the Antarctic Plateau. Equation 3 expresses specific humidity q as a function of e and pressure p.

On top of that, measurements of temperature done at -5cm, -15cm and -30cm in the snow as well as radiative atmsopheric fluxes (ISAC/CNR, Rome, Italy) are available for this period and are used to calibrate some parameters of the model (described in 3.2) and to evaluate the capability of the model to retrieve a correct thermal vertical profile in the first centimeters. Observed longwave and shortwave downward radiation terms are used as an input of the model and are shown in fig 2. During winter time, the solar radiation equals to zero and the exchanges of energy between the surface and the atmosphere are driven by the infrared radiation term.

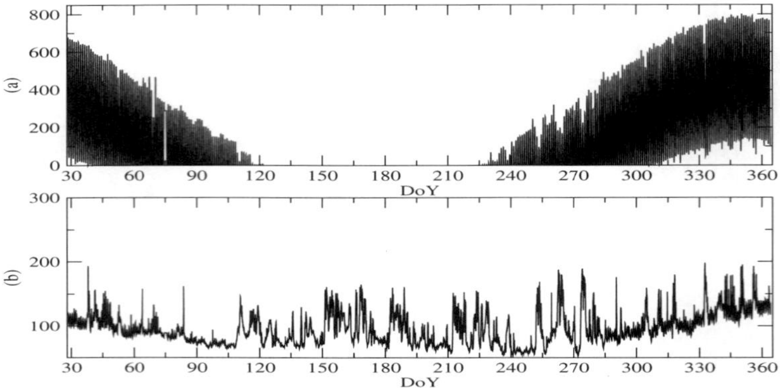

Fig. 2. Radiative forcing at DomeC: (a) solar radiation (W/m2), (b) infrared radiation (W/m2)

3. Numerical Aspects

3.1. *ISBA Model*

A simple version of the surface scheme, with only two layers to represent the heat transfer, is used to simulate the surface temperature representative of the first centimeters and a deep temperature that corresponds to the daily average of the surface temperature. Deep temperature is representative of a thickness of a few tens of centimeters. On top of that, a climatological temperature is used to avoid a thermal drift of the system. This region of the Antarctic continent is essentially covered by snow. For that reason, the soil of the model is treated as frozen, which means that there are no transfer of liquid water and characteristics of ice are used to describe the soil column. The thermal coefficient defined in the model is expressed following *Boone, 2000*[9] like:

$$C_{ice} = 2\sqrt{\frac{\pi}{\lambda_i c_i \tau}} = 9.47 \times 10^{-6} \quad \mathrm{m^2KJ^{-1}} \tag{4}$$

τ is the relaxation time of the force restore model for the temperature, it's equal to 1 day, c_i is the ice thermal capacity and λ_i is the ice thermal conductivity. Thermal capacity is deduced from ice specific heat C_i and density ρ_i like it is shown in equation 5.

$$c_i = \rho_i C_i \tag{5}$$

Table 1. Thermal characteristics

symbol	Unit	Value	Meaning
ρ_i	kg m^{-3}	350	density
λ_i	W m^{-1} K^{-1}	2.2	thermal conductivity
C_i	J K^{-1} kg^{-1}	2106	specific heat

According to *Dang, 1997*,[10] density ρ_i is close to 350 kg m^{-3} in the first two meters of ice column and then increases with depth.

3.1.1. *Surface characteristics*

Main surface parameters needed by the model are summarized in table 2

Table 2. Main surface parameters

Name of surface parameter	Value	Unit
albedo	0.84	[-]
emissivity	0.98	[-]
roughness length	0.001	[mm]
number of layers	2	[-]
1st layer depth	0.05	[m]
2nd layer depth	0.57	[m]

3.1.2. *Model equation*

The time evolution of deep temperature T_2 in ISBA model involves the surface temperature T_s and a climatological temperature T_c associated to a relaxation time γ:

$$\frac{\partial T_2}{\partial t} = \frac{1}{\tau}(T_s - T_2) + \frac{1}{\gamma\tau}(T_c - T_2) \tag{6}$$

The bigger γ is, the less T_c influences T_2. Inversely, if γ is small (1 for instance), then the role of T_c will be as important as the surface temperature's one.

3.2. *Calibration*

The calibration of the climatological temperature and its relaxation time have been done on a monthly basis by minimizing the root mean square error between the simulated deep temperature and the temperature observed at -30cm. The calibrated parameters are given in table 3 and the figure 3 shows the results of the simulation with and without calibration, compared to the observed temperature at -30cm in snow. There's a strong improvement in the computation of deep temperature, which leads to an improvement of surface temperature.

Table 3. Monthly calibration

month	Jan	Feb	Mar	Apr	May	Jun	Jul	Aug	Sep	Oct	Nov	Dec
$T_c(K)$	236	236	220	209	206	211	214	210	207	212	220	209
γ	4	4	4	3	1	2	3	1	1	1	1	2

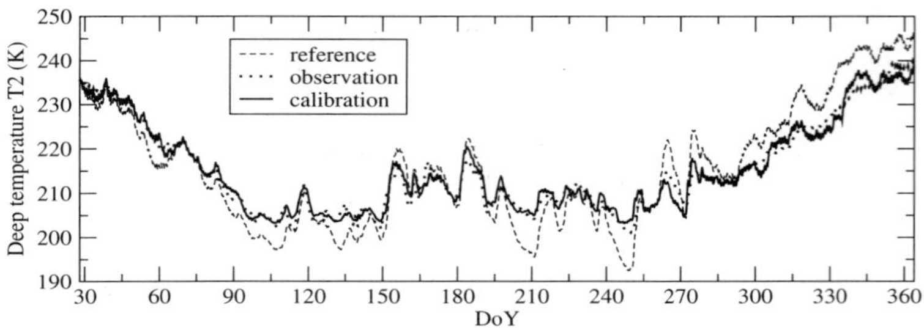

Fig. 3. Results of calibrated model for deep temperature

During April (between DoY 90 and 120), the reference run shows a cold bias of about 10 degrees compared to observations. Deep temperature ranges from $195K$ to $204K$ and the climatological temperature is much warmer ($209\ K$) and effectively corrects the bias.

4. Results

4.1. *Turbulent Fluxes*

In terms of surface turbulent fluxes, the impact of the calibration is rather strong during summer time for sensible heat flux, when solar radiation is present, but this impact is small during winter time because of the very strong air stability. From February to mid-September, the calibration increases the sensible heat flux because the reference run is colder than the calibrated one, while it decrases it for the rest of the year because reference run is warmer (figure 4(a)). Snow sublimation is usually small over snow, and the calibration tends to decrease it (figure 4(b)).

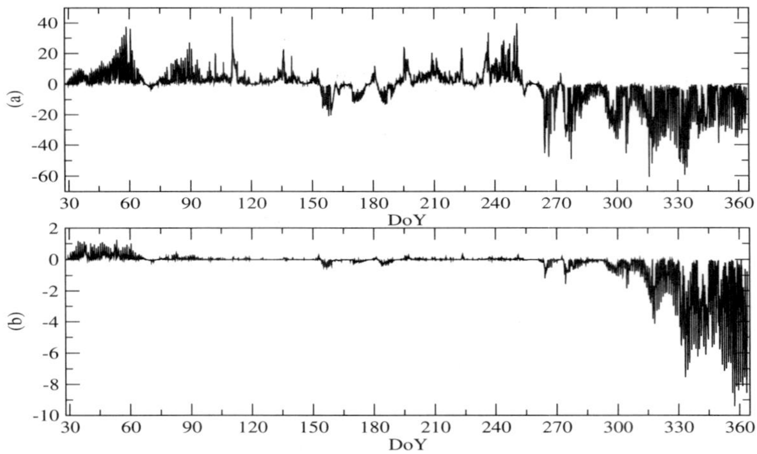

Fig. 4. Difference between reference and calibrated model for: (a) sensible heat flux (W/m2), (b) latent heat flux (W/m2)

5. Conclusion

A simple model, usually used to simulate the exchanges of water and energy over natural areas has been adapted and used over Antarctica. Force restore ISBA model has been used in a stand-alone mode at Concordia where time series of meteorological quantities have been observed or calculated. The main parameters that have been modified are firstly the thermal coefficient of the soil (which is directly used in the surface temperature equation) and secondly a climatological temperature that has been calibrated monthly to avoid a thermal drift of the system and to correct the differences between the observed temperature at -30cm and the simulated ones. In terms of results, it has been shown that the representation of ISBA deep and surface temperatures have been improved. The perspective of this work will be to evaluate the impact of the modifications when surface is coupled to the atmospheric meso-NH model, especially for optical turbulence purposes.

References

1. R. E. Dickinson, *Journal of Climate* **1**, 1086 (1988).
2. J. Noilhan and S. Planton, *Monthly Weather Review* **117**, 536 (1989).
3. J. Noilhan and J.-F. Mahfouf, *Global and Planet Change* **13**, 145 (1996).
4. J.-P. Lafore *et al.*, *Journal of Climate* **16**, 90 (1988).
5. C. Lac, in *OTAM08 Proceedings*,
6. P. Brousseau, in *OTAM08 Proceedings*,
7. Bubnova *et al.*, *Monthly Weather Review* **123**, 515 (1995).
8. J. King and J. Turner, *Antarctic Meteorology and Climatology*, 2nd edn. (Cambridge University Press, 2007).
9. A. Boone, Modelisation des processus hydrologiques dans le schema de surface isba: Inclusion d'un reservoir hydrologique, du gel et modelisation de la neige, PhD thesis2000.
10. H. Dang *et al.*, *Annals of Glaciology* **25**, 170 (1997).

MESOSCALE NWP OVER ANTARCTICA: AMPS AND SUPPORT FOR GROUND-BASED ASTRONOMY

JORDAN G. POWERS†

Mesoscale and Microscale Meteorology Division, Earth & Sun Systems Laboratory, National Center for Atmospheric Research, P.O Box 3000, Boulder, Colorado, USA

The Antarctic Mesoscale Prediction System (AMPS) effort is one by the United States to provide real-time, high-resolution numerical weather prediction (NWP) over Antarctica. AMPS is a mesoscale modeling system run at the U.S. National Center for Atmospheric Research (NCAR) that generates twice-daily forecasts for the U.S. Antarctic Program. As AMPS and its products have been adapted over the years to provide NWP guidance over Antarctica, supporting ground-based astronomy may be a possible extension of the system. Presented are an introduction to AMPS, its capabilities, and its potential for support of weather forecasting for ground-based astronomy in Antarctica. Limitations are also discussed.

1. Background

The Antarctic Mesoscale Prediction System (AMPS) [1] is an experimental mesoscale numerical weather prediction (NWP) capability providing support to the weather forecasting for the United States Antarctic Program (USAP). The AMPS effort is supported by the U.S. National Science Foundation Office of Polar Programs, and the principal partners are the National Center for Atmospheric Research (NCAR), The Ohio State University (Byrd Polar Research Center— BPRC), and the University of Colorado. Since 2000, AMPS has served both the forecasting and research communities. For the USAP, AMPS has provided forecasters with model guidance at the highest spatial and temporal resolution available for Antarctica. Moreover, the model products have been tailored to their needs, reflecting close interaction of AMPS with SPAWAR (Space and Naval Warfare Systems Center, Charleston), the organization that supplies the meteorologists. In addition to supporting the USAP forecasters, over the years AMPS has served scientific field campaigns and high-profile emergency rescue missions. In addition, to researchers in

† Work supported by U.S. National Science Foundation Office of Polar Programs.

Antarctic meteorology and climatology, AMPS has provided a capability for process and event studies, a platform to test parameterizations in a polar region, and a database of high-resolution numerical output. Lastly, AMPS has offered NWP support for foreign nations as resources have allowed. Given its history of generating weather products for science on the Ice, AMPS may be able to provide items useful for those doing optical turbulence forecasting for ground-based astronomy.

The mesoscale model in AMPS is the Weather Research and Forecasting (WRF) Model [2]. Developed since the late '90s as an NWP tool for both research and operations, WRF currently has over 8500 registered users in 114 countries worldwide. For AMPS, "Polar WRF' is used, a version of the model that has been tailored for the high latitudes [3] (See Sec. 2). At this time, AMPS has not been used to generate forecast products targeted at optical turbulence or the needs of ground-based astronomy. This paper looks ahead to such possibilities.

A number of modeling efforts discussed at the Optical Turbulence: Astronomy Meets Meteorology conference have addressed the current applications of mesoscale models in forecasting for ground-based astronomy. The main example is that of the use of the Meso-NH model by the ForOT Team [4]. Recent notable results of the ForOT* group include the application of Meso-NH to Dome C [5]. Other efforts have been those of the University of Hawaii (UH) for the Mauna Kea Observatory using the MM5 and WRF models; of Meteo-France for Antarctica and Dome C using Meso-NH; and of the University of Valparaiso using the MM5 for simulations over candidate telescope sites in Chile. The UH's forecasting in the Mauna Kea Weather Center represents experience in real-time NWP of optical turbulence parameters.

2. The Antarctic Mesoscale Prediction System

AMPS runs at NCAR and produces two forecasts per day. A host of products are derived from the model output and distributed to users through various mechanisms, primarily the AMPS web page (described below). As noted, AMPS uses a version of WRF modified for use in the polar regions. The modifications have been developed primarily by The Ohio State University (Polar Meteorology Group, BPRC), with other contributions from NCAR and the National Centers for Environmental Prediction. The modifications better

---

* Information on ForOT may be found at http://forot.arcetri.astro.it.

capture conditions unique to the high latitudes and over extensive ice sheets. In contrast, the physics and configuration for the standard version of WRF, as with most mesoscale models, reflect basic development and predominant application in the middle latitudes.

The polar modifications to WRF mostly involve adjusted lower boundary and land surface characteristics. First, Polar WRF accounts for fractional sea ice coverage in grid cells. Grid cells are quantified as fractionally covered if not fully ice-covered or fully ice-free, as indicated by daily sea ice analyses obtained from the (U.S.) National Snow and Ice Data Center. In contrast, in the default WRF grid cells are either fully open ocean or fully sea ice covered. Many of the polar changes are to the Noah land surface model (LSM) that is available to WRF. These are: the use of the latent heat of sublimation for calculations of latent heat fluxes over ice surfaces (permanent ice, sea ice, and snow cover); the assumption of ice saturation when calculating surface saturation mixing ratios over ice; an increase in the values of snow albedo and emissivity; adjustment of snow density; adjustment of snow heat capacity and thermal diffusivity for the subsurface layer; and a modification of the skin temperature calculation (enforcing a sensible heat/latent heat/radiation energy balance at the surface). Other modifications outside of the Noah LSM are: a stability-dependent formulation for the thermal roughness length and a modified initialization of low-level air temperatures (using an interpolation of the previous forecast's subsurface temperature and the first-guess field's low-level air temperatures). In addition, soil temperatures are cycled in polar WRF; that is, the previous forecasts' subsurface temperatures are used in the initializations of new forecasts. Lastly, the shortwave radiation scattering parameter is reduced when running Polar WRF over Antarctica to account for less clear-air scattering. Polar WRF modifications are described in Powers and Manning (2008) [6] and Hines and Bromwich (2008) [3].

AMPS has six computational/forecast grids, shown in Fig. 1. It features an outermost grid of 45-km horizontal spacing, with nesting to a finest grid of 1.67-km spacing. The 45-km grid extends to the middle latitudes of the Southern Hemisphere, a 15-km nest covers the Antarctic continent, and three 5-km grids cover the western Ross Sea, the South Pole, and the Antarctic Peninsula (Figs. 1(a),(b)). Finally, a 1.67-km grid (Fig. 1(b)) covers the critical Ross Island and McMurdo Station region. All nesting is two-way interactive. The model top is at 10 hPa, with vertical resolution reflecting 44 computational levels (η-levels) (see [1]) between the ground and the top.

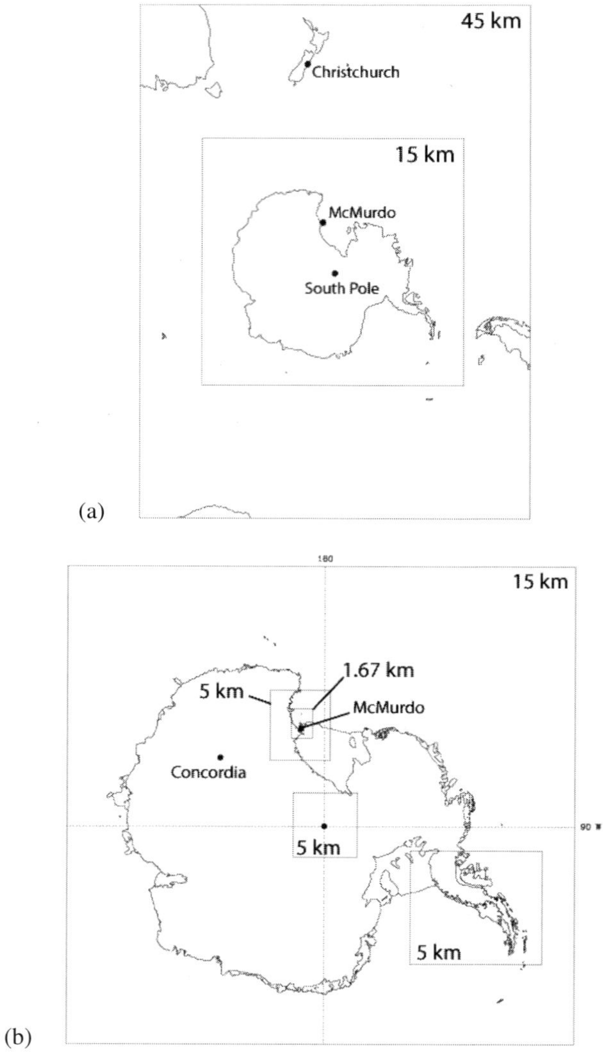

Figure 1. AMPS forecast grids. (a) 45-km Southern Hemisphere (outer frame) and 15-km Antarctica (inner frame) domains. Christchurch, New Zealand, McMurdo Station, and South Pole indicated. (b) 15-km Antarctica, 5-km western Ross Sea, 5-km South Pole, 5-km Antarctic Peninsula, and 1.67-km Ross Island grids. Concordia Station (Dome C) and McMurdo Station indicated.

Two forecasts per day are produced, initialized at 0000 and 1200 UTC. The forecast lengths are 120 hr for the 45- and 15-km grids and 36 hr for the 5- and 1.67-km grids. For WRF initialization a three-dimensional variational data assimilation (3DVAR) system is used (WRF-Var). This blends observations into the first-guess field that comes from NCEP's GFS (Global Forecasting System) model. The observations used for reanalysis include standard surface and upper-air observations (*e.g.*, surface stations, AWSs (Automatic Weather Stations), ships, buoys, radiosondes, aircraft), satellite cloud-track winds, COSMIC (Constellation Observing System for Meteorology, Ionosphere, and Climate) radio occultations, and MODIS (Moderate-Resolution Imaging Spectroradiometer) winds. AMSU-A (Advanced Microwave Sounding Unit-A) radiance measurements have been tested in AMPS and shown to have a slight positive impact, but further development of optimal assimilation of these by WRF is in progress.

As resources allow, AMPS provides forecast products to the international Antarctic community. The main users have been: Italy, the UK, Germany, Australia, South Africa, and Chile. AMPS output is used by the German Weather Service (Deutscher Wetterdienst) in support of Dronning Maud Land Air Network (DROMLAN) forecasting and activities, thus assisting the DROMLAN nations: Belgium, Finland, Germany, India, Japan, the Netherlands, Norway, Russia, South Africa, Sweden, and the UK. While the AMPS effort attempts to help non-USAP entities when possible, their assistance is secondary in priority to AMPS's support of USAP forecasting and needs.

The model output is delivered to users through various means. The primary venue is the AMPS web page: www.mmm.ucar.edu/rt/amps. This is available to all, with no password required. The page features selectors for forecast initialization, forecast hour, grid, and products. In addition, products (forecast plots, tables, etc.) are posted to a server for downloading via the web. Some users, with limited internet connectivity or bandwidth, have AMPS forecast products e-mailed directly to them. Lastly, gridded model output, such as temperature and wind fields, is put on the web page and the Antarctic Internet Data Distribution system (Antarctic IDD) [7]. The Antarctic IDD provides a data stream of Antarctic observations and model output that may be acquired with local data manager (LDM) software from Unidata (See www.unidata.ucar.edu/software/ldm).

3. Possibilities for Astronomy and Related Issues

The predominant use of AMPS is for flight forecasting. The key routes for the United States are Christchurch, New Zealand to McMurdo Station and McMurdo to Pole (Fig. 1(a)). In addition, the forecasters support flights from McMurdo to field camps across the Ice. AMPS also supports the meteorologists responsible for issuing marine forecasts for USAP research vessels operating in Antarctic waters, as well as for the annual supply convoy that goes to McMurdo in mid-summer (austral). AMPS has also been used for field campaign forecasting and to support forecasting for Antarctic rescues, typically medical evacuations. As noted, AMPS has not yet been applied for forecasting for ground-based astronomy. Given its flexibility in addressing various forecasting needs over the continent, however, it has potential to contribute to this.

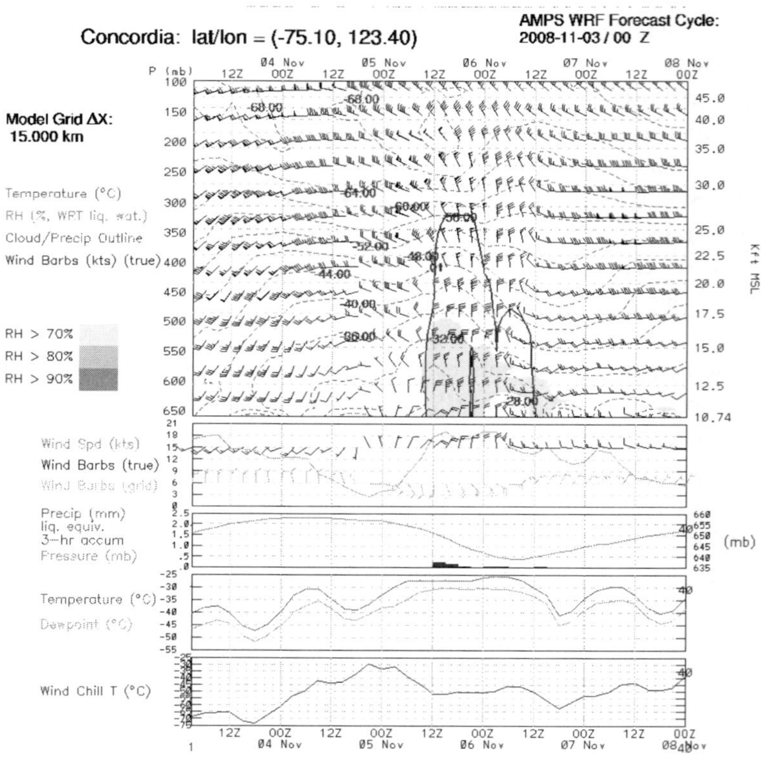

Figure 2. AMPS meteogram for Concordia Station, based on 00 UTC 3 Nov 2008 forecast. Forecast dates/times shown along bottom. Top panel: Temperature (°C), RH (%), cloud outline, and wind barbs. Second panel: Wind speed (kt) and wind barbs (deg true and grid). Third panel: 3-hourly accumulated precipitation (mm) and surface pressure (mb). Fourth panel: Temperature (°C) and dewpoint temperature (°C). Fifth panel: Wind chill temperature (°C).

Examples of AMPS's products indicate how it might be useful for forecasters for astronomy in Antarctica. AMPS's output ranges from standard weather charts to specialized items like "pseudo-satellite" plots, which depict model-produced clouds in satellite imagery format. Of import to astronomers may be products such as meteograms generated for individual sites on the ice. Figure 2 provides an example, for Concordia Station. This depicts surface and upper-air parameters (e.g., wind, temperature, pressure, cloud) at this location in times series format. Time series depictions could represent the variation of optical turbulence or seeing conditions over a given forecast period.

Figures 3 and 4 provide examples of vertical views through the troposphere and stratosphere from AMPS. Figure 3 shows a 24-hr forecast sounding for South Pole, with the sounding plotted on a skew-T diagram. For vertical temperature and dewpoint information in a different format, Fig. 4 presents a profile plot. Profiles may be generated for other parameters of interest, such as wind speed, cloud, Richardson number, etc. Moreover, these should be extendable to depict vertical profiles of optical turbulence strength. Thus, the real-time capability of AMPS might be able to offer various optical turbulence or seeing products for scheduling telescope operations on the Ice.

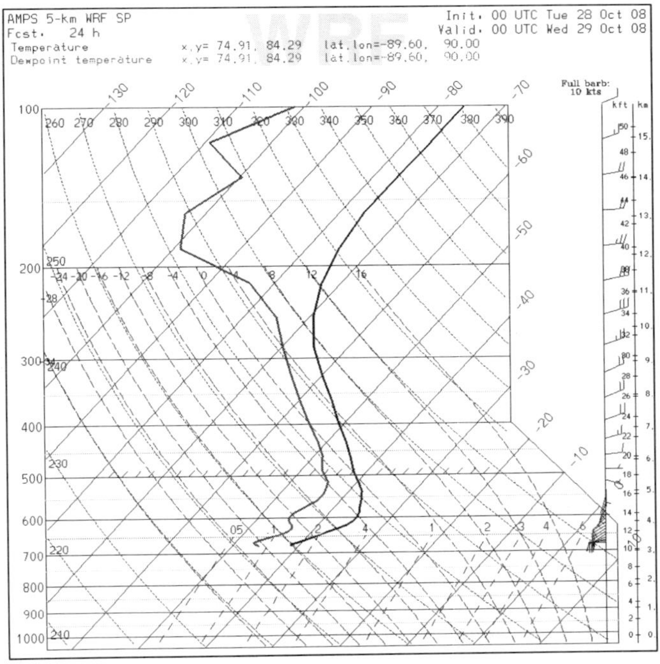

Figure 3. (color on-line) South Pole sounding for hour 24 of AMPS forecast initialized at 00 UTC 28 Oct 2008 (valid 00 UTC 29 Oct 2008). Temperature (°C), blue (right); dewpoint temperature (°C), red (left).

A relevant resource related to AMPS is the AMPS archive: a repository of the gridded forecast output accumulated since 2001. The archive can offer model datasets for investigations of past time periods or areas. For example, there may be interest in examining AMPS's forecasts of atmospheric conditions in a selected area of the Antarctic (*e.g.*, around Dome C) for periods in which optical turbulence or seeing observations have been made.

AMPS's employment of a polar-modified model is a plus for its possible use in ground-based astronomy over Antarctica. The polar modifications have been found to have a positive impact on WRF's simulations in polar regions (see, *e.g.*, [3] and [8]), and they have improved predictions of atmospheric conditions over Antarctica [9]. The polar modifications for WRF are scheduled for release to the public in 2009. This will make Polar WRF available to others who may want to test and apply the model to astronomical forecasting over the high latitudes.

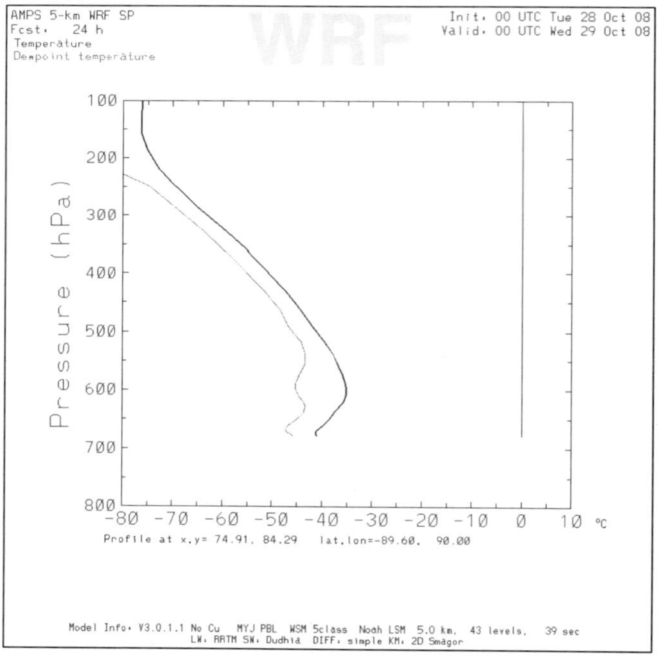

Figure 4. (color on-line) South Pole temperature (°C) and dewpoint temperature (°C) profiles for hour 24 of AMPS forecast initialized at 00 UTC 28 Oct 2008 (valid 00 UTC 29 Oct 2008). Temperature, blue (right); dewpoint termperature, red (left).

Its potential notwithstanding, there are limitations on AMPS's capabilities for supporting astronomical forecasting. The fundamental ones are mission priority and resources. As noted, the primary role of AMPS is to serve the USAP: thus, attempts to support the needs of astronomers would have a lower priority. This limitation is similar to AMPS's provision of other weather products to the international community. In addition, work would be necessary to develop and implement the calculations and generation of products needed for optical turbulence forecasting. The level of effort may require that support be provided.

A second caveat on the use of AMPS is that the system is untested for astronomical forecasting. No products for optical turbulence or seeing have yet been produced from AMPS. It is unknown how they would verify, and, as noted below, verification and iteration would be expected in any product development.

Third, WRF in AMPS is limited for the foreseeable future to the current vertical and horizontal resolutions. Thus, any optical turbulence or seeing products will reflect the WRF grid setups described above. It is unknown how well the vertical resolution, particularly in the surface layer and PBL, will suffice to accurately depict and forecast the astronomical parameters. The calculations and the accuracy will be resolution-dependent. Output from the 15-km grid, as over Dome C where a telescope might be sited, might not suffice for generating the optical turbulence forecast information of the precision and accuracy required. Increasing the model resolution for the sake of astronomical forecasting is unlikely, however, given that AMPS's primary mission is to the USAP and that a key constraint of time-to-forecast of the model must still be met.

Fourth, one aspect that would be necessary for the AMPS program in an optical turbulence forecasting effort is the collection of relevant observations for verifying the output. Verification of forecasts and estimation of the accuracy of model-based parameters are always important to system understanding and development, so measurements of optical turbulence or seeing would be needed. The verification process, furthermore, may reveal where modifications are needed in the model physics, and measurements would be used to assess forecast improvements from such changes. Thus, future use of AMPS for forecasting support for ground-based astronomy should involve verification using optical turbulence or seeing data from the forecast sites.

4. Summary

A number of numerical weather prediction (NWP) models operate in real-time over Antarctica. At this time, however, these are mainly represented by global models covering the continent. An exception is the Antarctic Mesoscale Prediction System (AMPS). AMPS is a real-time mesoscale modeling system for Antarctica providing twice-daily predictions with the highest-resolution forecasts available over the Ice. Its primary purpose is the support of meteorologists of the United States Antarctic Program, although it has supported a range of activities from research to field campaigns to rescue missions. AMPS uses a polar-optimized version of the Weather Research and Forecasting (WRF) model configured with six domains ranging in horizontal grid spacing from 45 km to 1.67 km. A large suite of model-based forecast products has been developed over the years and is provided mainly through a web site.

AMPS has the potential to offer real-time forecast products for ground-based astronomy over Antarctica. Its gridded output may be used for generating prognostic optical turbulence or seeing products for specific sites on the Ice, including the Dome C area, a favorable locale for telescope operations. The predicted parameters could be used for telescope planning out to periods of five days. Optimally, any forecasts or products of astronomical parameters coming from AMPS would be verified with in-situ observations.

It is recognized that AMPS, while applied to a range of logistic and scientific activities thus far, has not yet been tested for ground-based astronomy. In addition, any application of the system for astronomy in Antarctica would have a secondary priority in relation to the system's mandate of USAP support.

References

1. Powers, J.G., A.J. Monaghan, A.M. Cayette, D.H. Bromwich, Y-H. Kuo, and K.W. Manning, 2003: Real-time mesoscale modeling over Antarctica: The Antarctic Mesoscale Prediction System (AMPS). *Bull. Amer. Meteor. Soc.*, **84**, 1533–1545.
2. Skamarock, W.C., J.B. Klemp, J. Dudhia, D.O. Gill, D.M. Barker, M.G. Duda, X.-Y. Huang, W. Wang, and J.G. Powers, 2008: A Description of the Advanced Research WRF Version 3. NCAR Tech. Note, NCAR/TN–475+STR, 113 pp.
3. Hines, K.M., and D.H. Bromwich, 2008: Development and testing of Polar WRF. Part I: Greenland ice sheet meteorology. *Mon. Wea. Rev.*, **136**, 1971–1989.

4. Masciadri, E., 2006: Optical turbulence forecasts: Towards a new era of the ground-based astronomy. *SPIE Orlando, Ground-based and Airborne Telescopes*, 6267, 62671C.
5. Lascaux, F., E. Masciadri, S. Hagelin, and J. Stoesz, 2008: Meso-NH simulations of the atmospheric flow above the internal Antarctic plateau. *Proc. SPIE*, 7012, 70124D.
6. Powers, J.G., and K.W. Manning, 2008: Comparisons of polar WRF and polar MM5 in Antarctica. *9th WRF Users' Workshop*, Boulder, CO, National Center for Atmos. Res.
7. Lazzara, M.A., G. Langbauer, K.W. Manning, R. Redinger, M.W. Seefeldt, R. Vehorn, and T. Yoksas, 2006: Antarctic internet data distribution (Antarctic-IDD) system. *22d Int'l Conf. on Interactive Information Processing Systems for Meteorology, Oceanography, and Hydrology*, Atlanta, GA, Amer. Meteor. Soc.
8. Hines, K.M., D.H. Bromwich, and L.-S. Bai, 2008: Polar WRF. *9th WRF Users' Workshop*, Boulder, CO, National Center for Atmos. Res. http://www.mmm.ucar.edu/wrf/users/workshops/WS2008/abstracts/6-03.pdf
9. Powers, J.G., and K.W. Manning, 2007: Polar WRF testing in Antarctica. *8th WRF Users' Workshop*, Boulder, CO, National Center for Atmos. Res.

ForOT: A NEW APPROACH FOR THE OPTICAL TURBULENCE STUDIES APPLIED TO THE GROUND-BASED ASTRONOMY

E. MASCIADRI*, F. LASCAUX, S. HAGELIN, J. STOESZ

INAF, Osservatorio Astrofisico di Arcetri, Firenze, Italy
**E-mail: masciadri@arcetri.astro.it*

The characterization of the optical turbulence (OT) done with meso-scale models for astronomical applications is an alternative approach to this science that intrinsically presents some interesting and complementary features/advantages with respect to the characterization done with measurements.

The most important advantages are namely: (1) the possibility to describe a 3D map of the C_N^2 in a region around a telescope, (2) the possibility to forecast the optical turbulence i.e. to know with some hours in advance the state of the turbulence conditions above an astronomical site and (3) the possibility to perform a climatology of the optical turbulence extended over decades. No other tool of investigation with comparable potentialities can be figured out at present to achieve these 3 scientific goals.

The forecast of the optical turbulence is a fundamental requirement for the optimization of the management of the scientific programs to be carried out at ground-based telescopes foci. Ground-based astronomy will remain an appealing option for astronomers with respect to the spaced-based one only if the telescopes management will be performed taking advantage of the best turbulence conditions. The future of new ground-based telescopes generation relies therefore upon the success of these studies.

ForOT is a scientific project but even more, it identifies a philosophic approach to studies related to the characterization of the optical turbulence in an astronomical context. In this contribution we will deal about the main success obtained so far in this discipline in the past, the new goals predetermined by ForOT and the main results we obtained so far. To conclude, we will trace a perspective at long time scale indicating where our research is addressed to and how the scientific community and the managers who lead the ground-based astronomical facilities can support our researchers to progress in these studies.

Keywords: Optical turbulence; turbulence modelling; site testing.

1. Optical Turbulence Modelling in Astronomy

The characterization of the optical turbulence is fundamental to exploit, in a suitable way, the potential angular resolution achievable with the large ground-based facilities. There are several reasons that get appealing the ground-based astronomy with respect to the space-based one, provided the turbulence effects on the wavefronts are known and corrected: **(1)** a longer typical life time of the telescopes, **(2)** a lower financial investment and **(3)** an easier management of the astronomical facilities. At present there exist many instruments able to quantify one or many astroclimatic parameters characterizing the turbulence. Measurements can in principle access to all the spatial and temporal frequencies (the whole spectrum) of the optical turbulence (OT). However, *'measurements'* can provide only local estimates that, depending on the typology of the instrument, can be function of the single point, as is the case of the in situ instruments, or function of a single line of sight, as is the case of the optical instruments. The instruments can not provide a 3D C_N^2 maps in a region around a telescope and 3D C_N^2 maps in a region around a telescope placed in the future. These are by far the two main potentialities for mesoscale atmospheric models in Astronomy in the context of an exhaustive characterization of the OT. We refer the readers to Masciadri (2009)[1,2] for an extended introduction to the topic of the optical turbulence modeling in the astronomical context.

The optical turbulence is quantified by the constant of the structure function of the refractive index (C_N^2), and can be coded with suitable algorithms in atmospheric models.[4] The C_N^2 is a so called *'diagnostic parameter'*, i.e. a parameter that does not affect the spatio-temporal evolution of the atmospheric flow. Among the main parameters that can be simulated by 3 dimensional maps by an atmospheric model we remind the wind speed ($\tilde{\mathbf{V}}$), the pressure (P), the temperature (T), the mixing ratio (q), the dynamic outer scale ($\mathcal{L}_0$) and the C_N^2 that depends on all the meterological parameters that have been just cited. The astroclimatic parameters that we can simulate with 2 dimensional maps derive all from an integration of the C_N^2 over the whole atmosphere weighted by a function F=F(h^a, $\mathcal{L}_0{}^b$, $\tilde{\mathbf{V}}^c$) as:

$$\int_0^{+\infty} F(h^a, V^b, L_0^c) \cdot C_N^2(h)\, dh \tag{1}$$

(with a, b and c constant). Among the most important integrated astroclimatic parameters we cite the seeing (ε), the wavefront coherence time

(τ_0), the isoplanatic angle (θ_0), the spatial coherence outer scale ($\mathcal{L}_0$), the optimal heights at which to conjugate the deformable mirrors (h_m) and the isoplanatic angle for the MCAO (θ_M).

The three main challenges for the optical turbulence estimates done with mesoscale models are:
(1) the possibility *to forecast the OT for the implementation of the flexible-scheduling on ground-based telescopes.* The management of the scientific programs to be carried out as well as the management of the instruments to be placed at the focus of telescopes pass necessarily by the forecast of the OT. Ground-based astronomy will remain competitive with respect to the space-based one only if telescopes management will be performed taking advantage of the best turbulence conditions. The future of the ground-base astronomy relies therefore upon the success of these studies.
(2) the possibility *to reconstruct 3D maps in a region around a telescope.*
(3) the possibility *to perform a climatology of the OT extended over time scales of the order of decades.* It is worth to highlight that only models (and not instruments) can access to the past, necessary condition to reconstruct, in a homogeneous way, the climatology of a parameter.
Climatologic studies of the OT are pretty useful, for example, to quantify effects such as the deterioration of the seeing observed on time scales of the order of a few years at Cerro Paranal,[3] site of the Very Large Telescope (VLT). It is worth to highlight that no other tool of investigation with comparable potentialities can be figure out, at present, to achieve these three scientific goals. The OT modeling represents, therefore, a fundamental challenging goal to be achieved in the next future to guarantee the success of the ground-based astronomy.

2. Optical Turbulence Parameterization

The principal concept on which the OT simulation with mesoscale models is based on is the 'parameterization' that is the possibility to numerically represent the spatial and temporal evolution of parameters not explicitely resolved by a model because their typical scales of spatial and temporal fluctuations are smaller than the model grid size (model resolution).

The turbulence parameterization is done according to the *turbulence exchange* method allowing to represent the flux of a microscopic quantity ξ as a function of the gradient of the same quantity averaged on a larger spatial size:

$$\overline{w'\xi'} = K_\xi \frac{\partial \overline{\xi}}{\partial z} \tag{2}$$

K_ξ is the exchange coefficient, a constant expressing the properties of the turbulent flux in the whole turbulent layer. In other words, we express the microscopic characteristics in a statistical way as a function of the gradient of the same quantity but space averaged on a larger region.

To be effective, the parameterization of these physical phenomena, such as the optical turbulence, can not be done on General Circulation Models that are extended on the whole earth but on models (so called *'mesoscale models'*) extended on a limited surface (a few tens of kilometers). The typical resolution at which the best results[4–7] have been obtained so far for the OT reconstruction is a few hundreds of meters (400, 500 m)[a]. In perspective, it will be interesting to consider also models with higher horizonthal resolution such as the LES. See Masciadri (2009)[1] for a more extended discussion on this topic. We need to distinguish between the parameterization of the dynamic and optical turbulence. The **dynamic turbulence** implies the presence of kinematic and dynamic fluctuation of the atmopsheric flow and it is expressed by the buoyancy term[9,10] of the TKE equation that, in the Meso-Nh model, is:

$$\overline{w'\theta'_v} = -0.16L\sqrt{e}\phi_3\frac{\partial\overline{\theta_v}}{\partial z} \qquad (3)$$

The factor K is proportional, therefore, to the mixing length (that is basically the dynamic outer scale), the square root of the kinetic energy and the function ϕ_3 that is a thermodynamic stability term introduced by Redelsperger & Sommeria in 1981.[8] Meso-Nh is conceived to be used with different mixing length and the one that is useful for our simulation is the BL89.[9] The **optical turbulence** (C_T^2) concerns the effects of the turbulence on a wavefront and this implies the interaction between light and matter. To have optical turbulence it is necessary to have, at the same time, the presence of a gradient of the wind speed (that means presence of the dynamic turbulence) and a gradient of the potential temperature different from zero. C_T^2 is equal, in Meso-Nh, to:

$$C_T^2 = 0.59\phi_3 L^{4/3}\left(\frac{\partial\overline{\theta}}{\partial z}\right)^2. \qquad (4)$$

and the C_N^2 is related to the C_T^2 by the Gladstone's law.[4]

[a]If the the grid-nesting configuration is used, the specification just reported correspond to the model having the highest horizontal resolution.

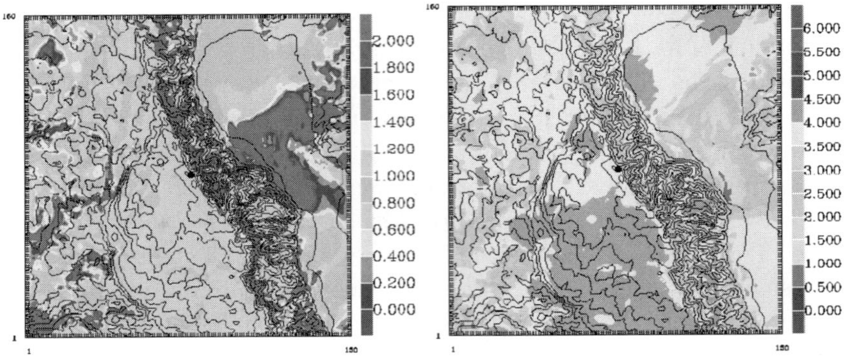

Fig. 1. Simulation of integrated astroclimatic parameters above the San Pedro Mártir Observatory done with Meso-Nh and the OT package. Horizontal maps (3 h simulation) of the seeing (ε, left side) and the isoplanatic angle (θ_0, right side) extended on 60 km x 60 km and an horizontal resolution of 400 m. The black lines are the ground levels of the orographic map. The black dot indicates the Observatory. The example shows a $\varepsilon \sim$ 0.6-0.8 arcsec and a $\theta_0 \sim 3$ arcsec.

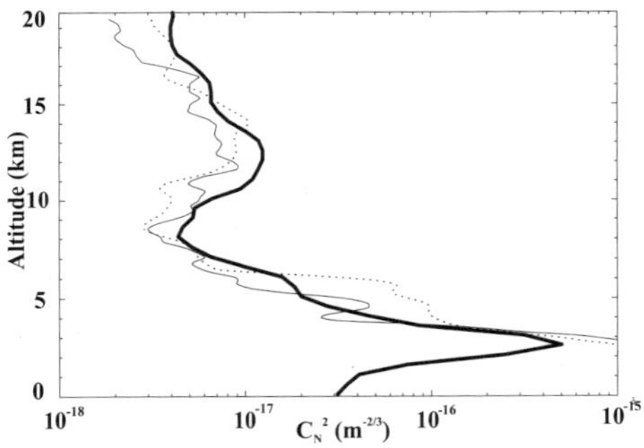

Fig. 2. Mean vertical C_N^2 profiles measured and simulated over the whole SPM2000 campaign. Bold line: Generalized Scidar. Thin line: radiosoundings. Dotted line: Meso-Nh model. The relative total seeing obtained from the integration of the C_N^2 profiles is: 0.79 arcsec (GS), 1.07 arcsec (balloons), 0.93 arcsec (Meso-Nh model).

3. OT Simulations: Main Results

In this contribution we will simply summarize the main results achieved in the past. We refer the readers to more extended reviews on this topic:[1,2,12] **(1)** We proved that a mesoscale model (Meso-Nh) can reconstruct the C_N^2 profiles on several astronomical sites (Cerro Paranal, Roque de Los Muchachos, San Pedro Mártir,..).[4–6] This means that the shape of the measured and simulated profiles is well correlated and the measured and simulated integrated values on the whole troposphere (i.e. the seeing ε) match.

(2) We proved that Meso-Nh can reconstruct horizontal maps of several integrated astroclimatic parameters characterized by values that are all coherent with a good astronomical site. Figure 1 shows, as an example, two horizontal maps of the seeing (ε) and the isoplanatic angle (θ_0) simulated above the San Pedro Mártir Observatory, Baja California, Mexico. See Masciadri (2009)[1] for a more complete analysis.

(3) We proposed a new and universal method to calibrate a mesoscale model that is independent from the model itself and aims to eliminate systematic errors.[13]

(4) We proved the reliability of the method and we validated the Meso-Nh model in statistical terms.[7,11] We compared all the measured C_N^2 profiles related to the SPM2000 site testing campaign done in San Pedro Mártir (Mexico) to the simulated profiles and we proved that the dispersion between the measurements and simulations ($\Delta\varepsilon_{GS,model}$) is comparable to the dispersion between measurements provided by different instruments ($\Delta\varepsilon_{GS,balloons}$) and this is of the order of 30 %. Also the vertical distribution of the optical turbulence, that is the shape of the average of the C_N^2 profiles measured and simulated is well correlated as shown in Fig. 2.[7] In an extended review (Masciadri, 2009)[1] it is shown that, at present, all comparisons done between measurements obtained with different instruments, above different sites and on different statistical samples, converge on a value that is not smaller then 20-30 %.[14,15]

(5) We completed the first seasonal variation study of the vertical distribution of the optical turbulence (C_N^2 profiles) and the integrated astroclimatic parameters (ε, τ_0, θ_0, h_M, θ_M,....) above the San Pedro Mártir on a sample of 80 nights uniformly distributed along one year. In this study an interesting seasonal trend of the position and strenghtness of the typical C_N^2 peak located in the free atmosphere has been observed (Masciadri & Egner, 2006, Fig. 4). More recent results obtained above different sites (Mt. Graham[1] and Roque de los Muchachos[16]) show similar trends and let us think that this might be an universal trend for the OT.

(6) We provided the first evidence of the finite horizontal extent of the turbulence layers at different heights.[17] Such a result has certainly very important consequences in applications to the ground-based astronomy and puts some limits on the assumption on horizontal uniformity of the turbulence layers.

(7) We proved that meso-scale models can discriminate and identify sites having weak wind seed near the ground while GCM can not.[18] Moreover, the relative error between the measured and reconstructed wind speed near the ground obtained by meso-scale models is 27-47 % smaller than what obtained by the GCMs.
It is worth to highlight that we are talking about statistical comparison i.e. median and mean values on one or a set of nights. No comparison between measurements and simulations is possible, at least at present, on time scale of minutes.

4. ForOT: Scientific and Technologic Goals

The validation of the Meso-Nh model and its application to several astronomical sites marked an important step in these studies in the decade 1996-2006. ForOT was born at the beginning of 2006 as an experiment to do a step ahead in this discipline. The OT modeling necessarily needs to pass now to a new dimension to completely exploit its potentialities. This leap ahead has to be done in two directions: research and technology. The ForOT Project moved from these motivations:
Research: ForOT is conceived to reach two main scientific goals: **(1)** to prove/test the ability of the non-hydrostatic meso-scale models in forecasting the optical turbulence and **(2)** to prove/test the ability of these models in discriminating the quality (from the turbulence point of view) of astronomical sites. To better focalize the objectives it has been decided to apply this technique in two different regions of the world.
Mt. Graham (Arizona), site of the Large Binocular Telescope.
Main Scientific goal: to test the model ability in forecasting the optical turbulence.
Antarctic Plateau (Dome C, South Pole and Dome A).
Main Scientific goal: to test the model ability in discriminating the turbulence characteristics on different sites above the Antarctic Plateau.
These three pre-selected sites are among the most interesting for applications to the astronomy. At South Pole a base exists since several years. At Dome C, an Italo-French base (Concordia) is being built and a set of projects of small-medium size telescopes and interferometers is planned by several international consortia. Dome A has not yet been explored and monitored during the night time so far but it might be even better than Dome C due to a higher altitude of the site. The main idea is to validate and calibrate the model above Dome C and South Pole using the measurements done so far above these two sites. After the calibration, the model should

be able to simulate the turbulence above Dome A with a high level of reliability.

Technology: astronomers need to appropriate themselves of a tool (atmospherical model) usually employed by meteorologists. This numerical technique applied to astronomy needs, to be effective, to achieve a technological development aiming to get operational a nightly forecasting of the optical turbulence above an astronomical site. It has been planned, therefore, to install the Meso-Nh model on an autonomous machine (instead of supercomputers of meteorologic centers) to be able to run in a systematic way the model above different astronomical sites. This allowed us to do a great leap in the direction of an implementation of the real flexible-scheduling system for astronomical facilities but still remaining in a research dimension.

From a more general point of view ForOT intends to be one of the possible answers to the missing link between Astronomy and Meteorology. This answer is based on two principles/constraints: **(a)** the multidisciplinary approach **(b)** the intention to maintain in this effort the character of 'research' in high angular resolution techniques. The ForOT manifesto (Fig. 3) represents schematically the final goal of ForOT. The High Angular Resolution (HAR) is a science developped traditionally in the context of the Astronomy and it is composed by several research lines such as the Adaptive Optics (AO), the interferometry, the image processing and the study of perturbations induced by the atmospheric turbulence on the wavefronts. ForOT aims to integrate into the traditional HAR a research line typical of the Meteorology: the physics of the atmosphere & hydrodynamic. Our final goal is to put the bases for a New Generation HAR discipline.

5. ForOT: Main Results

The main activities of ForOT addressed so far in three directions: **(1)** Optical turbulence measurements, **(2)** Optical turbulence simulations with mesoscale models, **(3)** Adaptive Optics applications.

In another paper (Ref. 19), in these Proceedings, results obtained in the topic (1) are summarized. We highlight here the main results obtained in the topics (2) and (3). An extended and detailed characterization of the atmosphere above the Internal Antarctic Plateau has been just concluded[20] in which a preliminary rank of the sites (Dome A, South Pole and Dome C) has been depicted. A complete validation of the ECMWF[b] analyses (provided by the General Circulation Models) in this region of the world has

[b]European Centre for Medium Weather Forecasts (ECMWF).

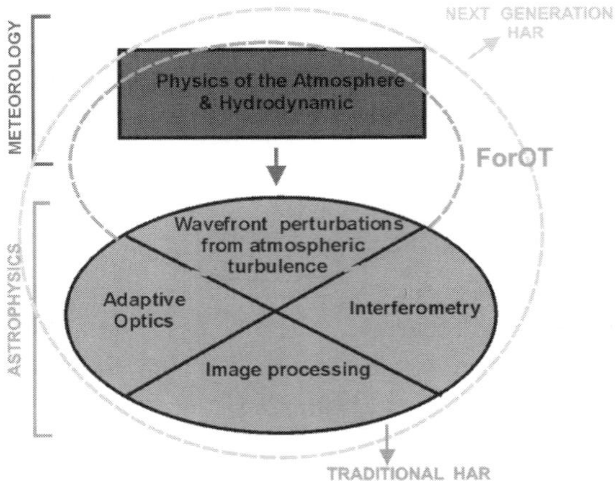

Fig. 3. From *http : //forot.arcetri.astro.it* Web Page. Schematic representation of the Manifesto of the ForOT Project. See text in Section 4.

been performed. This is a fundamental result for at least two reasons. The ECMWF analyses are the initialization product of the mesoscale models and they can provide therefore an indirect information on the mesoscale models performances. Secondly, it is fundamental to identify precisely the limits at which the GCM models can provide reliable information. This can tell us in which context mesoscale models represent the unique solution to answer to some precise questions. A few among the main results are highlighted[20] here:

(1) it has been proved that the probability to trigger turbulence in the free atmosphere above the Internal Antarctic Plateau is lower than above mid-latitude sites.

(2) it has been proved that Dome C is not necessarily the best site on the Plateau. It has been found a large median wind speed in the free atmosphere in winter time (due to the presence of the polar vortex) and a strong wind shear near the surface.

(3) Dome A shows the highest thermal stability near the surface. We expect a surface turbulent layer size comparable or thinner than the one developed above Dome C but characterized by a higher level of optical turbulence.

For what concerns the progress in the field of the mesoscale models simulations it is worth to highlight that the Meso-Nh model runs now on an autonomous machine with the optical turbulence package. The model can run now in a more sophisticated configuration with respect to what done

so far including (1) the grid-nesting and (2) a 6 hours external forcing performed with ECMWF analyses.
The most important results obtained so far[22] are listed here down:
(1) A statistic analysis done on 47 nights permitted us to put in evidence that the mesoscale model Meso-Nh reconstructs the meteorological parameters responsible of the development of the optical turbulence in the surface and boundary layer in a more reliable way than what can be done by the General Circulation Models conceived for the weather forecast.
(2) The first C_N^2 simulations extended on the whole 20 km ever done above the Antarctic Plateau have been performed.
(3) A first statistical comparison of the C_N^2 simulated by Meso-Nh and measured by balloons[23] has been done on all the 16 nights of the winter time for which measurements are available. The mean simulated thickness of the surface layer appears $\sim$ 10 m greater than the measured one (h_{surf}=35 m[23]). Figure 4 shows the temporal evolution of the C_N^2 simulated above Dome C and referring to a night in the Antarctic winter time in the 2005. The C_N^2 temporal evolution is extended on 5 hours and in the selected night is well visible a surface layer well confined in the first 30 m. We refer to Ref. 22 for a more extended discussion of these results.
(4) The seeing in the free atmosphere simulated by Meso-Nh is already pretty well correlated with measurements. On the same 16 nights it has been calculated a simulated[22] median seeing ε_{obs}=0.46 arcsec versus a measured[23] ε_{sim}=0.30.
(5) It has been proved that the grid-nesting configuration better reconstructs the optical turbulence vertical distribution. The grid-nesting configuration revealed to be more efficient than the monomodel configuration in spite of the flat topography of the Antarctic Plateau.

If the surface turbulence layer is confirmed to be thin above Dome C, is this site competitive with respect to mid-latitude sites in the near-infrared regime ? We proved[21] that, provided the projection of the single actuator on the pupil of the telescope is of the order of 38 cm or less (in J band), a telescope placed at Dome C can concretely provide better corrections on large field of view ($\theta \geq 4$ arcmin) than what is achievable at a good mid-latitude site. For actuators having a greater size the mid-latitude sites are still competitive. Translated in simple words our results indicate that small pupil size telescopes are astronomical facilities with interest for Dome C provided they are equipped with an AO system in agreement with the constraints indicated. However, large astronomical facilities are less interesting

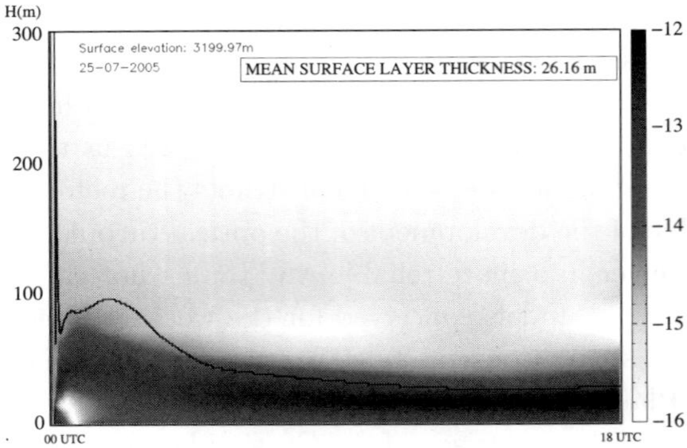

Fig. 4. Temporal evolution (18 h) of the C_N^2 above Dome C in a night of the Antarctic winter time (25/07/2005). The thickness of the surface layer ($h_{surf.}$=26.16 m) is calculated respecting Eq. 1 (Ref. 22) averaging between 11 h and 17 h from the starting of the simulation.

for Dome C because a less evident gain can be predicted with respect to what the same facility located in good mid-latitude sites might obtain. As an example, a LBT equipped with ARGOS would produce worse results in Antarctica than at Mt. Graham. These conclusions are therefore certainly fundamental in the evaluation process that the international astronomical community is doing in these years concerning potential perspectives of ground-based astronomy in Antarctica.

6. Conclusion

In this contribution we have dealt about a new strategy for the studies of the optical turbulence in astronomical applications based on a multidisciplinary approach that has been concretized by the ForOT project. Simulations done with mesoscale models are a unique tool of investigation to answer to some precise questions and necessities for astronomers. The numerical approach can achieve some fundamental and challenging goals namely the forecasts of the optical turbulence, the reconstruction of 3D C_N^2 maps and the reconstruction of a climatology of the C_N^2. The implementation of the flexible-scheduling in the ground-based facilities of next generation is strictly correlated to the success of these studies and this justifies a serious attention to this topic. In this paper we summarized the main results obtained in the last decade in this discipline and we dealt about the more recent results obtained by ForOT. To guarantee the development of this discipline in the future it is certainly important to conceive a benchmark site testing campaign expressly conceived for the validation of the mesoscale models in which several instruments measuring the optical turbulence and other atmospheric parameters are run simultaneously. In the next years we intend to achieve this goal that will required the contribution of the whole international community.

Acknowledgments

This work is funded by a Marie Curie Excellence Grant (ForOT), MEXT-CT-2005-023878 - FP6 Program.

References

1. Masciadri, 2009, Book on Seeing, Ed. Institute for Astronomy, Hawaii, US.
2. Masciadri, E., 2006, SPIE, 6267, 62671C
3. Sarazin, M., Lombardi, G., Navarrete, J., this Conf. Proceed.
4. Masciadri, E., Vernin, J. and Bougeault, P., 1999a, A&ASS, 137, 185
5. Masciadri, E., Vernin, J. and Bougeault, P., 1999b, A&ASS, 137, 203
6. Masciadri, E., Vernin, J. and Bougeault, P., 2001, A&A, 365, 699
7. Masciadri, E., Avila, R. and Sánchez, L.J., 2004, RMxAA, 40, 3
8. Redelsperger & Sommeria, 1981, Bound. Layer Meteor., 21, 509
9. Bougeault & Lacarrere, Mon. Wea. Rev., 117, 1872
10. Cuxart, J., Bougeault, P., Redelsperger, J-L., 1995, 11th Symp. Boundary Layer & Turb., 13, 409
11. Masciadri, E. & Egner, S., 2006, PASP, 118, 849
12. ForOT Web Page: *http://forot.arcetri.astro.it*
13. Masciadri, E. & Jabouille, P., 2001, A&A, 376, 727
14. Azouit & Vernin, 2005, PASP, 117, 831, 536
15. Tokovinin, A., Vernin, J., Ziad, A., Chun, M., 2005, PASP, 117, 395
16. Garcia-Lorenzo, B., Fuensalida, J.J., Castro-Almazan, J., et al., 2009, this Conf. Proceed.
17. Masciadri, E., Avila, R. and Sánchez, L.J., 2002, A&A, 382, 378
18. Masciadri, E., 2003, RMxAA, 39, 249
19. Masciadri, E., Stoesz, J., Hagelin, S., Lascaux, F., this Conf. Proceed.
20. Hagelin, S., Masciadri, E., Lascaux, F., Stoesz, J., 2008, MNRAS, 387, 4, 1499
21. Stoesz, J., Masciadri, E., Lascaux, F., Hagelin, S., 2008, SPIE, 7012, 70121W
22. Lascaux, F., Masciadri, E., Hagelin, S., Stoesz, J., 2008, SPIE, 7012, 70124D
23. Trinquet, H., Agabi, K., Vernin, J., Azouit, M., Aristidi, E., Fossat, E., 2008, PASP, 120, 864, 203

THE MAUNA KEA WEATHER CENTER: A CASE FOR CUSTOM SEEING FORECASTS

T. CHERUBINI, S. BUSINGER and R. LYMAN

Department of Meteorology, University of Hawaii, 2525 Correa Rd, Honolulu HI, 96822, USA

This paper briefly describes the current methods employed at the Mauna Kea Weather Center in the development of optical turbulence and seeing forecasts for the observatories at the summit of Mauna Kea. This paper reports the content of two talks (Businger et al. and Cherubini et al. – see 'Scientific Program, Session 5') given to the International Conference "Optical Turbulence – Astronomy meets Meteorology".

1. Introduction

The Mauna Kea Weather Center (MKWC) is a joint project between the Department of Meteorology and the Institute for Astronomy at the University of Hawaii. The main task of the MKWC is to provide weather guidance to the community of astronomers who operate the observatories on the summit of Mauna Kea. Since January 1999, the MKWC has issued two daily (10am and 5pm) custom weather forecasts for the Mauna Kea Observatories (Businger et al. [2]). Each detailed forecast extends over a 5-day period, but also includes an outlook 6-10 days into the future. In constructing the forecasts, the MKWC forecaster uses as guidance the output from global models from the national operational centers, the synoptic and a-synoptic (local) observations available in real-time, and the output from a mesoscale model, run at the MKWC. The custom forecasts, along with satellite and radar data, model output, and summit weather observations are made available to the public and the astronomers via the MKWC web site (http://mkwc.ifa.hawaii.edu). The MKWC forecasts include a range of meteorological variables of importance to the operation of the summit observatories, such as trends in summit temperature, wind speed, cloudiness, and precipitable water. However, the focus of this short paper will be the development of optical turbulence and seeing forecasts.

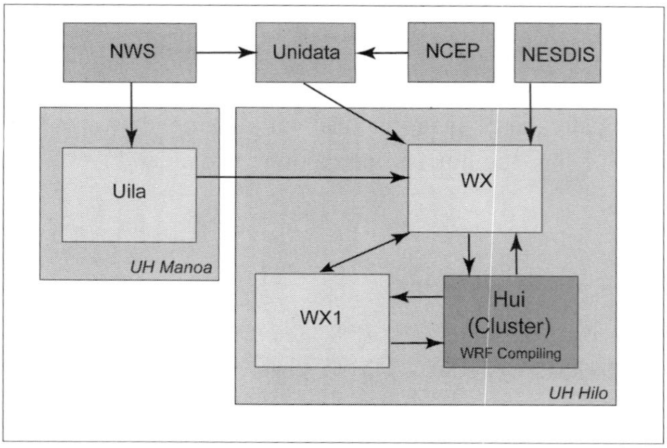

Fig. 1. (color on-line) Schematic flow chart of the hardware and data flow distribution for the MKWC, which comprises two locations; UH Manoa and UH Hilo. Green boxes indicate data sources. Light blue boxes indicate the three servers involved in data management, web serving, and pre and post-processing for WRF. The red box indicates the compute cluster that runs WRF.

2. Overview of Hardware and Data Flow

MKWC staff maintains a complex hardware and software configuration that handles the data flow, model input/output, and serves products to the web pages (Fig. 1). Our primary data source, including model output from the National Centers for Environmental Prediction (NCEP), is Unidata, which is located in Boulder, Colorado, and is funded by the National Science Foundation. Unidata distributes geosciences data sets and software applications to handle these data. One such application is the Local Data Manager (LDM), which is run on all three MKWC servers (Uila, WX and WX1) to handle the Internet data stream in real time (Fig. 1).

In addition to the data stream provided by Unidata, special satellite data sets are received directly from the NWS and NASA. The diverse observational data and model output are combined in the Local Analysis and Prediction System (LAPS; McGinley [17]), which runs on WX to produce and provide initial and boundary conditions for our mesoscale model as described in Cherubini et al. [3]. The mesoscale model, which is described in the next section, runs on a dedicated Silicon Mechanics cluster with 16 nodes (128 cpu/2 GB RAM each) with Infiniband switches for enhanced performance.

To make the operation of the MKWC more robust, WX and WX1 are mirrored servers that duplicate data management and web serving functionality. When one server has a hardware or software issue, the other server automatically takes over the task of making data available to users via the Internet, thus minimizing disruption in availability of MKWC web products.

3. Overview of Mesoscale Model

From 1999 to the beginning of 2008, the MKWC implemented the fifth-generation Mesoscale Model (MM5) that was originally developed at Pennsylvania State University and National Centre for Atmospheric Research (Grell et al. [6]; Cherubini et al. [4,5]). Early in 2008, the MKWC staff transitioned from MM5 to the Weather Research and Forecasting (WRF) model (Klemp et al. [12], http://www.wrf-model.org). By this time WRF, the next-generation mesoscale numerical weather prediction system, had become sufficiently stable and mature for our operational needs.

WRF features different dynamical cores and the Advanced Research WRF (Skamarock et al. [19]) core was chosen for implementation at the MKWC. WRF is a non-hydrostatic model with multiple nesting capabilities. The MKWC configuration encompasses 3 two-way nested domains, with horizontal resolution spacing of 15, 3 and 1 km, corresponding to a time step of 75, 25 and 5 seconds respectively (Fig. 2). The outermost domain covers a wide portion of the central Pacific area, the 3-km resolution domain spans all of the larger Hawaiian Islands, and the innermost domain is centered over the summit of Mauna Kea. Forty levels in the vertical are used. The vertical spacing is on the order of tens of meters for the levels nearest the ground and gradually increases with height as shown in Fig. 2. The model top is fixed at 10 mb, which correspond to an height of about ~25 km above the ground level.

The WRF physics package used in the operational configuration includes (i) a 3-class simple ice scheme for the microphysics scheme (Hong et al. 2004), (ii) the Kain-Frietsch cumulus convection scheme (Kain and Fritsch, [9,10]) for the 15- and 3-km domains, (iii) the Mellor-Yamada-Janjic (MYJ) planetary boundary layer scheme (Janjic [8]), which solves the prognostic equation for the TKE, (iv) the RRTM longwave-shortwave radiation scheme (Mlawer et al. [18]).

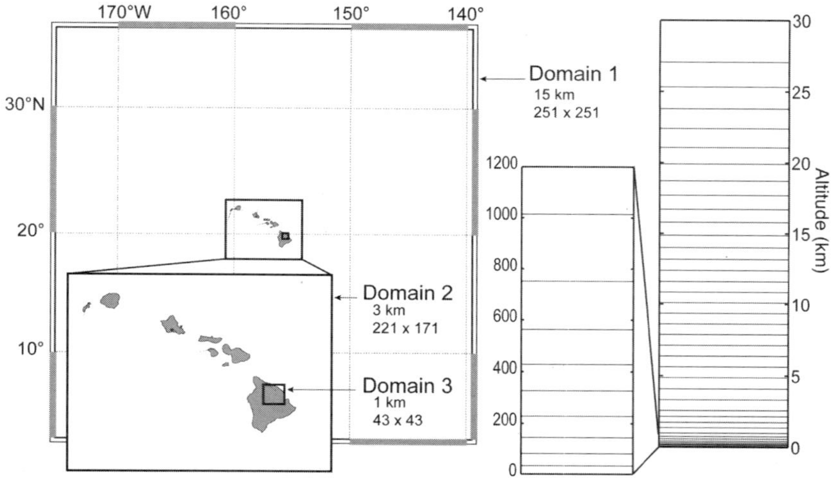

Fig. 2. MKWC configuration of the nested grids in WRF. Inset map shows an expanded view of the main island in the Hawaiian chain. The vertical resolution of the model is depicted in the diagram at right, including an expansion of the lowest 1200 meters of the model domain.

The WRF model is run twice daily with 0000 and 1200 UTC initial conditions produced by LAPS as discussed in section 2. Boundary conditions are updated every six hours using model output from the National Centers for Environmental Prediction Global Forecasting System.

4. The Optical Turbulence Algorithm

The prognostic algorithm for atmospheric turbulence previously implemented in the MM5 model calculates the refractive index structure function C_n^2, which describes the turbulent fluctuations of the atmospheric refractive index (Cherubini et al. [4]). The integral of C_n^2 along the line-of-sight of a telescope is proportionally correlated to a parameter referred to as the telescope's *seeing*. Seeing describes the image quality of a telescope.

The optical turbulence algorithm implemented within WRF differs somewhat from the one implemented in MM5. The boundary layer schemes in WRF and MM5 differ and these schemes provide information on the turbulent kinetic energy (TKE) profiles that is critical for the calculation of the C_n^2. The optical turbulence implemented in WRF is calculated following Masciadri et al. [14,15], with modifications introduced by Kemp et al. [11]. In particular, the C_n^2

profile is calculated using the following relation (Bean and Dutton [1]; Tatarskii [20]; Masciadri et al. [14], Masciadri and Jabouille [16]):

$$C_n^2(z) = \left(\frac{80 \times 10^{-6} p}{\vartheta^2} \right)^2 C_\vartheta^2(z) \tag{1}$$

where the temperature structure coefficient C_ϑ^2 is calculated according to:

$$C_\vartheta^2(z) = a^2 \left(\frac{K_H}{K_M} \right) L^{4/3}{}_0 \left(\frac{\partial \vartheta}{\partial z} \right)^2 . \tag{2}$$

In equation (2), K_H and K_M are the exchange coefficients for heat and momentum, a is an empirical constant and L_0 is the outer length scale of turbulence (the upper bound of the inertial sub-range), which, in stable conditions can be calculated as:

$$L_0 = \frac{0.76\sqrt{e}}{N}, \tag{3}$$

where N is the Brunt-Vaisala frequency and e is the TKE. For unstable conditions L_0 is calculated as follows:

$$L_0 = (l_{up} l_{down})^{1/2} \tag{4}$$

where l_{up} and l_{down} are calculated by solving:

$$\int_z^{z+l_{up}} \frac{g}{\vartheta} \left(\vartheta(z) - \vartheta(z') \right) dz' = e(z)$$

$$\int_{z-l_{down}}^{z} \frac{g}{\vartheta} \left(\vartheta(z') - \vartheta(z) \right) dz' = e(z)$$

The scheme implemented within WRF also includes a correction in the calculation of the ratio K_H/K_M. This ratio is about 1 in the original MYJ scheme. This does not agree with the observation in thermally stable conditions as pointed out by Walters and Miller [21]. The ratio has been modified using the Kondo et al. [13] relation as follows:

$$\frac{K_H}{K_M} = \frac{1}{7R_i} \qquad \text{for} \quad \text{Ri} \geq 1$$

$$\frac{K_H}{K_M} = \frac{1}{6.873R_i + \dfrac{1}{1+6.873R_i}} \qquad \text{for} \quad 0.01 < \text{Ri} \leq 1$$

where R*i* is the Richardson number. This correction decreases the effect of the buoyancy term, which acts as a sink in thermally stable conditions. Therefore it is easier for layer with wind shears to generate TKE.

The MYJ scheme includes a background TKE of 10^{-1} m^2 s^{-2}. The turbulent closure requires in fact a non-zero background TKE. Following Kemp et al. 2008, to produce realistic values of C_n^2 profile in the upper troposphere, this background value has been changed to a preliminary value of 10^{-5} m^2 s^{-2}. Future studies will be focused on calibrating the background value for TKE throughout the atmosphere.

5. A Case Study

The WRF model has been rerun for several nights during the 2002 Mauna Kea monitoring campaign site (Cherubini et al. [4]). The results from the run for the night of the 23 October 2002 are presented here. Seeing and C_n^2 simulated data are derived from the WRF domain-3. Turbulence data from MASS and G-SCIDAR were collected between 0600 and 1600 UTC.

The model run starts at 00 UTC on 23 October, therefore the forecast output presented is from the 6th to the 16th hour of WRF forecast.

Nightly averaged profiles of C_n^2 observed by G-SCIDAR shows two turbulent layers centered at about 4 and 12 km above the site, and a boundary layer below 2 km with higher values of C_n^2 nearer the ground (Fig. 3). WRF's simulated profile also

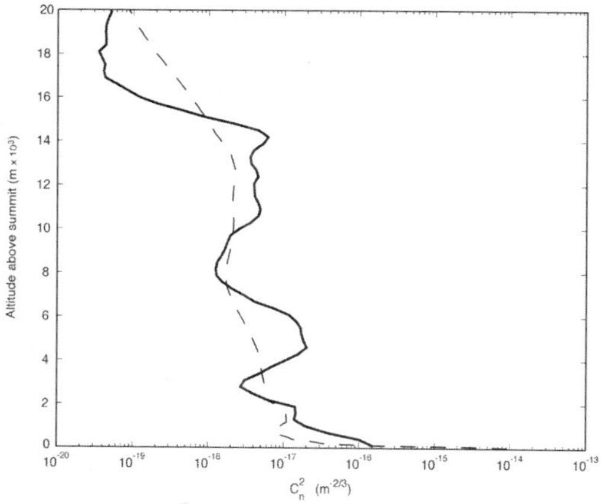

Fig. 3. Average C_n^2 profiles from G-Scidar (black solid line) and from the WRF (dashed line). Data are averaged from 0600 to 1600 UTC on 23 October 2002.

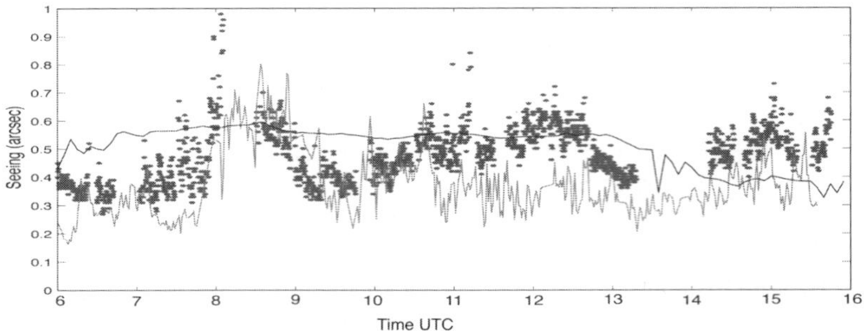

Fig. 4. Time series of seeing data as from the G-Scidar (gray dots), MASS (gray thin line) and WRF (dark solid line). The G-SCIDAR seeing represents the integrated optical turbulence upwards of ~80 m above the ground, while the MASS seeing represents the integrated optical turbulence upwards of ~250 m above the ground.

shows two levels of turbulence aloft, but these significantly underestimate the observed intensity. Both profiles show a significant contribution to the optical turbulence from the atmosphere nearest the ground.

Figure 4 shows the variation of the atmospheric seeing through the night as measured by the G-SCIDAR (gray dots) and by MASS (gray solid line). WRF-derived seeing is from the closest model level to the first G-SCIDAR level considered for seeing (~60 m above the ground). WRF simulated seeing results in a smoother time series than those of the SCIDAR and MASS. The magnitude and trend of the WRF simulation is in the right range to provide useful forecast guidance during much of the night (Fig. 4). However, predicted values over estimate those observed during the first 90 minutes, and tend to underestimate those observed during the last hour of the plot.

Output from WRF's optical-turbulence algorithm can be displayed in a terrain-following map of seeing (Fig. 5). The map shows increasing seeing values on the lower slopes of Mauna Kea associated with clouds and background turbulence in the planetary boundary layer beneath the tradewind inversion.

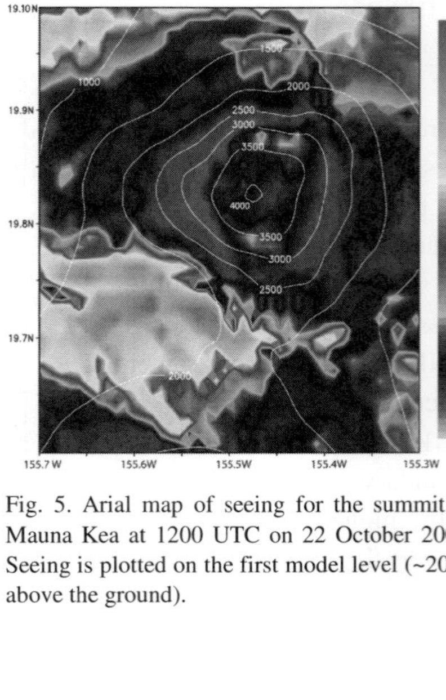

Fig. 5. Arial map of seeing for the summit of Mauna Kea at 1200 UTC on 22 October 2002. Seeing is plotted on the first model level (~20 m above the ground).

Although improvement is needed and possible, the preliminary performance of optical-turbulence algorithm is encouraging. Calibration of the minimum turbulence kinetic energy within the scheme is a necessary next step to adjust the algorithm to the Mauna Kea site and improve the simulation results.

5.1. *Preliminary Results from the Sensitivity Study to Vertical Resolution*

The results from a preliminary sensitivity test of the optical turbulence algorithm to the model vertical resolution are presented here. Two more simulations of the 23 October case applying 80 and 120 levels in the vertical have been carried out. The WRF vertical structure in our configuration is such that the first 10 model levels are at fixed heights within the boundary layer. Increasing the number of the total vertical levels results in an overall increase in resolution in the free atmosphere, where it is most needed. As expected, the nightly averaged C_n^2 profiles show a consistent increase in ability to resolve the vertical structure for C_n^2 as the number of vertical levels increases (Fig. 6).

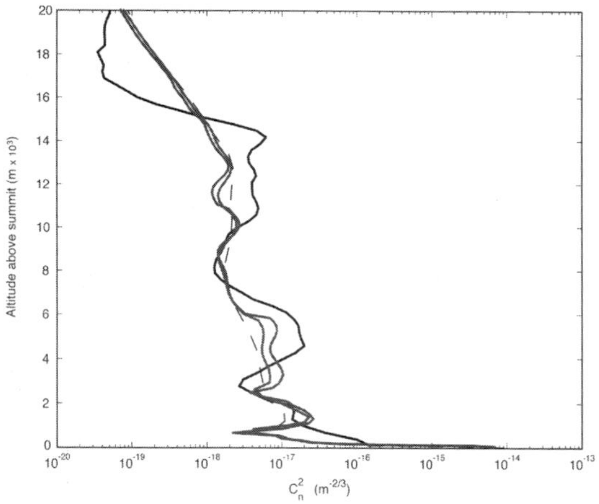

Fig. 6. (color on-line) Average C_n^2 profiles from G-Scidar (black solid line), WRF 40 vertical levels control run (dashed black line), WRF 80 vertical levels run (red solid line) and WRF 120 vertical levels run (blue solid line). Data are averaged from 0600 to 1600 UTC on 23 October 2002.

Differences in the variation of seeing through the night of the 23 October between the three model runs are noticeable and suggest that the vertical resolution of WRF has a significant impact on the algorithm performance. The nightly mean values for the seeing is 0.47 arcsec as observed from the G-SCIDAR data, 0.36 arcsec from the MASS data, and 0.51, 0.45 and 0.47 arcsec as calculated during WRF runs with 40, 80 and 120 vertical levels, respectively.

6. "The MKWC Daily Forecast for Optical Turbulence" or "Learning from the Every Day Experience: the Human Added Value"

Synthetic C_n^2 profiles along with a seeing forecast trends are available on the MKWC web pages and updated twice daily. Since July 2006 a quantitative seeing forecast has been part of the MKWC bulletin. MKWC forecasters rely on the output from the turbulence algorithm for guidance and add this prediction to their observational experience with synoptic and local weather conditions and astronomical observing conditions to construct a final seeing forecast.

Somewhat surprisingly, forecaster experience suggests that the two extreme conditions of very good and very bad observing conditions can be reasonably well anticipated. The reason is that excellent seeing is very often correlated to a strong and large-scale ridge parked over the Hawaiian Islands, producing light, uniform winds and subsiding dry air above the summit. In contrast, very bad seeing, is well correlated to poor weather conditions above the summit, usually emanating from a trough to the west of the Island of Hawaii that produces a moist south or southwesterly flow over the summit. Both of these large-scale weather patterns are reasonably well handled by numerical weather prediction models, giving the forecaster better guidance from which to construct the final forecast.

The middle part of the seeing spectrum tends to be more difficult to forecast, resulting in greater errors. There are numerous factors that impact the turbulent state of the atmosphere, including for example, horizontal and vertical wind shear at various levels induced by upper-level troughs and jet streaks, remnant tropical cyclones, gravity waves ahead of cold fronts, just to name a few. Additionally, summit-level winds will produce boundary layer turbulence and advect in moisture and/or clouds. Each of these circumstances impacts the observing conditions. Not surprisingly, turbulence near the surface and/or aloft is very difficult to anticipate solely by human or instrument observation alone. Therefore, guidance from an optical turbulent scheme within a high-resolution weather prediction model represents a powerful approach in forecast development.

7. Conclusions and Future Work

The primary goal of the Mauna Kea Weather Center (MKWC) is to provide weather forecast guidance to astronomers to optimize their ability to make astronomical observations from the summit of Mauna Kea (Businger et al. [2]). Custom Mauna Kea weather forecasts, along with satellite and radar data, model

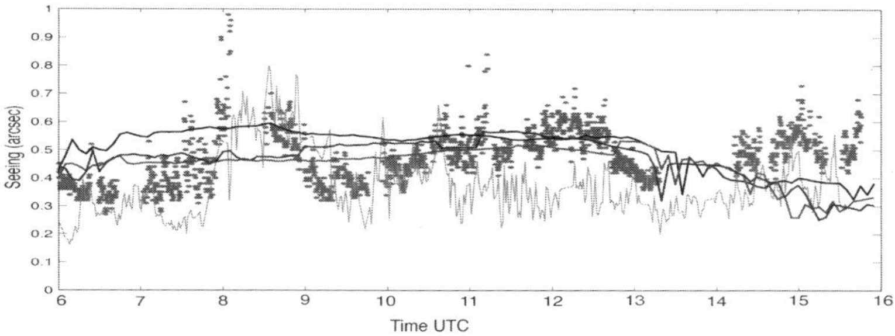

Fig. 7. (color on-line) As in Fig. 4 but the seeing time series for the WRF run with 80 (red solid line) and 120 (blue solid line) vertical levels are also shown.

output, and summit weather observations are made available to the public and the astronomers via the MKWC web site (http://mkwc.ifa.hawaii.edu). The focus of this short paper has been the development of optical turbulence and seeing forecasts with guidance provided by an optical turbulence algorithm run as part of the WRF model.

The WRF simulated seeing data results in a smoother time series than those of the SCIDAR and MASS. The magnitude and trend of the WRF simulation is in the right range to provide useful forecast guidance during much of the night (Fig. 7). Preliminary sensitivity tests of the optical turbulence algorithm to the model vertical resolution shows in a consistent increase in the ability of the turbulence algorithm to resolve the vertical structure of C_n^2 as the number of vertical levels increases (Fig. 6). Although improvement is needed and possible, these preliminary performance results of optical-turbulence algorithm are encouraging.

In the future, calibration of the minimum turbulence kinetic energy within the scheme is a necessary next step to adjust the algorithm to the Mauna Kea site and improve the simulation results. Calibration of the background turbulence kinetic energy will be accomplished using field observations taken by the Thirty Meter Telescope and Ground Layer Adaptive Optics campaigns. Additional studies of the sensitivity of turbulence algorithm to vertical and horizontal resolution will be carried out.

Acknowledgments

We would like to thank the Eric Kemp and Randall Alliss (Northrop Group, TASC) for their collaboration and willingness to share their research results with us. We also thank the Mark Chun (Institute for Astronomy) for providing the G-SCIDAR and MASS data.

References

1. Bean, B. R., and E. J. Dutton, 1966: *Radio Meteorology*. NBS Monogr. 92, 435 pp.
2. Businger, S., R. McLaren, R. Ogasawara, D. Simons, and R. J. Wainscoat, 2001: Starcasting. *Bull. Amer. Meteor. Soc.*, **83**, 858-871.
3. Cherubini, T., S. Businger, C. Velden, and R. Ogasawara, 2006: The impact of satellite-derived atmospheric motion vectors on mesoscale forecasts over Hawaii. *Mon. Wea. Rev.*, **134**, 2009-2020.
4. Cherubini, T., S. Businger, R. Lyman, and M. Chun, 2008: Modeling Optical Turbulence and Seeing over Mauna Kea. *J. Appl. Metor. Clim.* **47**, 1140-1155.
5. Cherubini, T., S. Businger, R. Lyman, and M. Chun, 2008: Modeling Optical Turbulence and Seeing over Mauna Kea: Verification and Algorithm Refinement. *J. Appl. Metor. Clim.* **48**, 3033-3043.
6. Grell, G. A., J. Dudhia, and D. R. Stauffer, 1985: A description of the fifth-generation PENN State/NCAR Mesoscale Model (MM5). NCAR/TN-398+STR. Technical report, National Center for Atmospheric Research, Boulder, CO.
7. Hong, S.-Y., J. Dudhia, and S.-H. Chen, 2004: A Revised Approach to Ice Microphysical Processes for the Bulk Parameterization of Clouds and Precipitation, Mon. Wea. Rev., **132**, 103-120.
8. Janjic, Z. I., 2002: Nonsingular Implementation of the Mellor–Yamada Level 2.5 Scheme in the NCEP Meso model, NCEP Office Note, No. 437, 61 pp.
9. Kain, J. S., and J. M. Fritsch, 1990: A one-dimensional entraining/ detraining plume model and its application in convective parameterization, J. Atmos. Sci., **47**, 2784-2802.
10. Kain, J. S., and J. M. Fritsch, 1993: Convective parameterization for mesoscale models: The Kain-Fritcsh scheme, The representation of cumulus convection in numerical models, K. A. Emanuel and D.J. Raymond, Eds., Amer. M eteor. Soc., 246 pp.
11. Kemp, E. M., B. D. Felton and R. J. Alliss, 2008: estimating the refractive index structure-function and related optical seeing parameters with the WRF-ARW. 2008 WRF Users Workshop *proceedings* [available at: http://www.mmm.ucar.edu/wrf/users/workshops/WS2008/abstracts/P9-28.pdf]
12. Klemp, J. B., W. C. Skamarock, and J. Dudhia, 2007: Conservative split-explicit time integration methods for the compressible nonhydrostatic equations. Mon. Wea. Rev., *in press*.

13. Kondo, J., O. Kanechika, and N. Yasuda, 1978: Heat and momentum transfers under strong stability in the atmospheric surface layer. *J. Atmos. Sci.*, **35**, 1012-1021.
14. Masciadri, E., J. Vernin, and P. Bougeault, 1999a: 3d mapping of optical turbulence using an atmospheric numerical model. i: a useful tool for the ground-based astronomy. *Astron. Astrophys. Supplement Series*, **137**, 185-202.
15. Masciadri, E., J. Vernin, and P. Bougeault, 1999b: 3D mapping of optical turbulence using an atmospheric numerical model. ii: First results at Cerro Paranal. *Astron. Astrophys. Supplement Series*, **137**, 203-216.
16. Masciadri, E., and J. P. Jabouille, 2001: Improvements in the optical turbulence parameterization for 3d simulations in a region around a telescope. *Astron. Astrophys.*, **376**, 727-734.
17. McGinley, J. A., 1989: The local analysis and prediction system. *12th Conf. on Analysis and Forecasting*, Amer. Meteor. Soc., Monterey, CA, 15-20.
18. Mlawer, E. J., S. J. Taubman, P. D. Brown, M. J. Iacono, and S. A. Clough, 1997: Radiative transfer for inhomogeneous atmosphere: RRTM, a validated correlated-k model for the longwave. J. Geophys. Res., 102 (D14), 16663-16682.
19. Skamarock, W. C., J. B. Klemp, J. Dudhia, D. O. Gill, D. M. Barker, W. Wang, and J. G. Powers, 2005: A description of the Advanced Research WRF Version 2. NCAR Tech Notes-468+STR [available at: http://www.mmm.ucar.edu/wrf/users/pub-doc.html]
20. Tatarskii, V. I., 1971: The effects of the turbulent atmosphere on wave propagation. Technical report, U.S. Department of Commerce, NTIS TT-68-50464. 472pp.
21. Walters, D. L. and D. K. Miller, 1999: Evolution of an upper-tropospheric turbulence-event Comparison of observations to numerical simulations. *Preprints, 13th Symposium on Boundary Layer Turbulence*, AMS, 157-160, Dallas, TX.

NUMERICAL SIMULATIONS OF THE WINTERTIME OPTICAL TURBULENCE IN ANTARCTICA WITH THE MESOSCALE MODEL MESO-NH

F. LASCAUX*, E. MASCIADRI**, S. HAGELIN and J. STOESZ

INAF/Osservatorio Astrofisico di Arcetri
Florence, Italy
*E-mail: * lascaux@arcetri.astro.it*
*** masciadri@arcetri.astro.it*

Mesoscale model such as Meso-NH have proven to be highly reliable in reproducing 3D maps of optical turbulence (OT).[1–3] These last years ground-based astronomy has been looking towards Antarctica, especially its summits and the continental plateau where the OT appears to be confined in a shallow layer close to the surface. However some uncertainties remain. That's why our group is focusing on a detailed study of the atmospheric flow and turbulence in the internal Antarctic Plateau. Our intention in this study is to use the Meso-NH model to do predictions of the atmospheric flow in the internal plateau. The use of this model permits us to have access to informations inside an entire 3D volume, which is not the case with observations only. Two different configurations of the model have been used: one with a low horizontal resolution ($\Delta X = 100$ km) and another one with higher horizontal resolution with the help of the grid-nesting interactive technique ($\Delta X = 1$ km in the innermost domain). The impact of the configuration on the meteorological parameters has already been studied.[4] We present here the results obtained with Meso-Nh of forecasted C_N^2 profiles, surface layer thickness (SLT) and seeing values at Dome C for the 16 winter nights, whose C_N^2 profiles have been measured by Ref. 5.

Keywords: Atmospheric Turbulence; Site Testing; Numerical Simulations.

1. Introduction

The internal Antarctic Plateau is, at present, a site of potential great interest for astronomical applications. Its extreme low temperatures, dryness and typical high altitude (more than 2500 m), joint to the fact that the OT seems to be concentrated in a thin surface layer whose thickness is of the order of a few tens of meters do of this site a place in which, potentially, we could achieve astronomical observations otherwise possible only by space. In spite of the exciting first results[5–7] the effective gain that astronomers might achieve from ground-based astronomical observation from this location still suffers from serious uncertainties and doubts that have been pointed out in previous work.[8–10] A better estimate of the properties of the OT above the internal Antarctic Plateau can be achieved with both dedicated measurements done in simultaneous ways with different instruments and simulations provided by atmospheric models. Simulations offer the advantage to provide volumetric maps of the OT (C_N^2) extended on the whole internal plateau and, ideally, to retrieve comparative estimates in a relative short time and homogeneous way at different places. In a previous paper[10] our group performed a detailed analysis of the meteorological parameters from which the OT depends on, provided by the General Circulation Model (GCM) of the European Center for Medium-range Weather Forecasts (ECMWF). In that work we quantified the accuracy of the ECMWF estimates for all the major meteorological parameters and, at the same time, we pointed out which are the limitations of the GCMs. In contexts in which the GCMs fail, mesoscale models can supply more information. The latter are indeed conceived to reconstruct phenomena (such as the OT) that develop at a too small spatial and temporal scale to be described by a GCM. In spite of the fact that mesoscale models can attain higher resolution than the GCM, parameters such as the OT are not explicitly resolved but are parameterized, i.e. the fluctuations of the microscopic physical quantities are expressed as a function of the corresponding macroscopic quantities averaged on a larger spatial scale (cell of the model).

For classical meteorological parameters the use of a mesoscale model should be useless if GCMs such as the one of the ECMWF could provide estimate with equivalent level of accuracy.

For this reason the Ref. 10 has been a first step towards the exploitation of the mesoscale Meso-Nh model. We retrieved everything that was possible from the ECMWF analyses (wind field, potential temperature, absolute temperature...) and we defined their limitations at the same time. We concluded that in the first 10-20 m, the ECMWF analyses show a discrepancy with respect to measurements of the order of 2-3 $m.s^{-1}$ for the wind speed and of 4-5 K for the temperature. Preliminary tests concerning the optimization of the Meso-NH model configuration and sensitivity to the horizontal and the vertical resolution have already been conducted by our team[4] for the internal Antarctic Plateau. Here we present further progress of that work. A more detailed description is found in Refs. 11, 12.

2. Meso-Nh Validation at Dome C

The purpose of this section is to verify the performances of the mesoscale Meso-Nh model above the internal Antarctic Plateau and to verify if such a mesoscale model can provide a better estimate of the atmospheric flow than the GCM from the ECMWF.

The numerical set-up of the Meso-Nh model was already presented in Refs. 11, 12.

An important number of winter nights (47) were simulated with the Meso-Nh mesocale model. We analyze here the key meteorologic parameters from which the OT depends on: the temperature and the wind speed. Two configurations (low horizontal resolution monomodel, and high horizontal resolution grid-nesting) are tested and evaluated. All the simulations started at 00 UTC and were integrated for 18 hours. Simulations outputs at 12 UTC can be compared with measurements as well as with the analysis from the GCM of the ECMWF.

A statistical study of the wind and temperature profiles at Dome C is performed. The 47 nights have been selected in winters 2005 and 2006. For all the 47 nights selected, we respected two criteria. First, a radiosounding must be available at the end of the simulation (at 12 UTC of the selected night) to perform comparisons between Meson-Nh outputs, ECMWF analysis and observations. And for the selected nights, the radiosoundings must cover the longest path along the z-axis (perpendicolar to the ground) before to explode. It was impossible to collect in winter time 47 nights in which all the balloons reached the 20 km. The mean altitude reached by the selected balloons was about 13 km. The mean vertical profiles of temperature and wind speed were computed for 47 winter nights from the two model configurations, the ECMWF analyses and the radiosoundings. The location of the profiles is Dome C as shown in Fig. 1, Ref. 11. The mean temperature profiles are very similar over the entire free atmosphere. Some light discrepancies appear in the first kilometer above the ground, where both the ECMWF analyses and the Meso-Nh model present a negative bias of 4-5 K up to 200 m . At this height, the grid-nested simulation and the low resolution simulation give similar results and are not necessarily better than the ECMWF. The temperature gradients provided by the two Meso-Nh configurations and the ECMWF analyses are not as pronounced as the one obtained with the radiosoundings. Considering the wind speed, from the ground up to 10 km, analyses and radiosoundings mean wind speed are well correlated. Above 10 km the wind speed reconstructed by the ECMWF analyses is larger than the one reproduced by Meso-Nh (monomodel and grid-nesting). It is hard to say whether the ECMWF analyses or the Meso-Nh simulation is the best since no mean value from the observations is available at this altitude. In the first kilometer above the ground, and especially around 150 m, it is well visible that Meso-Nh better reconstructs the strong wind shear than the ECMWF analyses. The wind speed provided by the ECMWF analyses is a bit too weak (11 m.s^{-1} instead of 13 m.s^{-1} in the observations). At the same altitude the Meso-Nh simulations give better results, with a mean wind profile perfectly correlated to the one measured by the radiosoundings. The improvement is even better in the case of the high-resolution model. The difference between low and high horizontal simulations is more important above 12 km, with an increase speed of the wind more important in the high-resolution simulation. The mean values of the wind speed and the temperature immediatly above the ground at Dome C were also computed for the 47 nights. The results are reported in the tables 1 and 2 (mean values for ECMWF analyses and radiosoudings, and Meso-Nh low and high horizontal resolution simulations, respectively). One can see that the ECMWF analyses are too warm at the

Table 1. Mean values on 47 winter days at Dome C of wind speed and temperature at the surface level, for radiosoundings and ECMWF analyses. Into brackets, the corresponding sigma.

	Radiosondes	ECMWF
Windspeed (m.s^{-1})	4.02 ($\pm$ 2.55)	6.51 ($\pm$ 2.51)
Temperature (K)	212.90 ($\pm$ 7.64)	216.64 ($\pm$ 5.83)

Table 2. Same as table 1 but for the monomodel (1-MOD) and grid-nested (Grid-N) numerical simulations.

	1-MOD	Grid-N
Windspeed (m.s^{-1})	4.23 ($\pm$ 1.77)	3.98 ($\pm$ 1.95)
Temperature (K)	214.92 ($\pm$ 4.64)	214.50 ($\pm$ 4.97)

surface (with a difference of almost 4 K, tables 1) in winter with respect to the observations. The mean surface temperature simulated by Meso-Nh after 12 hours is closer to the observations than the ECMWF analyses. The difference between ECMWF analyses and observed mean temperature is 3.74 K. The mean surface temperature in the low-resolution simulation is $\sim$2 K higher than in the observations. It is only 1.6 K higher for the grid nested simulation. Therefore Meso-NH can improve up to a factor of 2 the accuracy of the reconstructed temperature. More over it can be seen that the surface temperature is better retrieved with the use of the high horizontal resolution (ΔX=1 km in the innermost domain) than the low horizontal resolution (ΔX=100 km). Tables 1 and 2 show the mean values of the wind speed at the first interpolated point of the profiles. The mean wind speed in the ECMWF analyses is higher than the observed wind speed ($\sim$6.5 m.s^{-1} against $\sim$4 m.s^{-1}, thus a difference of 2.5 m.s^{-1}). Both low and high horizontal simulations reproduce better the surface wind than the ECMWF analyses. With a mesh-size of 100 km in Meso-Nh, the difference in simulated mean wind speed and observations is of $\sim$0.2 m.s^{-1}. The grid nested simulations give even better results.

3. Forecasting of OT at Dome C with the Astro-Meso-Nh Package

We have seen in the previous section that the Meso-Nh model allows for a better forecasting of meteorological parameters as wind speed and temperature than the analyses from the GCM of the ECMWF. They are very important for OT forecasting, since the astronomical parameters derives from these thermodynamic parameters. In order to verify how well simulations match with measurements we start with computing the typical SLT using the same criterion as in Ref. 5. They defined the thickness h_{sl} of the surface layer as the slab containing 90% of the first kilometer BL optical turbulence:

$$\frac{\int_{8m}^{h_{sl}} C_N^2(z)dz}{\int_{8m}^{1km} C_N^2(z)dz} < 0.90 \qquad (1)$$

where C_N^2 is the refractive index structure parameter. The observed SLT for 16 winter nights are reported on table 3. We also computed h_{sl} from Meso-Nh for the same 16 nights, using criterion 1 (table 4). Independently from the time interval chosen for computing the mean (see Ref. 12), and using the criterion 1, the grid-nested simulations give a mean SLT of around 47 m and the monomodel a mean thickness of around 60 m (table 4). Both configurations leads to a higher mean forecasted thickness with respect to the observed one (34.5 m) but the grid-nested technique is closer to the observations ($\Delta h\sim$12 m). In spite of the computing cost, the grid-nested configuration seems to be necessary to better reconstruct the concentration

Table 3. SLT for 16 winter nights. Units in m. The criterion used is the one from Eq. 1. The mean value is also reported.

Date	Observed SLT (m)	Date	Observed SLT (m)
13/06/05	23	08/08/05	30
04/07/05	30	12/08/05	22
07/07/05	21	29/08/05	47
11/07/05	98	02/09/05	41
18/07/05	26	05/09/05	20
21/07/05	47	07/09/05	39
25/07/05	22	16/09/05	24
01/08/05	40	21/09/05	22
		MEAN	34.5

Table 4. Mean SLT for the same 16 winter nights than in table 3, deduced from Meso-Nh computations using the criterion in Eq. 1. Units in m. The mean value is also reported.

date	Grid-N	1-MOD	date	Grid-N	1-MOD
13/06/05	23.1	20.5	08/08/05	40.8	64.1
04/07/05	26.2	42.1	12/08/05	28.5	37.9
07/07/05	58.9	54.2	29/08/05	97.1	94.9
11/07/05	81.9	110.4	02/09/05	65.6	85.5
18/07/05	54.1	84.9	05/09/05	28.2	45.7
21/07/05	65.9	76.0	07/09/05	85.3	117.3
25/07/05	16.9	25.7	16/09/05	22.2	32.5
01/08/05	34.6	44.5	21/09/05	22.0	27.7
			MEAN	46.8	60.2

of the turbulence in a thin layer near the surface. To highlight the potentiality of Meso-Nh in discriminating between different turbulent conditions in different nights, the temporal evolution of the C_N^2 profiles near the surface is displayed for 2 nights (figure 1). These 2 nights had different turbulent conditions: the night 1 (11 July 2005) had a well developed surface layer (h_{obs}=98 m); the night 2 (25 July 2005) is less turbulent, with the turbulence concentrated in a thin surface layer (h_{obs}=22 m). During the night 1 (Fig. 1(a,b) for monomodel and grid-nested configurations) the predicted SLT remained constant in time, with a mean forecasted value of ∼82 m for the grid-nested simulation. The monomodel simulations gave a value of ∼110 m. Both are closed to the observation made for this night.[5] The night 2 (figure 1c,d), with less turbulence, has a mean forecasted SLT of 17 m in grid-nested mode (∼26 m in monomodel). The observed value was 22 m.[5] This shows the ability of Meso-Nh to react and predict the evolution of dif-

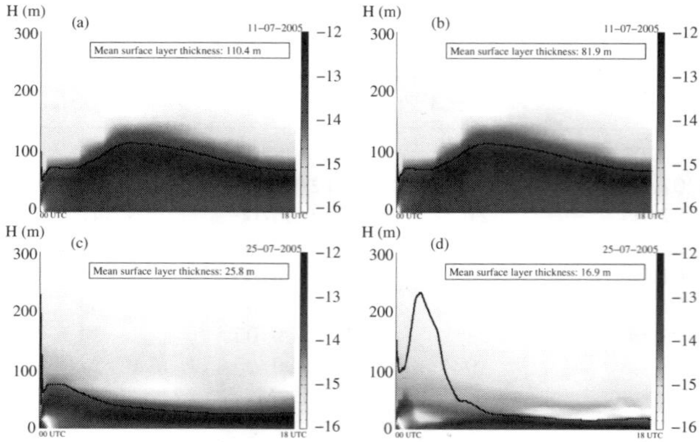

Fig. 1. Temporal evolution of the C_N^2 profile at Dome C obtained with Meso-Nh, between 00 and 18 UTC for 2005 July 11, (a)low horizontal resolution simulation and (b)grid-nested simulation, and 2005 July 25, (c)low horizontal resolution simulation and (d)grid-nested simulation, The mean value of the SLT (criterion 1) between 11 and 17 UTC is written for each night. The thin black line represents the evolution of the SLT during the night.

ferent surface layers. What also is interesting to see is that the grid-nested configuration tends to concentrate the turbulence in a thinner layer than the monomodel configuration. This is especially visible for night 2.

The mean value of the free atmosphere seeing was also computed for the 16 nights forecasted with the grid-nested configuration of the model. The free atmosphere seeing was defined as the seeing above the mean altitude of the top of the computed SLT (46.8 m) and up to 13 km high (in order to perform valuable comparisons with the observations[5]). We obtained a value of 0.46", to be compare to the 0.3" deduced from the observations.

4. Conclusions

We studied the performances of the Meso-Nh mesoscale meteorological model in reconstructing wind and temperatures profiles above Dome C, a site in the internal Antarctic Plateau. Two different configurations were tested. First, monomodel simulations using a low horizontal resolution (ΔX=100 km) were performed for 47 winter nights. Then the same nights were simulated using a grid-nesting high horizontal resolution configuration (ΔX=1 km for the innermost domain). The results were compared with the analyses from the ECMWF Global Circulation Model and radiosoundings. Several conclusions can be drawn: (1) we showed that near the surface, Meso-Nh retrieved better the wind vertical gradient than the

ECMWF analyses, thanks to the use of a highest vertical resolution; (2) the analysis of the first vertical grid point permits us to conclude that the Meso-Nh model surface temperature is closest to the observations than the GCM of ECMWF which is too warm; (3) the outputs from the grid-nested simulations are closer to the observations than the monomodel simulations.

This study highlighted the necessity of the use of high horizontal resolution to reconstruct a good meteorological field in Antarctica, even if the orography is almost flat over the internal Antarctic Plateau. The computations estimates from a previous study[13] are probably affected by the low horizontal resolution they used in their simulations.

The Meso-Nh mesoscale model, even in a remote site as the Antarctica known for very extreme weather conditions, already can give valuable estimates of the SLT and free atmosphere seeing in winter. We planned to extend this study to other astronomical parameters. We will especially look at the boundary layer seeing.

Acknowledgments

This study has been funded by the Marie Curie Excellence Grant (FOROT) - MEXT-CT-2005-023878.

References

1. E. Masciadri, J. Vernin and P. Bougeault, *A&ASS* **137**, 185 (1999).
2. E. Masciadri and P. Jabouille, *A&A* **376**, 727 (2001).
3. E. Masciadri and R. A. amd L.J. Sanchez, *RMxAA* **40**, 3 (2004).
4. F. Lascaux, E. Masciadri, J. Stoesz and S. Hagelin, Mesoscale simulations above antartica: how to choose an optimal configuration of the meteorological model to characterize turbulence above a potential astronomical site ?, in *Proc. Symposium on Seeing (SOS)*, (Kona, Hawai'i, 2007).
5. H. Trinquet, K. Agabi, J. Vernin, M. Azouit, E. Aristidi and E. Fossat, *PASP* **120**, 203 (2008).
6. J. Lawrence, M. Ashley, A. Tokovinin and T. Travouillon, *Nature* **431**, 278 (2004).
7. K. Agabi, E. Aristidi, M. Azaouit, E. Fossat, F. Martin, T. Sadibekova, J. Vernin and A. Ziad, *PASP* **118**, 344 (2006).
8. E. Masciadri and S. Egner, *PASP* **117**, 1604 (2006).
9. K. Geissler and E. Masciadri, *PASP* **118**, 1048 (2006).
10. S. Hagelin, E. Masciadri, F. Lascaux and J. Stoesz, *MNRAS* **387**, 1499 (2008).
11. F. Lascaux, E. Masciadri, J. Stoesz and S. Hagelin, *Proc. SPIE* **70124D** (2008).
12. F. Lascaux, E. Masciadri, S. Hagelin and J. Stoesz, *to be submitted.*
13. M. Swain and H. Gallée, *PASP* **118**, 1190 (2008).

METEOROLOGICAL STUDY OF AN EVENT OF OCCURRENCE OF OPTICAL TURBULENCE AT THE TMT CANDIDATE SITES

O. CUEVAS*, M. CURE and A. CHACÓN

AstroMeteorology Group, Department of Physics and Astronomy,
University of Valparaíso,
Gran Bretaña 1111, Casilla 5030, Valparaíso, Chile
**E-mail: omar.cuevas@uv.cl*
www.dfa.uv.cl/astromet

S. ELS[1,3], M. SCHÖCK[2,3], R. RIDDLE[3], W. SKIDMORE[3] and T. TRAVOUILLON[3]

[1] *Cerro Tololo, Inter-American Observatory, La Serena, Chile*
[2] *TMT Observatory Corporation, Victoria, Canada*
[3] *TMT Observatory Corporation, Pasadena, USA*

In the north of Chile three of the five pre-selected candidate sites to host TMT are located; namely, these were Cerro Armazones, Cerro Tolar and Cerro Toloncha. The TMT site testing program conducted measurements of the meteorological conditions and the optical turbulence strength throughout the atmosphere (the turbulence profiles) above these sites. A peculiar case of turbulence occurrence above these sites happened during the night of August 13, 2006. We will show a study on the meteorological conditions above these sites adressing the origin of this particular turbulence behaviour.

A synoptic analysis for this night shows anticyclonic predominance (AP) and Jet Stream (JS) activity at the high levels of the atmosphere. Simulations using the MM5 mesoscale model are performed and compared to the measured meteorological site conditions. Turbulence above Tolonchar seems be associated to the Jet Stream, but the turbulence above Tolar can not clearly be associated to the synoptic conditions. Also trajectories were calculated for each site at altitude levels of the turbulence profiles. The low level trajectories for Tolar show the origin of air during this night being South-East of the central valley, following the orography. For Armazones the low level trajectories originate from North-East. The flow which affects Tolonchar is strongly influenced by the western circulation which is present at these intermediate altitudes at around 500 hPa. Simulations based on MM5 show clear signs of a local circulation for each site, but to improve the results it will be necessary to increase the spatial resolution of the models in order to trace turbulent processes, follow their evolution and investigate their relation with good and bad seeing.

Keywords: Optical turbulence; meteorological simulations; synoptic patterns; trajectories.

1. Introduction

In the north of Chile three candidate sites for the construction of TMT (Thirty Meter Telescope) telescope had been selected for on-site testing; Tolar (21.964°, 70.009°, 2290 m.a.s.l.), Tolonchar (23.933°, 67.975°, 4477 m.a.s.l.) and Armazones (24.58°, 70.189°, 3064 m.a.s.l.). The TMT site testing program performed at these sites measurements of meteorological standard variables by automated weather stations and of the optical turbulence profiles $C_n^2(h)dh$ by MASS instruments (Multi Aperture Scintillation Sensor, [1]). More details on the TMT site testing program can be found in Schöck et al. (this conference) and Riddle et al. (this conference).

The three Chilean candidate sites formed a triangle with sides of approximately 400 km, which could allow to trace the onset and evolution of turbulence on such larger spatial scales. A peculiar case of the occurrence of optical turbulence above these sites happened during the night of August 13, 2006. The analysis of the MASS based $C_n^2(h)dh$ data shows a strong increase of the turbulence strength above Tolonchar between 3 and 4 UTC, at altitudes between 2 and 16 km. At the location of Tolar a similarly strong increase of turbulence strength occured between 5 and 8 UTC and at altitudes between 1 and 4 km above the site. In contrast, above Armazones no increase of turbulence was registered during that night (Figure 1).

In the following we will investigate the meteorological conditions during that night in the North of Chile to look for patterns which could have caused this increase in optical turbulence strength only above two out of the three locations.

2. Atmospheric Simulations

The MM5 system model was applied to simulate the atmospheric conditions during the night of August 13th, 2006. Three domains were setup, centered on 22.8°S and 68.8°W. The domain 1 has 80x80 grid points and 27 km of horizontal resolution, the domain 2 has 82x97 grid points and 9 km of horizontal resolution. The high resolution domain 3 has 130x160 grid points, 3 km of horizontal resolution and it covers all tree mountains of interest (Tolar, Tolonchar and Armazones). The boundary conditions were taken from the GFS (Global Forecast System, [2]) using its operational data at 00 UTC. The model was configured with 30 sigma levels between 1000 hPa to 100 hPa and to forecast 24 hrs. The output of data was performed every 10 minutes. For this case we used the following parameterization of the MM5 system: Schultz for explicit moisture scheme, Kain-Fritsch 2

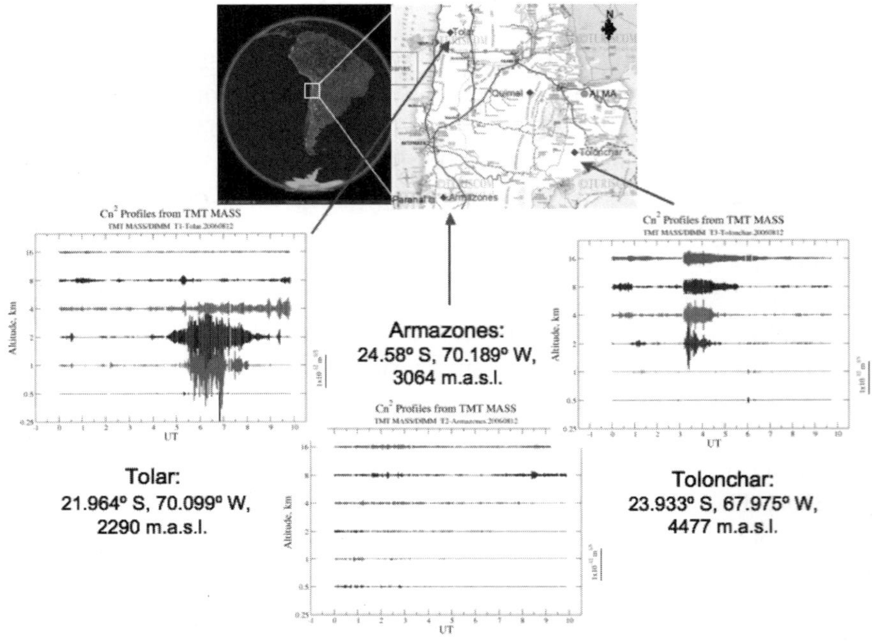

Fig. 1. MASS measurements taken during the night of 13-08-2006 at the locations of Tolar, Tolonchar and Armazones in the North of Chile.

for cumulus scheme, Eta for planetary boundary layer and RRTM (Rapid Radiative Transfer Model) for atmospheric radiation. For details on the numerical schemes employed we refer to [3,4,5].

3. Analyses

The climatic conditions within the zone of interest are arid (Atacama desert) and the regular synoptic condition is an Anticyclonic Predominance (AP) during the year. During the night of August 13 the predominant condition at low and middle atmospheric levels was also AP. The infrared satellite image shows a gray cloud cover above the sea, typical for these synoptic conditions (AP, Figure 2). The Antofagasta radiosonde launched during that day (23.43°S, 70,43°W, 12 UTC) shows the clouds at low levels (900 HPa), above which the AP is strong. The dewpoint line (left black line) shows 2 peaks. The temperature (right black line) shows also two peaks, one at 500 hPa (5830 m.a.s.l.) and another one in at hPa (9800 m.a.s.l.). This indicates that two humidity intrusions and thermal inversions were present. Such conditions are generally being associated with atmospheric instability and thought to be related to optical turbulence. Another condition related to optical turbulence is the Jet Stream (JS) at the 200 hPa level. The MM5 simulation for this night shows that the Jet Stream, with strong wind speeds, passed over the zone of interest. Its intensity increased during the night (Figure 3).

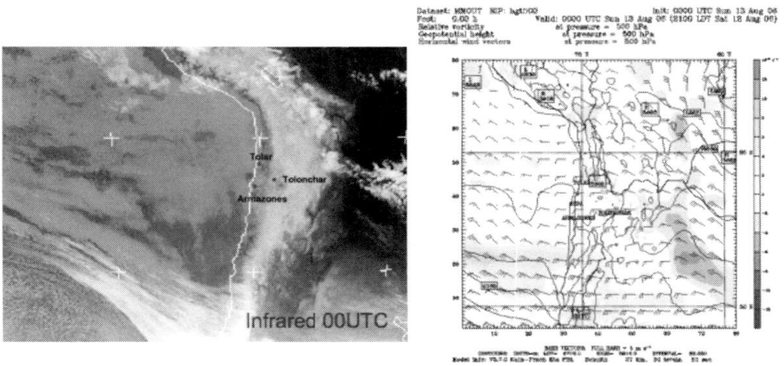

Fig. 2. (color on-line) Infrared satellite image (left). Chart of 500 hPa level (right), the shadow color is a relative vorticity, black lines is a potential height and blue vectors indicate the wind. Image from 13-08-2008 at 00 UTC.

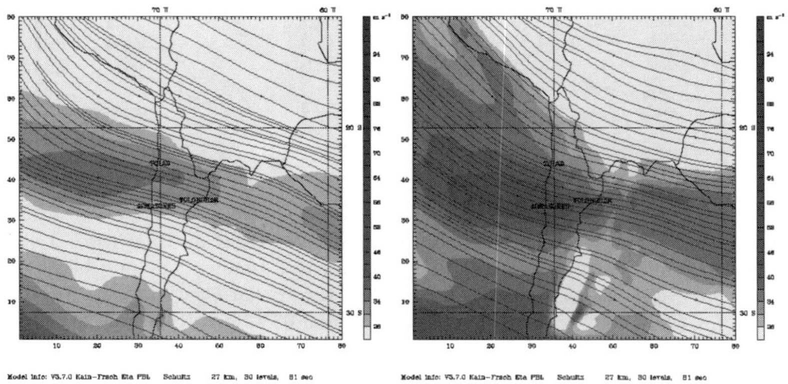

Fig. 3. (color on-line) Chart of 250 hPa. The color indicates the wind intensity and the blue lines are the streamlines. Compued for 13-08-2008 00 UTC (left) and 06 UTC (right).

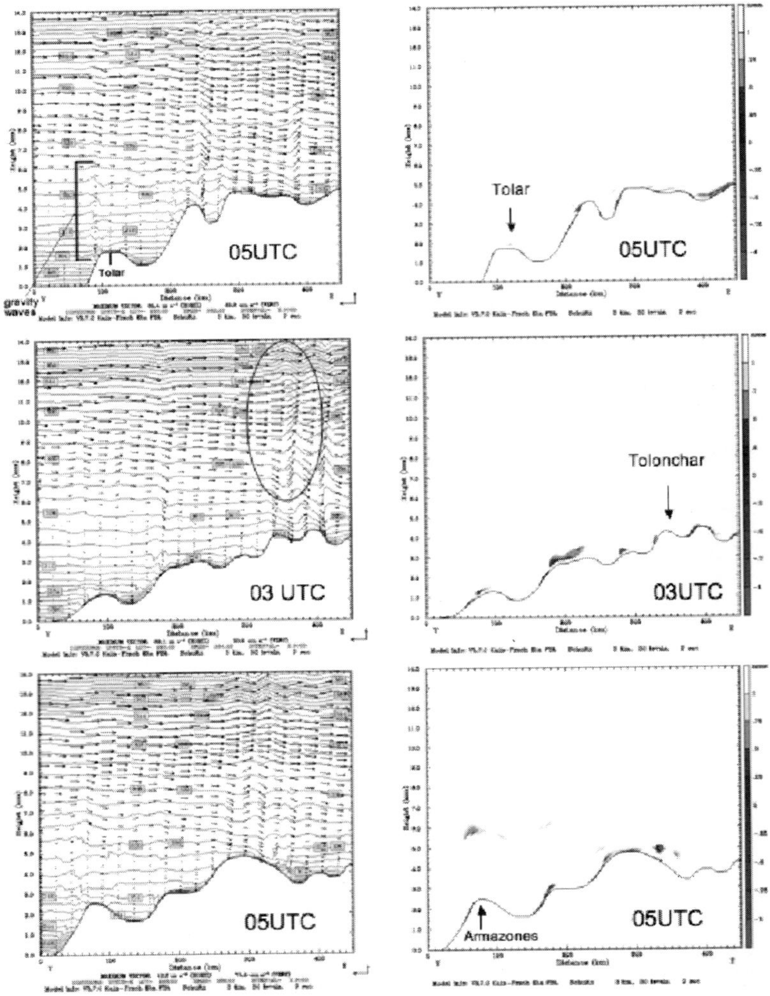

Fig. 4. (color on-line) *Left panels:* Cross section (West-East) at Tolar (up), Tolonchar (center) and Armazones (down). Red lines are potential temperature and black vectors indicate wind. *Right panels:* The color indicates the Richardson number (Ri). Images computed for 13-08-2008 at 03 UTC (Tolonchar) and 05 UTC (Tolar and Armazones).

3.1. *Vertical Analyses*

The West-East cross section at Tolar (Figure 4) shows a thermal break (potential isotherm) produced b gravity waves formed by the orographic obstacle (Tolar). This thermal break was present between 2 km at 7 km and between 05 UTC to 08 UT, coinciding with the domain when the optical turbulence was most intense. The Richardson number *Ri* was calculated as atmospheric turbulence indicator but does not show turbulence during those hours. In the Tolonchar case, the gravity waves caused by the shock of the flow with the orography is the dominant and constant condition during the night (Figure 4). The *Ri* did not indicate turbulent conditions during the hours when the optical turbulence was seen with MASS. At Armazones (Figure 4) gravity waves were a dominant condition, similar to Tolar, with the thermal break being at 7 km during the night. In this case the *Ri* indicates turbulence up to 6 km height. Starting at 01 UTC and increasing during the night. This condition is inconsistent with the MASS measurements, which did not show an increase of optical turbulence during that night. It is likely that the spatial resolution of our models is not high enough to properly trace turbulence by *Ri*.

3.2. *Trajectories*

The trajectories were calculated at 4 levels above each site. The altitudes of these levels were chosen to be identical to four levels of the MASS turbulence profiles: 1 km, 2 km, 4 km and 8 km above the sites. The backward trajectories (to know where the air comes from) were calculated at 09 UTC at 00 UTC. The trajectories at 1 km and 2 km above Tolar are in South-East direction (Figure 5), however, at Armazones these trajectories originate from the North-East direction. In the Tolonchar case the air comes from the West. This is because the altitude of the summit of Tolonchar is similar to the middle atmosphere, where the circulation is already synoptic and thus from the West. The same happens with the trajectories at 4 km and 8 km above Tolar and Armazones, where the West circulation is dominant at these altitudes.

4. Conclusions

In this work a meteorologic study was made for a particular case of optical turbulence registered by MASS instruments located on three mountains in the North of Chile and which were part of the site selection for the TMT telescope. During the night of August 13 2006, Tolar and Tolonchar registered an increase of optical turbulence during different hours during the night and at different altitude levels (Section 1.1). The differences in time and height suggest that these were due to different meteorological patterns at sub-synoptic scales. Another site analysed was Armazones, but here no

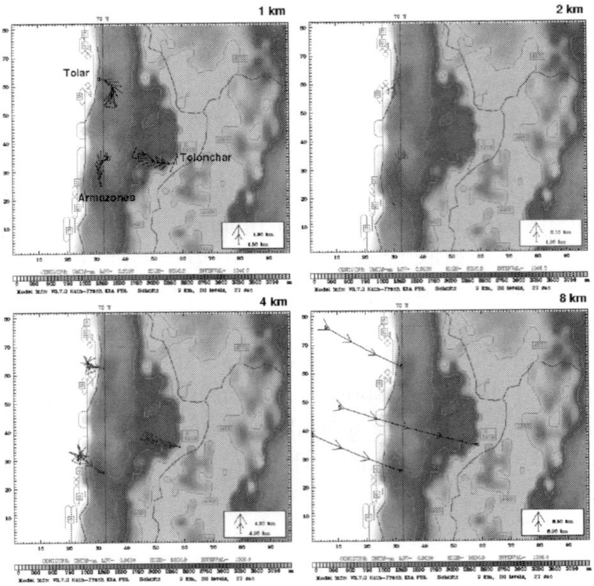

Fig. 5. Trajectories from 00 UTC to 09 UTC at 1 km (left up), 2 km (right up), 4 km (left down) and 8 km (right down). Images valid to 13-08-2008 at 03 UTC.

increase of optical turbulence was observed during that night. The synoptic conditions during this night were AP and also the jet stream increased its intensity during that night (section 2). These conditions suggest stability in the low and middle troposphere (500 hPa) but instability in the high troposphere (250 hPa). Local analyses show a thermal break above Tolar, being more intense during the hours when the optical turbulence appeared to be strong (Figure 4). Similar conditions shows the Armazones site (Figure 4) but the in-situ observations did not find optical turbulence during that night. In the Tolonchar case the origin of the optical turbulence remains unclear as gravity waves remain steady in time due the proximity to the Andes mountain range (Figure 4). As turbulence index the Richardson number (Ri) was analysed. However, it did not show a clear relation with the observed $C_n^2(h)dh$ as in the case of Armazones the Ri suggests a turbulent flow (Figure 4), whereas the MASS instrument did not observe an increased optical turbulence. The clearest correlation is observed between the potential temperature and $C_n^2(h)dh$. The trajectories show that the direction of the air at Tolar is opposite to Armazones at 1 km and 2 km above these sites and suggest a relation of the path of the airflow and C_n^2.

During future work the configuration of the MM5 model should be improved by increasing the spatial resolution to 1 km and 100 meters in horizontal and vertical direction, respectively. Future work will have to investigate other cases using Ri and TKE (Turbulent Kinetic Energy) as indicator for atmospheric turbulence. Also a sensitivity study of MM5 to input parameters like NCEP compared with ECMWF globals models should be performed.

Acknowledgment

We thank Marc Sarazin (ESO) for his comments and support. The authors gratefully acknowledge the support of the TMT partner institutions. They are the Association of Canadian Universities for Research in Astronomy (ACURA), the California Institute of Technology and University of California. This work was supported as well by the Gordon and Betty Moore Foundation, the Canada Foundation for Innovation, the Ontario Ministry of Research and Innovation, the National Research Council of Canada, the Natural Sciences and Engineering Research Council of Canada, the British Columbia of Canada Knowledge Development Fund, The Association of Universities for Research in Astronomy (Aura) and the U.S. National Science Foundation.

References

1. Kornilov, V., Tokovinin, A., Shatsky, N., Voziakova, O., Potanin and Safonov, B.: Combined MASS-DIMM instrument for atmospheric turbulence studies. 2007, Mon. Notices of the Royal Astron. Soc., Vol. 382, 343, pp. 1268-1278.
2. Grell, G. A, Dudhia, J. and Stauffer, D. R., 1994: A description of the fifth-generation of Penn State/NCAR mesoescale model (MM5). NCAR Technical Note, NCAR/TN-398+SRT, pp. 117.
3. Cuevas, O., Chacón, A. and Cure, M. 2008: Analysis of local meteorological conditions in Macon using the MM5 modeling system. *Proceedings of The International society for Optical Engineering (SPIE), 7016-73*, Marseille, France.
4. White, B. G., J. Paegle, W. J. Streenburgh, J. D. Horel, R. T. Swanson, L. K. Cook, D. J. Onton and J. G. Miles, 1999: Short-Term forecast validation of six models. *Wea. Forecast, 14*, 84-108.
5. Whiteman, D. C., 2000: Mountain Meteorology: Fundamentals and Applications. *Oxford University Press.*

THE GROUND LAYER TURBULENCE OBSERVED BY MASS-DIMM DURING THE TMT SITE TESTING AND ITS RELATION TO METEOROLOGICAL PARAMETERS

S.G. ELS

Cerro Tololo Inter-American Observatory,
Casilla 603, La Serena, Chile
E-mail: sels@ctio.noao.edu

M. SCHÖCK

TMT Observatory Corporation, 5071 West Saanich Road,
Victoria, British Columbia, Canada

R. RIDDLE, W. SKIDMORE, T. TRAVOUILLON

TMT Observatory Corporation, 2632 E. Washington Blvd.
Pasadena, CA 91107, USA

A study of the ground layer turbulence as seen by the combined measurements by MASS and DIMM during the TMT site survey is presented. The TMT site survey data include also the necessary meteorological parameters to infer important micrometeorological quantities. Several cases are being presented demonstrating the regime in which small scale local effects dominate the seeing within the first few hundred meters of the atmosphere.

Keywords: Optical turbulence; MASS; DIMM; ground layer turbulence.

1. Introduction

The site testing for the Thirty Meter Telescope (TMT) has monitored over a period of more than 2.5 years the optical turbulence properties above five candidate sites. These were 13N ontop of Mauna Kea, San Pedro Mártir in Mexico and Cerro Tolar, Cerro Armazones and Cerro Tolonchar. The latter three being located in northern Chile. We refer to Riddle et al. and Schoeck et al. (this conference) for overviews of the TMT site testing program. At all sites 35 cm telescopes were deployed, mounted on 6 m tall towers and equipped with a combined MASS-DIMM instrument [1]. The Differential Image Motion Monitor (DIMM) measures the integrated

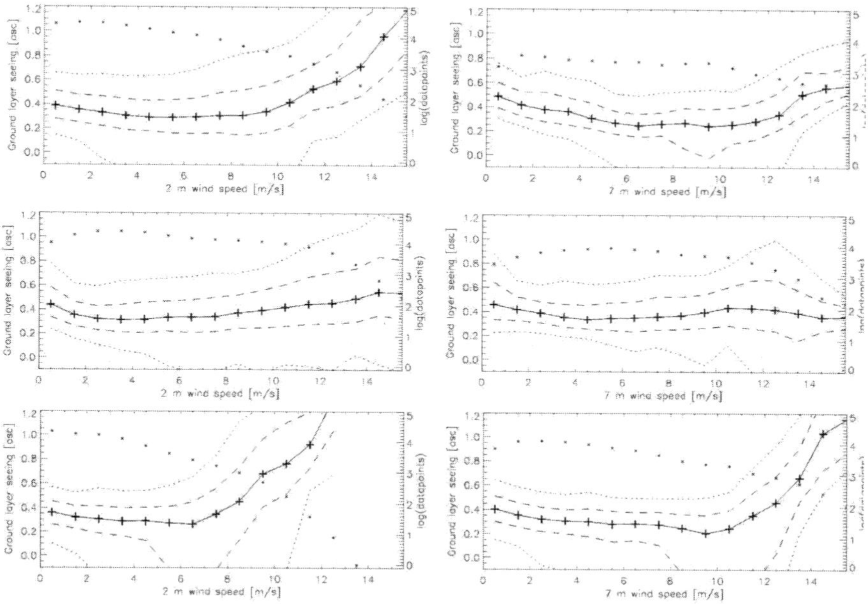

Fig. 1. MASS-DIMM based ground layer seeing vs. wind speed at Tolar (top panels), Armazones (middel panels) and Toloncha (lower panels). The left panels show the wind speeds measured at 2 m, the right panels the measurements taken 7 m above ground. The lines indicate the GL seeing percentiles within 1 m/s wind speed bins (dotted: 10% and 90%, dashed: 25% and 75%, solid and crosses: median). The dots indicate the amount of data within each bin (logarithmic scale on the right of each panel).

turbulence strength through the atmosphere; the total seeing. The Multi Aperture Scintillation Sensor (MASS) obtains a low resolution turbulence profile $C_n^2(h)dh$ at six altitudes h above the site monitoring telescope. From such a profile, the seeing can be computed which an observer would see who would be located 500 m above the telescope. By using the difference between DIMM and MASS seeing it is possible to compute the optical turbulence strength within the first 500 m above the telescope, the ground layer (GL). The general atmospheric conditions are monitored by automated weather stations. This extensive data set allows to investigate the dependence of optical turbulence on various atmospheric conditions. Here the connection between wind speed, the ground energy budget and ground layer turbulence will be investigated. Due to the much easier terrain situation (no trees, no close-by large surface structures) this study focusses on the Chilean sites only.

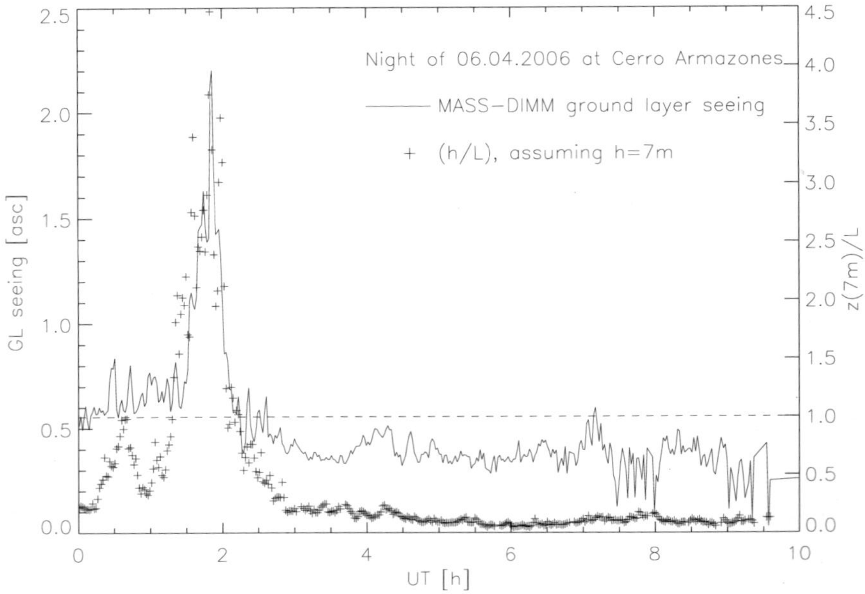

Fig. 2. Ground layer seeing (solid line) as observed by MASS-DIMM during the night of April 6, 2006 at Cerro Armazones. The crosses (scale on the right) represent the stability parameter (h/L), with Monin-Obukhov length L, computed from the weather station measurements and assuming h to be the telesocpe height. The dotted horizontal line indicates $(h/L) = 1$. The lowest wind speeds during this night were $v_{2m} \approx 2$ m/s and the main cause for the large (h/L) values.

2. Ground Layer Seeing Behaviour

Simultaneos MASS and DIMM observations were used to compute the integrated turbulence strength within the first 500 m, providing a measurement approximately every 70 sec to 90 sec. Wind speeds were recorded at each site by means of cup anemometers mounted 2 m above ground and by a CSAT-3 sonic anemometer located at the telescope level 6 m above ground. Both systems provide averaged measurements over 1 minute time intervalls. To investigate the dependence of the GL seeing on wind speed, we computed the statistics over all GL data in 1 m/s bins between 0–15 m/s, the typical operational limit for the site testing telescopes. This limit was defined by the wind speed at 2 m above ground and was set at Tolonchar to be slightly lower than at the other sites. The resulting viewgraphs of GL seeing versus wind speed are shown in the panels of Figure 1. The GL seeing calculation took into account data for which the MASS seeing was larger than the DIMM seeing, resulting in negative GL seeing values. The aim is to avoid biases in the final statistics due to times during which the GL strength was so low that MASS and DIMM measurment were reaching their accuracy limit. Figure 1 shows that the lowest GL seeing values are reached when the wind speed is non-zero and, for the sites in question, in a range between approximately 2 m/s and 6 m/s. Above such wind speeds the GL seeing increases again, in the case of Tolonchar very rapidly. That this increase is not an instrumental artefact indicates the very weak increase of the Armazones data, e.g., the curves are almost flat, and that similar behaviour has been observed by other site monitoring programes [2].

3. Low Wind Speeds

The increase of the turbulence strength towards very low wind speeds ($\lesssim$ 2 m/s) is suspected to be associated with changes in the stability of the atmosphere [3]. In Figure 2 data of a single night taken at Armazones are shown. During that night a strong increase of the GL seeing occured simultaneously with very low wind speed conditions. Also in this Figure the behaviour of the stability parameter h/L is shown. L is the Monin-Obuhov length and h the altitude above ground. $|L|$ indicates the altitude up to which turbulence production is dominated by shear. We set $h = 7$ m, equal to the telescope height. The GL seeing increases strongly when $|L| \lesssim$ 7 m, e.g., the telescope looks through a stable atmosphere in which the turbulence production is dominated by buoyancy.

At low wind speeds it could be expected that the local summit terrain surrounding the telescope will dominate the turbulence production within the air low above the site. Therefore, it can be hoped that in this regime the Monin-Obukhov similarity theory (MO), developed for the flows over flat terrain, can be, at least qualitatively, applied. We used equations (2.47) and (2.49) for $C_T^2(h)$ for the surface and mixed layers, provided in [4], to compute theoretical turbulence profiles. Using the median night time conditions (temperature, pressure, ground heat flux, net radiation) and various wind speeds (and surface roughnesses) at Armazones, vertical turbulence profiles $C_T^2(h)$ were computed; then converted to $C_n^2(h)$ and convolved with the MASS-DIMM GL response function to obtain the GL turbulence strength. The results for the stable stratification are disappointing and reproduce the observed turbulence strength only within orders of magnitude, even when altitude-less scaling is applied in the domain $h/L > 0.5$. The decrease of turbulence strength with increasing wind speeds in the results employing

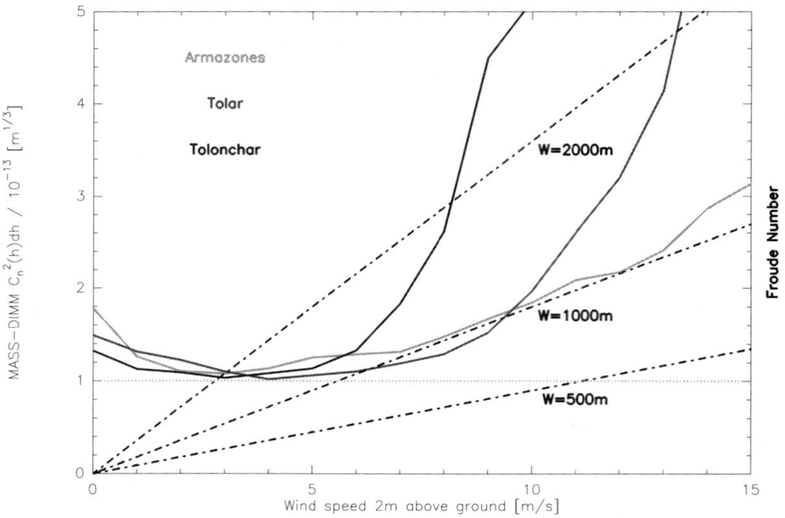

Fig. 3. The Froude number Fr(right scale and dashed-dotted lines) under median Armazones conditions as function of wind speed, for different mountain widths W. The horizontal dotted line indicates $Fr = 1$.

the unstable atmospheric models are somewhat better and do show a rate of decrease of the turbulence strength closer to the observed values.

4. High Wind Speeds

At higher wind speeds, the turbulence production due to wind shear will become more important and is affected by a larger area surrounding the summit. If the wind speed is strong, separation of the flow might be introduced by terrain features at the summit and/or by the geometry of the mountain itself. In the later case, this will also trigger the formation of gravity (mountain) waves. The parameter commonly used to decide whether the wind flows over the mountain or around it, is the Froude number $Fr = v/(WN_{BV})$, where W is the width of the mountain and N_{BV} is the Brunt-Väisälä frequency. If $Fr \gtrsim 1$ gravity waves will form due to the friction between the flow and the mountain. For $Fr < 1$ the streamlines will not separate and buoyant forces are dominating. We are interested at which wind speeds $Fr > 1$ becomes true. Again we use the median conditions at Armazones and compute Fr at different wind speeds and for various mountain widths. Figure 3 shows the results and overplotted again the observed GL turbulence strength. Three width parameters W were

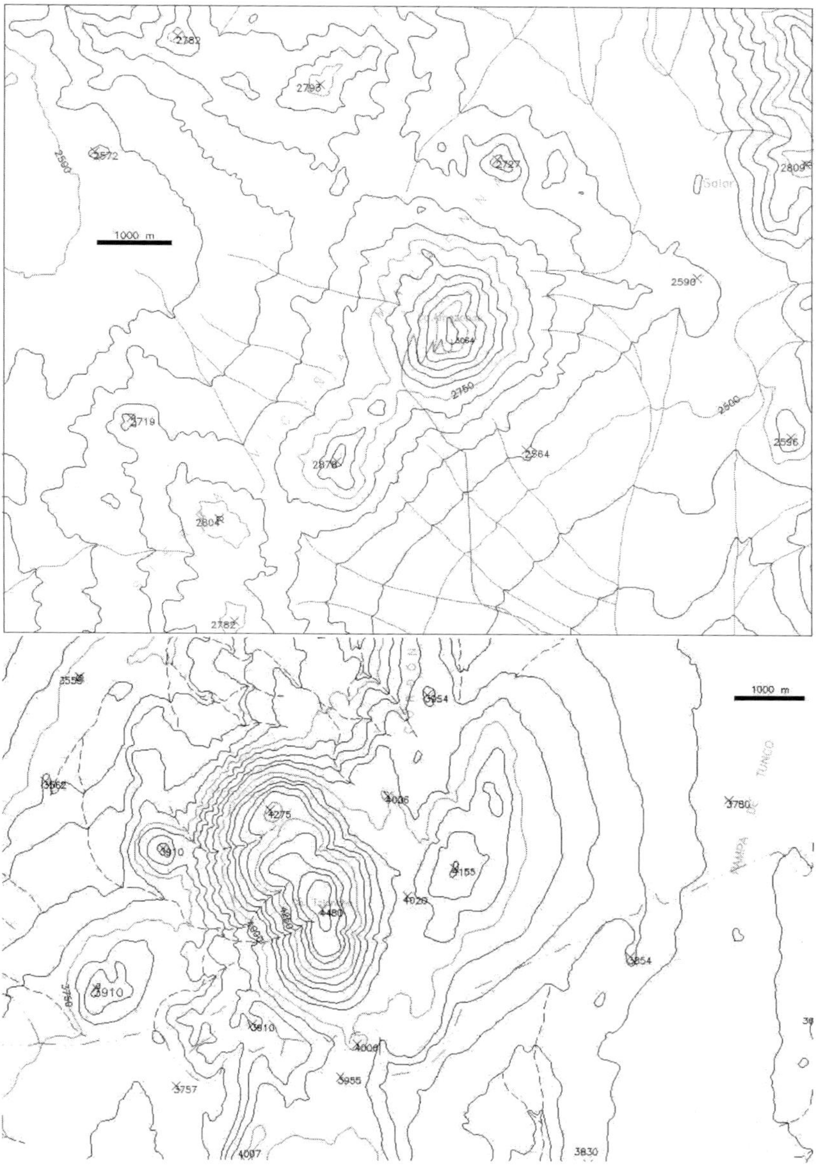

Fig. 4. (color on-line) Topographic map of Cerro Armazones (top panel) and Cerro Tolonchar (bottom panel). The thick bar represents a length of 1 km. Altitude lines are spaced by 50 m (black) and 250 m (purple); North is up. During the TMT site monitoring the prevailing night time wind directions were between approximately 340° and 030° at Armazones and 300° and 340° at Tolonchar. Map source: Instituto Geográfico Militar de Chile (`http://www.igm.cl`).

chosen, which shall cover the cases of the candidate mountains. We define W as the width of the mountain in the prevailing wind direction and at the altitude line 250 m below the summit. Armazones and Tolar we assume to be similar and associate $W = 500 - 1000$ m with them (Figure 4, top panel). Toloncha r is a very extended mountain range, rising steeply out of the surrounding area and represents rather $W = 2000$ m (Figure 4, bottom panel). As can be seen in Figure 3, $Fr > 1$ is fullfilled at lower wind speeds in the case of a larger mountain width W. The Froude number can also be defined and similarly interpreted by the height of the mountain H_m. Also in this case Tolonchar shows the largest scaling parameter H_m. We note, that the difference between the wind speeds measured at 2 m and 7 m do not change these basic qualitative conclusions.

5. Conclusions

The behaviour of the optical turbulence strength of the ground layer, as measured by MASS-DIMM, has been investigated in the cases of the three Chilean TMT candidate sites. At low wind speeds ($\lesssim 2 - 3$ m/s) buoyancy seems to be the dominant source for optical turbulence in this layer. Applying analytical formulae from the literature for $C_T^2(h)$, which were developed for flows over flat and uniform terrain, and computing the MASS-DIMM turbulence strength, we do not find a good absolute match with the observations. However, these models do show a similar trend towards lower turbulence strength at higher wind speeds. That these models do not represent our data well, is not surprising as no analytical form for $C_T^2(h)$ have been yet found for the layer above a stable surface layer and these models were developed a very different environment. At low wind speeds drainage flows are likely to occur and new models will have to be developed to take such effects properly into account.

At larger wind speeds friction becomes the dominant driver of turbulence. The Froude number with its geometric condition on the mountain shape (in the direction of the prevailing wind), appears to indicate the correct behaviour: long, steep mountains show an increase of their GL seeing at lower wind speeds as compared to lower and smoother mountains. Whether this is due to the formation of mountain waves or by separation bubbles forming locally at the summit edge has to be investigated by future studies.

To properly model the GL turbulence is very complex and 'might be impractical' [5]. Still, employing astronomical turbulence profilers allows to obtain measurements of atmospheric layers with high temporal sampling, which otherwise would be impossible to reach. They have the potential to become powerful tools for boundary layer meteorological studies.

Acknowledgment

The authors gratefully acknowledge the support of the TMT partner institutions. They are the Association of Canadian Universities for Research in Astronomy (ACURA), the California Institute of Technology and University of California. This work was supported as well by the Gordon and Betty Moore Foundation, the Canada Foundation for Innovation, the Ontario Ministry of Research and Innovation, the National Research Council of Canada, the Natural Sciences and Engineering Research Council of Canada, the British Columbia of Canada Knowledge Development Fund, The Association of Universities for Research in Astronomy (AURA) and the U.S. National Science Foundation.

References

1. Kornilov V., Tokovinin A., Shatsky N., Voziakova O., Potanin and Safonov B., "Combined MASS-DIMM instrument for atmospheric turbulence studies", 2007, Mon. Notices of the Royal Astron. Soc., *382*, 343, 1268
2. Thomas-Osip J.E., Prieto G., Johns M., Phillips M., "Giant Magellan Telescope Site Evaluation and Characterization at Las Campanas Observatory", in 'Ground-based and Airborne Telescopes II', Stepp L.M. and Gilmozzi R. (eds.), 2008, Proc. of SPIE, Vol. 7012, 70121U
3. Els S.G., Vogiatzis K., "Revealing the onset of free convection in terrestial planet atmospheres", in 'Convection in Astrophysics', Kupka F., Roxburgh I.W., Chan K.L. (eds.), 2007, Proc. IAU Symp. No. 239, 172
4. Kaimal J.C., Finigan J.J., "Atmospheric Boundary Layer Flows", 1994, Oxford University Press
5. Stull R.B., "An Introduction to Boundary Layer Meteorology", 1988, Kluwer Academic Publishers

ASTROCLIMATOLOGICAL ANALYSIS OF GROUND BASED OBSERVATORIES*

GIANLUCA LOMBARDI

Department of Astronomy, University of Bologna, Via Ranzani 1, I-40127 Bologna, Italy
INAF Bologna Astronomical Observatory, Via Ranzani 1, I-40127 Bologna, Italy
ESO European Southern Observatory, Av. Alonso de Cordova 3107, Santiago 19, Chile
E-mail: glombard@eso.org

VALENTINA ZITELLI

INAF Bologna Astronomical Observatory, Via Ranzani 1, I-40127 Bologna, Italy
E-mail: valentina.zitelli@oabo.inaf.it

SERGIO ORTOLANI

Department of Astronomy, University of Padova, Vicolo dell'Osservatorio 5, I-35122 Padova, Italy
E-mail: sergio.ortolani@unipd.it

This contribution shows some results obtained analysing a long time series of meteorological data taken at two of the most important astronomical sites in the world: the site of Observatorio del Roque de Los Muchachos (ORM, Canary Islands) and the site of Paranal Observatory (Chile). The aim of this analysis is to derive correlations between synoptic parameters and the physical properties of the atmosphere in both short and long time baseline in order to have the best scientific result from the astronomical observations.

1. Introduction

A key parameter in determining how good is an astronomical site is the image quality that we know to be linked with the local environmental conditions. The knowledge of correlations between synoptic parameters and the local properties of atmospheric turbulence in short and long time scale gives important

* This activity is supported by the European Community (Framework Programme 7, Preparing for the construction of the European Extremely Large Telescope, Grant Agreement Number 211257).

information in order to improve the global efficiency of the system telescope-instrumentation. This kind of analysis needs long series of data in order to avoid bias due to the short time fluctuations, and the same statistical approach, when comparing different sites, in order to avoid induced statistical errors. Due to their different characteristics, two of the most important astronomical sites are here taken into account: the site of Paranal Observatory (Chile) in the southern hemisphere, and the Observatorio del Roque de Los Muchachos (ORM) located in La Palma Islands (Canaries, Spain) in the northern hemisphere. Paranal Observatory (about 2600 m a.s.l.) is characterized by a continental desert area, at 12 km from the Pacific Ocean, and it is under the influence of the El Niño phenomenon (quantified by the Southern Oscillation Index, SOI). The ORM (about 2300 m a.s.l.) is 400 km far from Morocco coast and is influenced by the almost periodical variation of the North Atlantic Oscillation (NAO). More than 20 years of data have been collected with the local meteorological towers. At ORM we have considered two different telescopes: the Galileo Telescope (TNG) and the Carlsberg Meridian Telescope (CAMC). The row data are reduced to hourly averages and then in monthly averages. The missing data are replaced with the average obtained by two corresponding months in the other years showing similar mean values [2]. In this analysis we have taken into account several meteorological parameters and the analysis is performed using different time baselines allowing complementary information. In particular, the long time scale gives indications about the climatic trend of each site through decades, the seasonal effects and possible correlations with wide scale phenomena as El Niño and North Atlantic Oscillation. The short time scale data instead give information about the physical atmospheric conditions that are strongly linked with the image quality. For instance the knowledge of the vertical and horizontal temperature gradient is a key point in the analysis of the Surface Layer, while the analysis of the fluctuations of pressure and temperature are important for the thermalization of the instruments.

2. Air Temperature

The temperature is an important parameter that affects the quality of the astronomical observations, in particular the temperature gradients are correlated with the astronomical seeing and therefore with the final quality of the astronomical observations. To this purpose it is very important to predict a few hours in advance the working air temperature of the telescope mirrors to have good performance of the whole instrumentation.

2.1. *Long Time Scale*

Figure 1 shows the annual mean temperature measured at Paranal and ORM (left panel) computed in the last 20 years. All trend are positive, in particular the best fit for ORM gives an increasing in temperature of about 1 deg/10years, while the slope computed for Paranal gives an increasing in temperature of about 0.4 deg/10year. This result is an evidence of a climate change in two different sites located in two different hemisphere and far from urban areas. At Paranal the year 2000 shows a breakdown with a clear change in slope. This suggests a possible correlation with the occurrence of wide scale climatic event such as La Niña and El Niño. A deeper analysis shows that 1999 and 2000 were characterized by strong La Niña and El Niño episodes that are linked with the coldest annual temperatures.

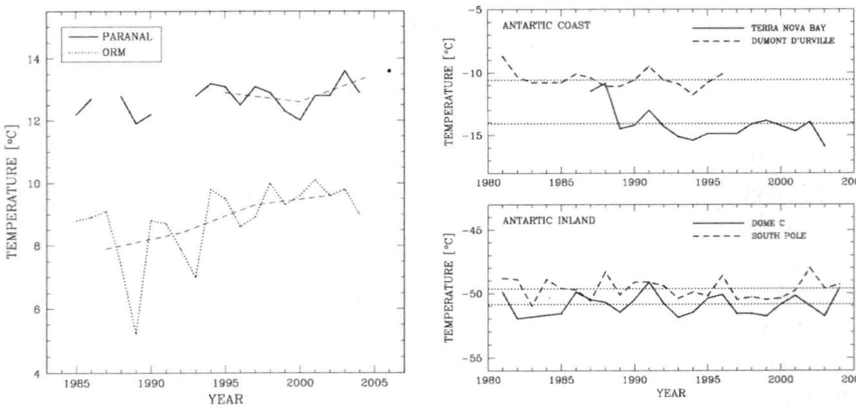

Figure 1. Left panel shows the annual mean temperatures at Paranal (solid line) and CAMC (dotted line). The short-dashed lines indicate the 5 years running means. The right panel shows, as comparison, the annual mean temperature at some Antarctic stations.

As comparison, the right panel of Figure 1 shows the plot of the annual mean temperature in the same time baseline obtained using the automatic meteo stations in the Antarctic Region from four different locations: Dome C and South Pole located in the inner land, and Terra Nova Bay and Dumont d'Urville located along the coast. The Antarctic climate appear quite stable not showing any clear positive trend during the interested years. But taking into account the more long baseline of the temperature shown in the web page of IPCC an increasing in temperature is evident since 1930. The analysis of the distribution of the meteo stations providing such data shows that they are not equally distributed around the Antarctic plateau with the majority of the stations located along the Peninsula. In conclusion we can argue that the warming has not uniform on the whole Earth.

2.2. *Short Time Scale*

Taking into account the short time scale, in terms of hours, it is well know that the image quality, computed as FWHM of the profile of the image reduced to the zenith, is strongly dependent on the temperature gradient between the main mirror and the external air. Since the years 1995 Zago [5] shown that a difference in temperature of 1 degree degrades the seeing about by 0.5 arcsec.

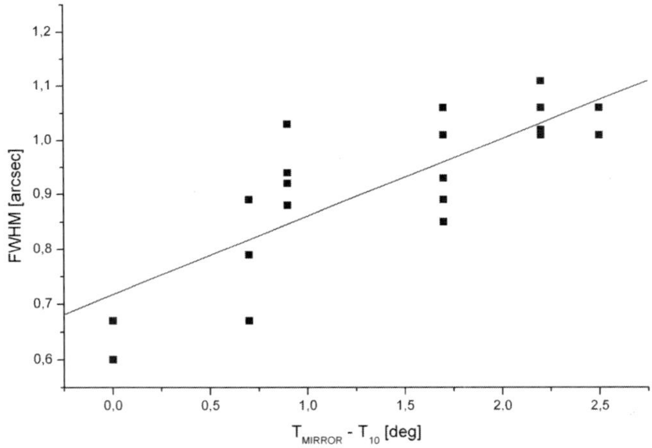

Figure 2. FWHM as a function of the difference $T_{mirror} - T_{10m}$. .

More recently Ortolani (see Figure 2) obtained the same results analysing images using the TNG telescope in June 2008. Figure 2 shows how good improvement of the image quality could be reached having care of the telescope temperature and how sensitive is the image quality to the temperature gradient. This result, obtained after a careful instrumental set up several nights long, puts the question if it is possible to predict a few hours in advance the observing temperature in order to optimize the telescope time. A collaboration with meteorologists should be welcome and very useful.

Also vertical temperature gradient just outside the dome may affect the image quality as shown in Figure 3. The plot shows the image quality in terms of FWHM as a function of the air temperature measured in the common period at two different levels. About 120 images obtained using TNG and in visual band are analysed. The image quality clearly deteriorates when ΔT >-0.6 deg.

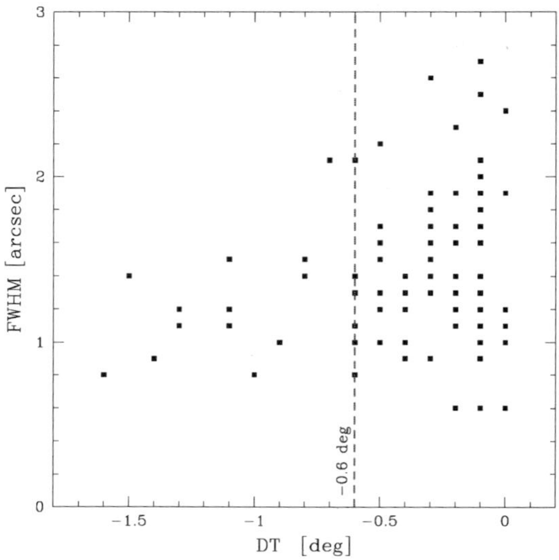

Figure 3. V-band FWHM recorded at TNG as a function of the vertical temperature gradient ΔT between the main mirror and 8 meters below.

3. Air Pressure

3.1. *Long Time Scale*

Figure 4 shows the mean monthly value of the air pressure measured at 2 meter above the ground at Paranal (upper panel). There is a clear decreasing of the dispersion during the years with a reduction of the dispersion of about 70% from 1989 to 2006 (bottom panel). The same analysis taken at ORM does not show the same effect. We can ask if these different behaviors can be explained as different links with wide scale climatic episodes. In fact the temperature annual average at Paranal shows to increase in connection with El Niño, vice versa minimum mean temperatures occur with La Niña episodes.

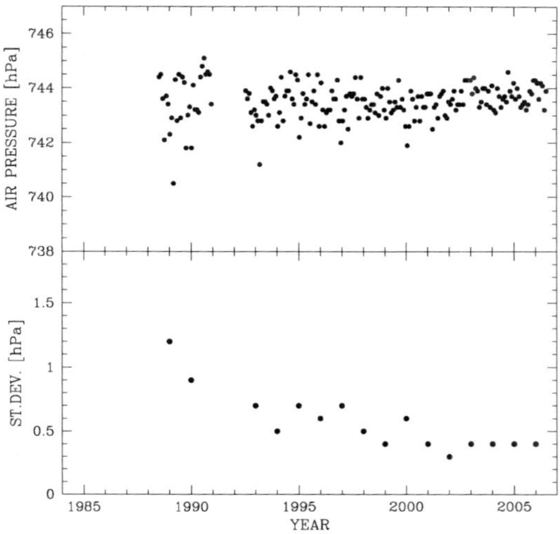

Figure 4. Mean monthly pressures at Paranal (top) and the annual standard deviations (bottom).

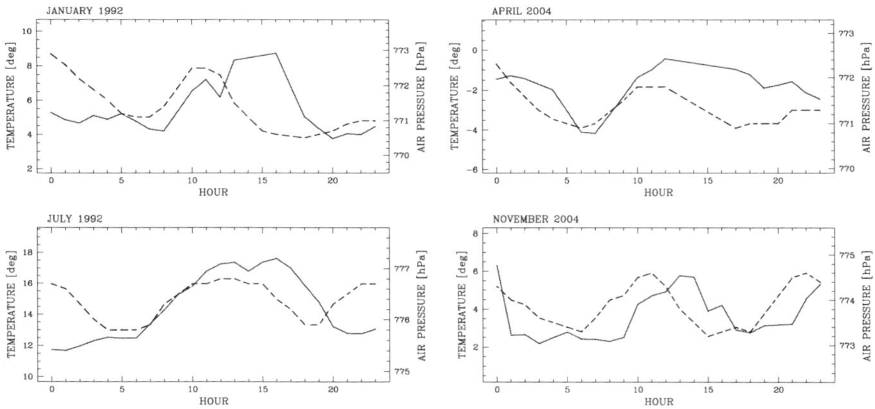

Figure 5. An example of distribution of hourly mean pressure and temperature at ORM.

3.2. *Short Time Scale*

Is it possible to foresee the changes in temperature a few hours in advance? The answer could be given analysing the short time scale relationship (hour-by-hour) between air pressure and temperature measured. Figure 5 shows that changes in temperature (solid line) follow changes in pressure (dashed line) at ORM. We have cross-correlated the hourly mean value of pressure and temperature in different years and we have obtained that the typical correlation range at ORM is between 50 and 90%, while at Paranal the correlation is between few percent to 70%. Although this effect is more clear at ORM in both observatory the delay decreases in warm season.

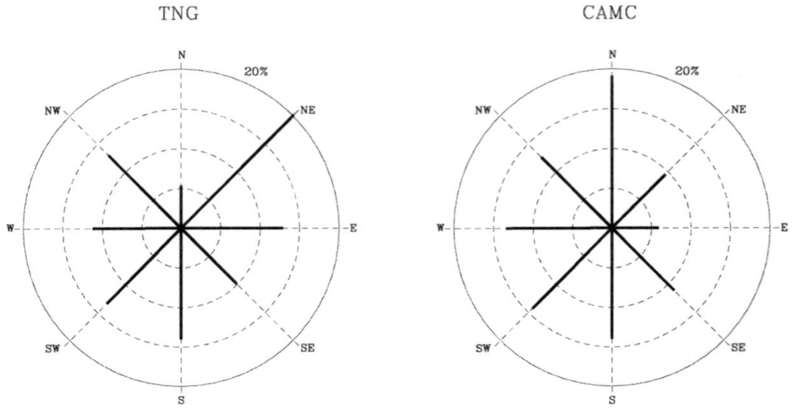

Figure 6. Nighttime wind direction at ORM during the years 1998-2004.

4. Wind Direction

The wind is an important parameter for the image quality, in particular wind speed is linked to the optical turbulence and to the wave front coherence time [1]. It is well known [4] that the effects of wind speed are negligible if the speed is included in a range that is site dependent. We found that the TNG has optimal seeing condition when the wind speed is between 3.3 and 12 m/s.

Another important parameter that may affect the image quality is the prevaling wind direction that can interact with the local orography. Figure 6 shows the mean nighttime wind roses for TNG and CAMC at ORM. Wind values are obtained using the local meteorological towers. TNG shows a dominant NE component, while CAMC shows a dominant N component. These different wind directions at ORM are due to the complex local orography. Recently Sarazin [3] found that under peculiar conditions Paranal is strongly affected by wind direction generating a local Surface Layer. Should be interesting to analyze the local Surface Layer at each different telescope at ORM to check if changes occur during the years.

5. Conclusions

We conclude that long baseline data give the history of the past evolution of the climate. Available atmospheric models may help to predict changes in the climate of the interested sites. Local environmental conditions affect local observations. We need models to predict short time scale environmental conditions in order to improve in real time the performance of the instruments and the observing techniques.

References

1. Geissler, K., & Masciadri, E., 2006, *PASP*, 118, 1048.
2. Lombardi, G., Zitelli, V., Ortolani, S., 2006, *PASP*, 118, 1198.
3. Sarazin, M., Melnick, J., Navarrete, J., & Lombardi, G., 2008, *The ESO Messenger*, 132, 11.
4. Sarazin, M, 1992, Update of environmental conditions on *ESO observatories, VLT-TRE-ESO-17400-0347*. European Southern Observatory.
5. Zago, L. 1995, *PhD Thesis*, Ecole Polytechnique Federal de Lausanne, Thesis No. 1394.

COMPARISON OF THE ATMOSPHERE ABOVE THE SOUTH POLE, DOME C AND DOME A: FIRST ATTEMPT

S. HAGELIN[1,2], E. MASCIADRI[1], F. LASCAUX[1] and J. STOESZ[1]

[1] *INAF - Osservatorio Astrofisico di Arcetri, Largo E. Fermi 5, I-50125 Firenze, Italy*
[2] *Uppsala Universitet, Dep. of Earth Sciences, Villavägen 16, S-75236 Uppsala, Sweden*

In this contribution we briefly summarize the results recently published in Ref. 1.

The atmosphere above three sites on the Internal Antarctic Plateau (Dome A, Dome C and the South Pole) is investigated for astronomical applications using the analysis-data from the ECMWF (European Centre for Medium-Range Weather Forecasts) for an entire year (2005). The monthly median of several meteorological parameters, such as wind speed and potential temperature, are calculated. Radiosoundings from Dome C and the South Pole are used to verify the reliability of the analyses and to study the wind speed in the first 100 m as the analysis-data are not optimized for this altitude-range. The wind speed in the free atmosphere is obtained from the ECMWF-analyses for all three sites. In this context a fourth site, Dome F, is also discussed. In the free atmosphere the stability is studied using the Richardson number, which is an indicator of the probability to trigger thermodynamical instabilities. We find that the free atmosphere over the Internal Antarctic Plateau is more stable than at mid-latitude sites. Dome C shows worse thermodynamic instability conditions than those predicted above the South Pole and Dome A in the same vertical slab. Finally we provide a ranking of the three sites with respect to wind speed, in the free atmosphere (ECMWF analyses) as well as in the surface layer (radiosoundings).

Keywords: Site testing; atmospheric effects; turbulence.

1. Introduction

The summits of the Internal Antarctic Plateau might be among the best places in the world for astronomical facilities. The free atmosphere is very clear with low humidity and low levels of turbulence. The turbulence appears to be confined into a narrow surface layer. The height of this surface layer is expected to be lower at the summits of the plateau than at the slopes. Measurements show that the surface layer at Dome C (a summit) is 33 m high[2] while at the South Pole (situated at a slope) it is 220 m.[3]

The largest source of turbulence in the surface layer is the near-surface winds. The surface winds are triggered by the sloping terrain in combination with a temperature inversion.[4] The temperature inversion forms when the incoming solar radiation is weaker than the radiative cooling of the snow surface. During most of the year a temperature inversion is present at the internal plateau. The calmest conditions are found at the summits since the principal cause triggering the surface winds, the slope of the terrain, is absent here.

The scientific goals of this study are:

(1) To carry out a detailed comparison of radiosoundings/analyses of the wind speed and the potential temperature (the main critical parameters defining the stability of the atmosphere) near the surface (the first 150 m) for winter as well as summer above Dome C and the South Pole.

(2) Using radiosoundings we will estimate the statistic median values of the wind speed in the first tens of metres above the South Pole and Dome C extended from April to November.

(3) Studying the wind speed in the free atmosphere we intend to quantify which site is the best for astronomical applications.

(4) We extend the analysis of the Richardson number done by Ref. 5 for Dome C to the three sites (South Pole, Dome A and Dome C) in order to quantify the regions and periods that are less favourable for the triggering of optical turbulence and to identify the site with the best characteristics.

In this contribution we will highlight just a few of the results obtained in this study. We refer to Ref. 1 for a complete analysis.

2. $\partial\theta/\partial z$ and $(\partial v/\partial z)^2$ at the South Pole, Dome C and Dome A

As shown by Ref. 1 there is an excellent agreement between the ECMWF analyses and the radiosoundings of both Dome C and the South Pole in the free atmosphere. However, in the first vertical grid-point there is a large offset in both wind speed and temperature. This difference can affect the calculation of the Richardson number so it would be quite risky to use the Richardson number retrieved from the ECMWF analyses (see Sec. 5) to study the stability of the surface layer since this number depends on the ratio of $\partial\theta/\partial z$ and $(\partial v/\partial z)^2$. We will therefore limit ourselves to retrieve a ranking of the three sites with respect to the thermal and dynamic stabilities in an independent way.

The dynamic properties are described by the shear of the wind speed and the thermal stability is represented by the vertical gradient of the potential temperature. A positive potential temperature gradient is defined as stable conditions, the vertical displacement of the air is suppressed and so is the production of dynamic turbulence.

The monthly median of the gradient of the potential temperature in the first 150 m at the synoptic hour closest to 00:00 Local Time at Dome A (thick lines), Dome C (thin lines) and the South Pole (dashed lines) is shown in Fig. 1. An inversion is always present during the winter. In the central months of the winter (June, July and August) Dome A shows the most stable conditions with a very sharp temperature inversion near the surface.

Regarding the median of the gradient of the wind speed we refer the reader to Fig. 5 of Ref. 1 which shows $(\partial v/\partial z)^2$ in the first 50 m.

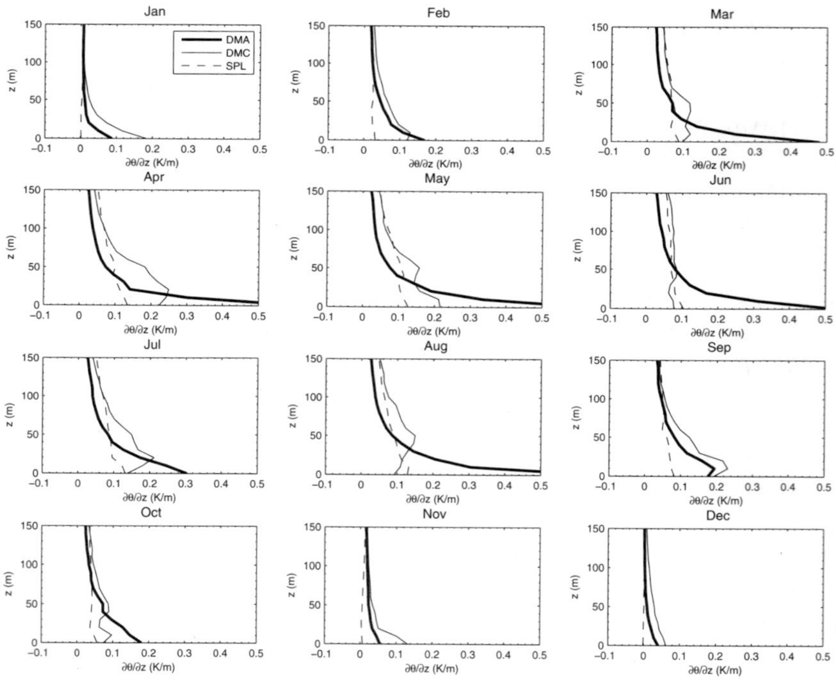

Fig. 1. The monthly median of the potential temperature 2005, ~00 LT for Dome A (thick lines), Dome C (thin lines) and the South Pole (dashed lines).

3. Radiosoundings: The Surface Wind Speed

Above the summits of the Internal Antarctic Plateau the surface winds are expected to be weaker than elsewhere on the plateau. Fig. 6 of Ref. 1 shows the median wind speed at the South Pole and Dome C from April to November. While it is true that the wind speed at the lowest level is weaker at the summit (Dome C) than at the slope (South Pole), it is clearly visible a sharp wind shear in the first 10/20 m at Dome C. Above this height the wind speed is either stronger or comparable to that of the South Pole.

In the core of the winter (June, July and August) the wind speed above Dome C reaches 8 m/s at 20 m and 9 m/s at 30 m. The sharp change in the wind speed in the first 10/20 m matches our expectations of a large wind speed gradient. This is a necessary condition to justify the presence of optical turbulence in the surface layer[2,6] in spite of very stable thermal conditions.

Similar measurements have also been published by Ref. 7 who, using sonic-anemoters mounted on tower, achieved similar results to ours, measuring the wind speed up to 30 m above the surface. Ref. 2, using a small sample of radoisoundings in winter, observed a wind speed of 5 m/s at 20 m altitude and 8 m/s at 70 m. Our result, obtained with a complete statistical sample in winter and substantially in agreement with those of Ref. 7, tells us that their estimate is too optimistic and we should expect a larger wind speed at low altitudes.

Mechanical vibrations in the first 10/20 m, that might be derived from the impact of the atmospheric flow, flowing at 8-9 m/s, on a telescope structure, are probably more critical at Dome C than at the South Pole. This should be taken into account carefully when designing astronomical facilities.

4. ECMWF Analyses: The Wind Speed in the Free Atmosphere

The monthly median of the wind speed, from the ground up to 25 km a.s.l. is shown in Fig. 2. During the summer the wind speed is quite weak and changes very little with height. The median wind speed from December to March never exceeds 15 m/s at any height for any site. As the winter approaches the wind speed above 10 km increases monotonically. The rate at which the wind speed increases is not the same above the different sites. The smallest increase rate is seen at the South Pole whereas the largest increase rate is found above Dome C. These differences are far from being negligible since at these heights the median wind speed at Dome C is almost twice that of the other three sites. Ref. 5 found a similar trend of the wind speed at Dome C for the years 2003 and 2004. The trend of the wind speed is therefore confirmed in different years.

In Ref. 1 we proved that the difference in the slope of the vertical profile of the wind speed in Fig. 2 is related to the special synoptic circulation over Antarctica. The jet-stream, that characterize the vertical wind speed profile at mid-latitude sites, does not form over Antarctica. Instead the large-scale circulation is dominated by the polar vortex, which creates a cone of strong high-altitude winds surrounding the polar high in winter.[4] The result observed in Fig. 2 can be explained by the fact that the South Pole is located near the centre of the polar vortex which means that the influence of the polar vortex is weak at this site. Dome C, for example, which is situated further from the centre of the continent, is experiencing stronger wind speeds. It appears that the farther from the centre of the polar vortex a site is, the greater is the strength of the wind speed in the upper atmosphere.

Knowing the distance of a site from the centre of the polar vortex would be extremely useful to identify, a priori, the site with the weakest wind speed in the upper atmosphere. However, as shown by Ref. 8, the polar vortex is not a perfect cone, the coordinates of the centre varies with both time and altitude.

5. The Richardson Number

The Richardson number is an indicator of the stability of the atmosphere, $Ri = \frac{g}{\theta}\frac{\partial\theta/\partial z}{(\partial v/\partial z)^2}$, where g is the gravitational acceleration (9.8 m/s^2), θ is the potential temperature and v is the horizontal wind speed. The atmosphere is considered to be stable if the Richardson number is larger than the critical value, typically 0.25. The smaller the Richardson number is the higher is the probability to trigger turbulence.

Comparing the Richardson number gives a relative estimate of the probability of the triggering of turbulence. The method proposed by Ref. 5, to calculate the inverse of the Richardson number to rank different sites qualitatively, has been definitely proved by comparing an Antarctic site (Dome C) with a mid-latitude one (Mt. Graham). Indeed, as seen in Fig. 9 in Ref. 1, Dome C is always more stable than Mt. Graham, except the high part of the atmosphere in September and October when the polar vortex creates strong high altitude winds at Dome C. As discussed in Ref. 1 this proves that the probability to trigger turbulence above mid-latitude sites is larger than above Dome C.

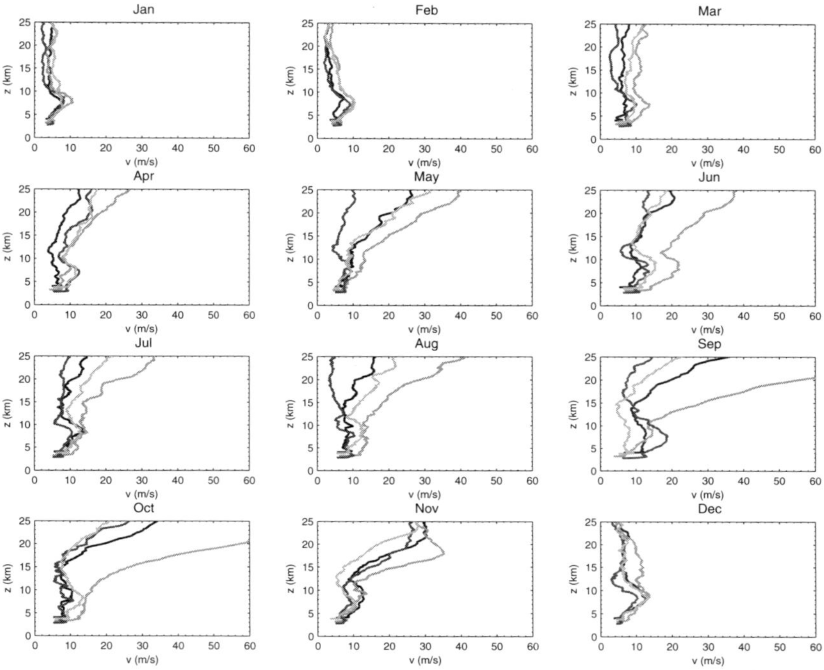

Fig. 2. (color on-line) The monthly median wind speed for 2005, 00 UTC for Dome A (dark blue lines), Dome C (green lines), Dome F (light blue lines) and the South Pole (red lines).

Comparing the three Antarctic sites with each other, see Fig. 10 of Ref. 1, it appears that the most stable conditions are found above the South Pole. Dome A is less stable than the South Pole and the least stable conditions are found at Dome C.

6. Conclusions

From a first comparison of the atmospheric characteristics of three different sites on the Internal Antarctic Plateau: Dome A, Dome C and the South Pole using ECMWF analyses and radiosoundings we can draw the following conclusions.

The gradient of the potential temperature indicates a stable stratification at all three sites, even though the ECMWF analyses generally underestimates this gradient when compared to measurements from radiosoundings. The stability is particularly strong in winter at Dome A.

During the period April to November the median wind speed near the ground is weaker at Dome C than at the South Pole. However, the wind shear in the first 10/20 m at Dome C is very sharp, above this height the wind speeds of the two sites are comparable. Such a strong wind shear in combination with a stable thermal stratification in this layer is most likely the cause of the intense optical turbulence measured in the first tens of meters at Dome C.[2,6] Such a strong wind speed, 8-9 m/s at 10 m, might be a source of vibrations produced by the impact of the atmospheric flow on a telescope structure and should therefore be taken into account in the design of astronomical facilities.

At higher altitudes in the free atmosphere, above 10 km, the polar vortex causes the wind speed to increase monotonically during the winter. The increase rate is proportional to the distance of the site from the centre of the polar vortex. This explains why the wind speed above Dome C is particularly large above 10 km in winter. At 15 km altitude the wind speed of Dome C can easily be almost twice the wind speed of Dome A or F and even thrice the wind speed of the South Pole. The wind speed at the South Pole is the weakest of the four in all seasons and at all heights.

The monthly median of the inverse of the Richardson number (1/Ri) indicate that the probability to trigger turbulence is larger above a mid-latitude site than above any site investigated by us on the Internal Antarctic Plateau, except the upper part of the atmosphere at Dome C in September and October. Knowing the typical values of the optical turbulence above Mt Graham, this is a definit proof that the method proposed by Ref. 5 is reliable. Moreover, using this method to compare the three Antarctic sites the 1/Ri indicates that the South Pole shows the most stable conditions, followed by Dome A, whereas Dome C is the least stable of the sites. This is most likely due to the polar vortex. Near the surface it is risky to use estimates of 1/Ri from the ECMWF analyses as we do not observe an equivalent smoothing effect for the gradients of potential temperature and wind speed.

This study allowed us to draw a first comprehensive picture of the atmospheric properties above the Internal Antarctic Plateau. In spite of the generally good conditions for astronomical applications, Dome C does not appear to be the best site with respect to the wind speed, in the free atmosphere as well as in the surface layer. All the other sites show a weaker wind speed in the free atmosphere. Above Dome A the gradient of potential temperature is particularly large near the surface, indicating extreme thermal stability associated to a strong the optical turbulence when a thermodynamic instability occurs. It is even possible that the optical turbulence here is larger than at Dome C. However, to predict the thickness of such a layer it is necessary to have either measurements or simulations with a mesoscale model with a higher spatial resolution. This is a part of our forthcoming activities.

At present, the best argument that makes Dome C a better place for astronomical applications than the South Pole is the extreme thinness of the optical turbulence surface layer. Dome A probably has comparable or larger values of the optical turbulence with respect to Dome C in the surface layer. We cannot conclude whether the surface layer at Dome A is thinner than what is observed at Dome C. However our study clearly indicates that Dome C is not the best site on the Internal Antarctic Plateau with respect to the wind speed (both in the surface layer and in the free atmosphere) nor is it the site with the most stable conditions in the free atmosphere. Both the South Pole and Dome A show more stable values of 1/Ri.

Acknowledgments

This study has been carried out using radiosoundings from the AMRC (Antarctic Meteorological Research Center), University of Wisconsin, Madison and from the Progetto di Ricerca 'Osservatorio Meteo Climatologico' of the PNRA (Programma Nazionale di Ricerche in Antartide). ECMWF products are extracted from the MARS-catalogue. This study has been funded by the Marie Curie Excellence Grant (ForOT) MEXT-CT 2005-023878, FP6.

References

1. S. Hagelin, E. Masciadri, F. Lascaux and J. Stoesz, *MNRAS* **387**, 1499 (2008).
2. H. Trinquet, K. Agabi, J. Vernin, M. Azouit, E. Aristidi and E. Fossat, *PASP* **120**, 203 (2008).
3. R. Marks, *A&A* **385**, 328 (2002).
4. W. Schwerdtfeger, *Weather and climate of the Antarctic, Developements in atmospheric science 15* (Elsiever, Amsterdam, 1984).
5. K. Geissler and E. Masciadri, *PASP* **118**, 1048 (2006).
6. A. Agabi, E. Aristidi, M. Azouit, E. Fossat, F. Martin, T. Sadibekova, J. Vernin and A. Ziad, *PASP* **118**, 344 (2006).
7. T. Travouillon, E. Aristidi, E. Fossat, J. S. Lawrence, D. Mekarnia, A. M. Moore, A. W. Skidmore and J. W. V. Storey, *Proc. of SPIE* **7012**, 70124B (2008).
8. F. Lascaux, E. Masciadri, S. Hagelin and J. Stoesz, *this conference* (2008).

CORRELATING SODAR TURBULENCE MEASUREMENTS WITH GROUND HEAT FLUX DATA

TONY TRAVOUILLON*, REED RIDDLE and WARREN SKIDMORE
TMT Observatory Corporation, 1200 E. California Blvd, Pasadena, CA, USA
**E-mail: tonyt@tmt.org*

MATTHIAS SCHÖCK
TMT Observatory Corporation, Victoria, BC, Canada

SEBASTIAN ELS
AURA-NIO, Cerro Tololo Inter-American Observatory, La Serena, Chile

The TMT site testing campaign includes the measurement of the ground layer turbulence profiles using SODARs. In this part of the atmosphere, turbulence is caused by the thermal exchange between the ground and the air above it. A complete knowledge of the energy balance can therefore be related to the turbulence profiles. With the measurement of the different heat fluxes, we were able to find a correlation between the kinetic heat flux and the integrated turbulence measured by the SODAR. In this paper we present these results and discuss their limitations.

Keywords: Turbulence; SODAR; Ground layer.

1. Introduction

The profiling of the ground layer (GL) turbulence is an important aspect of the Thirty Meter Telescope (TMT) site testing program.[1,2] Indeed, measuring the seeing at a height of 7m does not give an accurate picture of what the telescope will see when built. With a dome height at 50m, TMT needs a measurement of the turbulence below this height not only to know what fraction of the GL it will miss but also to allow the modeling of the dome seeing itself.

To measure accurately the turbulence profile of the GL of TMT candidate sites, 3 pairs of Scintec SODARs (SOund Detection And Ranging) were purchased. Two different models (namely the XFAS and the SFAS) were necessary to cover the vertical range and resolution required to obtain a full profile extending to the lowest altitude covered by the MASS (Multi Aperture Scintillation Sensor) which we use to profile the free atmosphere turbulence. The SFAS is a low-power unit that samples the first 200m of the GL with a resolution of 5m. This resolution is suitable for measuring the turbulence around the height of the future TMT dome. The XFAS unit is a high-power model that has a vertical resolution of 20m and can measure turbulence up to 800m. This fills the gap left between the SFAS and the MASS which has its first layer sensitivity peaking at 500m.

Such a complete setup is almost ideal to fully understand the turbulence properties of a site. However, its price tag is out of reach of small projects and difficult for testing simultaneously a large number of sites. For a comparative study of the GL condition of multiple sites, one may want a more pragmatic way to determine which site is likely to be the best without having to resort to a detailed study. For example, a 2m class telescope would be likely located close to the ground an hence do not require a high resolution profiling of each of its candidate sites. In this case it is the overall seeing that drives the selection and since the free atmosphere works on large scales, two close mountains can be discriminated by their GL contribution to the seeing. This is where it becomes important to understand the processes that contribute to the GL turbulence. To some extent, the GL and the free atmosphere are decoupled and poorly correlates when it comes to turbulence. The free atmosphere turbulence is driven by thermal exchanges, in such locations as the tropopause, on meso to global scales. The GL, however, is mainly sensitive to thermal exchanges with the ground. We can therefore expect that sites with the smallest thermal exchange during the night would have the smallest GL turbulence. How much knowledge of the GL turbulence can then be infered from the heat transfer balance of a site? Such measurements are possible thanks to heat flux sensors available on TMT's weather stations. In conjunction with the SODAR measurements, it can be determine whether a good correlation exists between heat transfer and the GL turbulence of a site.

2. Theory and Instrumentation

2.1. *SODAR*

Through the measurement of σ_0, the backscattering cross section of turbulence cells, SODARs determine the temperature fluctuation constant C_T^2

through the relation:

$$\sigma_0 = 0.0039k^{-1/3}\frac{C_T^2}{T^2} \tag{1}$$

where T is the temperature and k is the wave number of the acoustic wave emitted by the SODAR. The refractive index C_n^2, which is the figure of merit of turbulence in astronomical site testing can then be simply calculated with the independent knowledge of the atmospheric pressure P:

$$C_n^2 = 79 \times 10^{-6}(\frac{P}{T^2})^2 C_T^2 \tag{2}$$

After careful calibration,[3] the pair of SODAR used in the TMT site testing were operated in an interleaved manner since two acoustic instruments cannot operate simultaneously. After multiple samplings, the data of both instruments were integrated over the same 30min period hence giving 2 overlapping turbulence profiles. After operating for a minimum of a few months at each sites, giving a statistically representative sample of the GL conditions. Fig. 1 is an example of the median conditions found on Mauna Kea 13N both expressed in units of C_n^2 and integrated seeing. These results, which are details in Ref. 4 are the basis of our knowledge of the GL of the TMT candidate sites.

2.2. *Radiation Balance*

Kinetic heat flux is part of the radiative balance of the surface layer and the source of GL turbulence. The energy exchange at the ground level can be express as the equality:

$$R = L + G + H \tag{3}$$

where R is the net solar radiation (Solar radiation minus albedo), L is latent heat flux responsible for evaporation of soil moisture , G is the ground heat flux and H is the kinetic heat flux in question.

Since astronomical sites are usually extremely dry, the amount of energy transferred to latent heat can be neglected, hence allowing for the kinetic heat flux to be strictly a function of net radiation and ground heat flux. These two components are measured by the TMT weather station using sensors CN1-R and CN3 respectively. Both sensors make one measurement every minutes and work equally well during the day and the night. Combining the two measurements gives the kinetic heat flux has shown in Fig. 2. In this example showing the average flux for different time of the day on

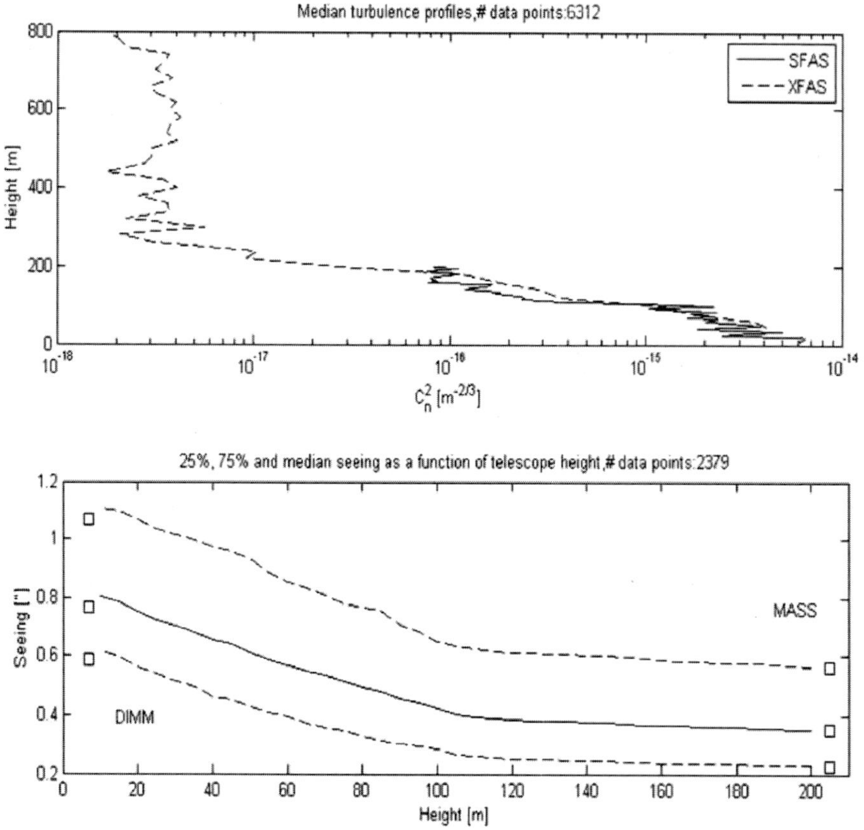

Fig. 1. Top: Median turbulence profiles on Mauna Kea 13N using both XFAS and SFAS. Bottom: 25%, median and 75% profiles of integrated seeing as function of starting height of observation. The squares and the matching DIMM and MASS data over the same period as the SODAR data were taken.

Cerro Toloncha r, we see that the kinetic or turbulence energy very stable during night time and caused by radiation coming from the ground.

Converting the turbulence heat flux to a more useful C_T^2 profile relies however on certain assumptions, the most important being the stability of the atmosphere. As shown in Ref. 5, the turbulence profile as function of elevation z can be related to the turbulence heat flux H through the following relationships:

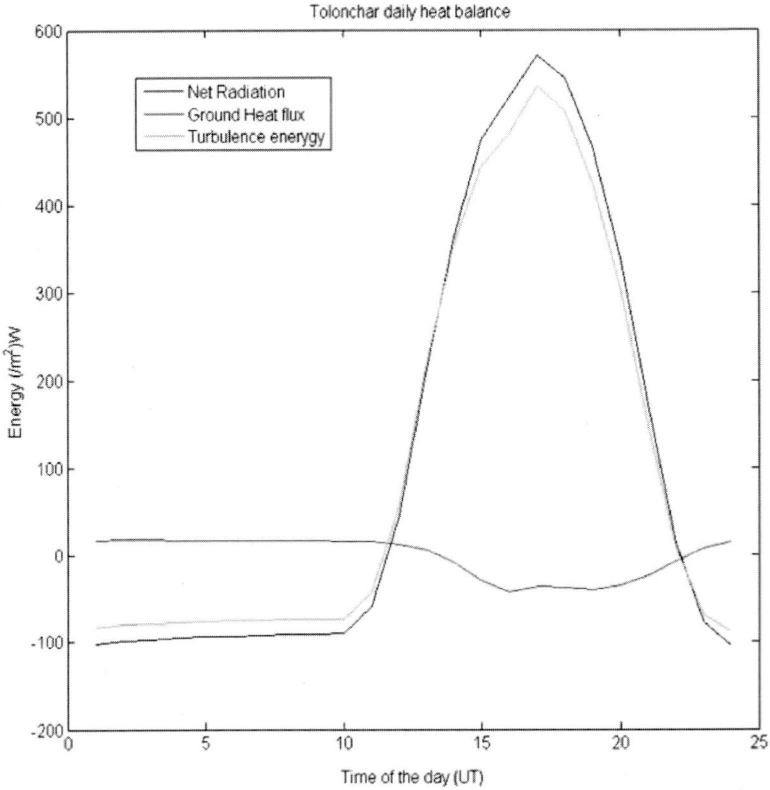

Fig. 2. Average of net radiation, ground heat and turbulence heat flux over the course of 24h.

- Unstable (convective) atmosphere:

$$C_T^2(z) \propto H^{4/3} z^{-4/3} \tag{4}$$

- Neutral atmosphere:

$$C_T^2(z) \propto H^2 z^{-2/3} \tag{5}$$

- Stable atmosphere:

$$C_T^2(z) \propto \frac{H^2}{u_*^2(z)} z^{-2/3} - \frac{H^3}{u_*^5(z)} z^{1/3} \tag{6}$$

The stable atmosphere, which applies to nighttime observation is also a function of the surface roughness u_* which itself is a function of wind speed.

3. Results

Comparing the turbulence heat flux and the turbulence profiles of the SODARs is not possible due to the complexity of the relationship and the absence of a measurement of surface roughness. Instead, the integrated GL seeing computed by integrating the SODAR turbulence profile over its entire range allows the dependance of the surface roughness to be treated as an integration constant and estimated by a fitting of the two parameters. The results shown in Tab. 1 demonstrate a correlation between the turbulence heat flux and the GL seeing. The Chilean sites share a very similar type of soil and hence radiate a comparable amount of turbulence heat flux. It is sensible to think that Armazones is slightly more turbulent due to its lower altitude (higher air pressure), although the difference are close to measurement error. A similar statement can be made regarding the respective altitude of Mauna Kea 13N and San Pedro Mártir, however, their soil and terrain are very different. Maunea Kea 13N rests on a volcanic terrain with a high amount of cinder on the ground. San Pedro Mártir, on the other hand is on a rocky mountain in the middle of a pine forest. The presence of vegetation suggest that the early assumption that latent heat flux is negligible may not apply for this site and may introduce an error of about $10W/m^2$. It must also be stressed that it is sometime difficult to measure in a representative soil as the relative composition of rocks and soft soil can be hard to estimate.

Table 1. Comparison of turbulence heat flux, GL seeing and BL height for all TMT candidate sites.

Site	Heat flux [W/m^2]	GL seeing ["]	BL Height [m]
Tolonchar	53	0.28	25
Armazones	58	0.33	60
San Pedro M.	64	0.61	100
Mauna Kea 13N	66	0.61	105

Tab. 1 also includes the GL height measured by the SODAR. It is defined at the altitude at which the turbulence double derivative is the lowest. It is not surprising that under stable atmosphere conditions the GL height is also correlated with the turbulence heat flux. It is however difficult to directly relate the two together due to their complex relationship. Simply integrating Eq. 6, however, allows us to fit the relationship between the turbulence heat flux and the GL seeing calculated with the SODAR. Such fit is shown in Fig. 3. With integration constants as free parameters, it is hardly

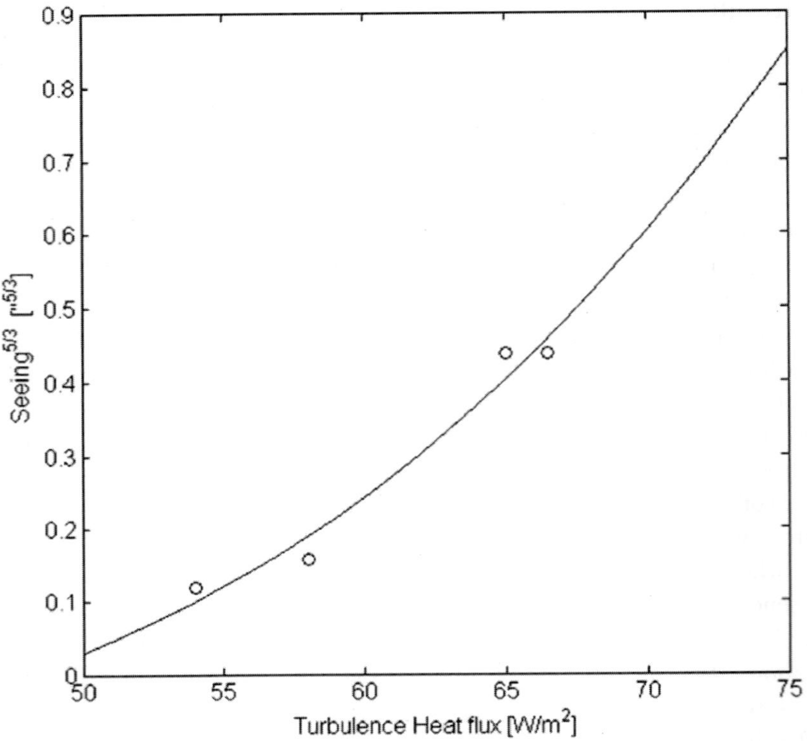

Fig. 3. GL seeing seeing as a function of Turbulence heat flux. The solid line correspond to the fit using Eq. 6.

conclusive that the relationship shown in Eq. 6 holds in our case. One may also wonder whether this relationship, which was developed experimentally for flat terrain holds for mountain peaks. The only certain conclusion is that using the data available from the TMT candidate sites, a correlation is found between the turbulence measured by the SODAR and the turbulence heat flux. This correlation is also confirmed by CFD simulations done on the same sites.[6] To that extent, it is safe to say that heat flux sensors can be used to compare the GL turbulence of astronomical sites.

4. Conclusions

Heat flux sensors have a modest cost and are easy to maintain and operate. While they are not enough to give a quantitative measurement of the GL turbulence they do offer a way to relatively rank the GL turbulence of different sites. This may be sufficient for site testing campaign with limited resources or a pre-selection campaign that aims at reducing the number of sites to be thoroughly tested.

SODARs on the other hand do offer a more complete and directly applicable knowledge of the GL. By also measuring the wind speed profile, it provides invaluable information for a future observatory.

References

1. R. Riddle, M. Schöck, S.G. Els, W. Skidmore and T. Travouillon, in *These proceedings* (2009).
2. M. Schöck, S.G. Els, R. Riddle, W. Skidmore and T. Travouillon, in *These proceedings* (2009).
3. T. Travouillon, S.G. Els, R. Riddle, M. Schöck and W. Skidmore, *App. Opt. In prep.* (2009).
4. S.G. Els, R. Riddle, M. Schöck, W. Skidmore and T. Travouillon, *P.A.S.P.*, *In prep.* (2009).
5. J.C Kaimal and J.J Finigan, Atmospheric Boundary Layer Flows, *Oxford University Press.* (1994).
6. S.G. Els, K. Vogiatzis, Proc. SPIE, Vol. 6267, 626723 (2006)

CLIMATOLOGY AT THE ROQUE DE LOS MUCHACHOS OBSERVATORY

ANTONIA M. VARELA[†]
Instituto de Astrofísica de Canarias, C./Via Láctea, S/N
La Laguna, Tenerife, 38200, Spain

CASIANA MUÑOZ-TUÑÓN
Instituto de Astrofísica de Canarias, C./Via Láctea, S/N
La Laguna, Tenerife, 38200, Spain

The Roque de los Muchachos Observatory (ORM) at La Palma (Canary Islands) is one of the two top pre-selected sites for hosting the future European Extremely Large Telescope (E-ELT), the other ones are Ventarrones (Chile), Macon (Argentine) and Aklim (Maroc). Meteorological and seeing conditions are crucial both for the site selection and for telescope design and feasibility studies for adaptive optics. The ELTs shall be very sensitive to wind behavior when operating in open air, therefore ground level wind velocity and wind gust are also required for the feasibility of the telescope construction. Here we analyze the wind speed and wind direction, the air temperature, the relative humidity and the barometric pressure statistical results obtained from data recorded at different sites at the ORM by several Automatic Weather Stations (AWS) since 1985, day and night time separately. Ground wind speed regimes (775mbar) are compared with those provided by satellites from 200 to 700mbar. There exists also observational evidence of the correlation between the seeing and the wind speed and wind direction that will be discussed in this work. [†]

1. Canary Islands Climatology

Both climatologic and geographical factors at the Canary Islands offer the conditions required for the astronomical observations, attracting some of the most modern telescopes.

The climate of the Canary Islands is dominated by a persistent area of high pressure in the North Atlantic (known as either the Azores anticyclone or the Bermuda High), which is normally situated over the North West of the archipelago.

[†] avp@iac.es
Work supported by European OPTICON and the FP6 Proposal for Site Selection for the E ELT.

In particular the trade wind scenario, the cold oceanic stream, the presence of a stable inversion layer (between 1200 and 1600m, thereby producing a clear dividing line between the lower and upper layers of air circulation), in combination with the local topography, play an important role in the retention of low cloud layers well below the north summits of the western islands (Tenerife and La Palma), leading a generally cloud-free, dry sky above it. This sea of clouds plays also an important role on the retention of light and aerosols that keep below the level of the observatories (2400m) (Varela et al., 2008).

2. Meteorological Campaigns at the Roque de los Muchachos Observatory

Several site-testing campaigns have been performed at the Roque de los Muchachos Observatory at La Palma Island (see Figure 1), in particular, tests at the TNG and WHT sites (since 1992 to 1994) are presented in Muñoz-Tuñón et al. (1997) and the seeing and meteorological results of the campaign to select the site for the Gran Telescopio de Canarias (during 1995 and 1996) are summarized in Muñoz-Tuñón et al. (1997), Muñoz-Tuñón et al. (1998); Mahoney et al. (1998). From 1997 to 1999 a site testing campaign was continued at the GTC site till starting telescope construction in October 1999. In February 2002 we have started a campaign at the Hoyo Verde site, towards the SW of GTC site, near la Caldera. The proximity of this site to the Caldera necessitates continuous statistical meteorological tests over long period in order to ensure also a seasonal sample. The Caldera could also produce local fluctuations in the thickness of the surface layer.

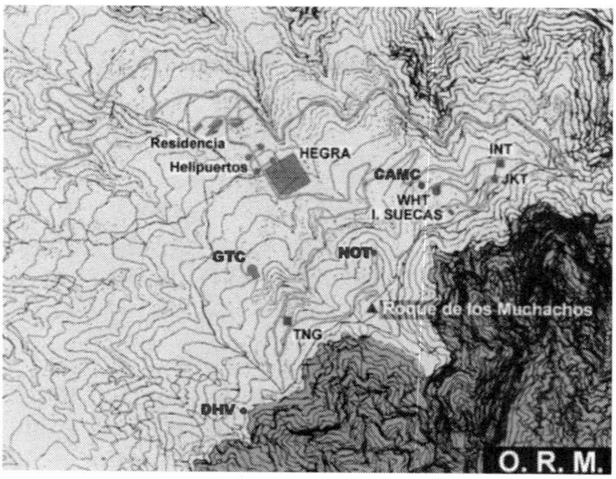

Figure 1. ORM map showing the different sites where the site-testing campaign was performed. These 4 stations can be representative of the entire ORM area.

Meteorological database has been recorded from several automatic weather stations (AWS) at the ORM indicated in Table 1. In the table is also indicated the sampling rate, the total number of data and the period to which the measurements correspond. Occasional technical problems can modify the number of data of a particular parameter, then the number of data corresponds to the available wind data.

Table 1. AWS meteorological database compiled for climatologic analysis. NOT corresponds to the AWS located at the Nordic Optical telescope, CAMC is the Carslberg Meridian Circle, GTC correspond to the data provide by the AWS at the Gran Telescopio de Canarias site, and the DHV is la Degollada del Hoyo Verde, towards the southwest of the GTC. Day and night time classification have been done by using solarimeter data monthly averages.

AWS	NDATA (day)	NDATA (night)	Period	Sample rate
NOT	415341	265496	feb'97-dec'03	1 data/5 min
CAMC	935838	606281	oct'85-jun'02	1 data/5 min
GTC	345624	300336	feb'95-dec'96	1 date/min
DHV	443862	289552	feb'02-feb'04	1 data/min

The IAC weather stations (at the GTC and Hoyo Verde sites) have been provided by Casella Ltd. The Casella stations consist on a metal framework holding the data acquisition unit (DAU) and, at a height of 2m, a combined hygrometer and air temperature sensor, a solarimeter, a surface-wetness probe and a temperature-compensated barometer. A 12-m lattice tower supporting and anemometer and wind vane is erected near the station. The system is supplied with solar power generator and a radio communication system.

3. Statistical Results

Meteorological data and statistics provided by the AWS are compiled by the Sky Quality Group of the IAC and are accessible at http://www.iac.es/site-testing/. The information about the NOT and the CAMC weather stations are available from their corresponding project web pages (http://www.not.iac.es and http://www.ast.cam.ac.uk respectively). Global statistical results for all meteorological parameters, daily and nocturnal time separately, provided by meteorological weather stations operating at the ORM are summarized in http://www.iac.es/site-testing/ -click on Data & Statistics- and in Varela and Muñoz-Tuñón (2009).

3.1. *Air Temperature and Relative Humidity*

Typical mean air temperatures ranges between 16°C and 18°C for daytime and between 13-16°C for night time. Differences between different locations are explained in terms of difference in altitude.

An important factor for determining the stability of observing conditions the air temperature gradient along the night which typically ranges between 2 and 4°C. The typical difference between the day and night time mean air temperature is about 2°C at any location at the ORM.

There is a reciprocal relation between both parameters: the warmest summer months (June-September) are marked by extremely low relative humidity values (10%-35%) whereas the coldest winter months give the lowest mean monthly air temperatures and higher relative humidity (> 40%).

In Figure 2 we show the cumulative frequency of the relative humidity (day and night time) at the CAMC. Smaller values than 80% are measured in a 85% of the cases.

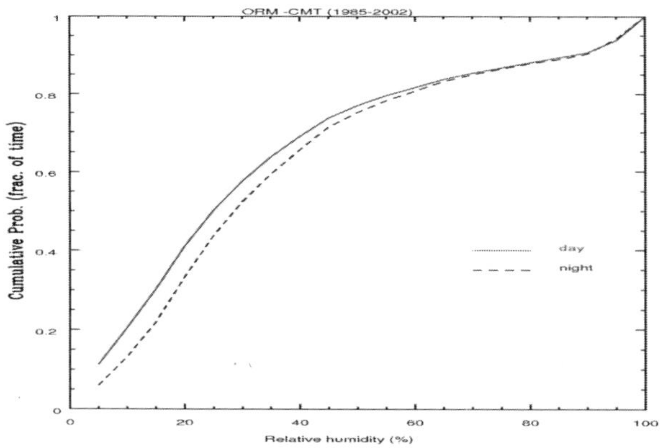

Figure 2. Cumulative frequency of the relative humidity (day and night time) at the CAMC since 1985 till 2002.

3.2. *Wind Speed and Wind Direction*

Wind speed statistical results from data recorded at different automatic weather stations (AWS) operating at the ORM provide typical mean and median wind speed at the ORM ranges from 5 to 6m/s, except at the CAMC due to its particularly leeward location. The mean and median wind speed decrease continuously with the height: ranging from 22.5m/s at 200mb to 8.6m/s at 700mb (day + night). If we would extrapolate to wind speed values at 775mbar (mean height where ground automatic meteorological stations are installed), we would expect typical 5-6m/s velocity values, which is coincident with the values provided by ground level automatic weather stations (Varela et al. 2006 and Lombardi et al. 2007).

The yearly average ground wind velocity at Paranal (750mbar) is 6,6m/s (http://www.eso.org/paranal/site/paranal.html) and the median wind speed at Mauna Kea (615mbar) is 7m/s (Sarazin, 2003). At the foothills of a major mountain chain as at Maidanak the ground mean velocity is 3m/s (Ehgamberdiev et al., 2000).

In Table 2 we summarize the frequency of the occurrence of different ranges of wind speed at the ORM. If 15m/s would be the practical limit of operation, the useful time ranges between 92% and 99%.

Table 2. Frequency of the wind speed values at different sites locations in the ORM. If 15m/s would be the practical limit of operation, the useful time is ranges from 92-99%.

AWS	NOT	GTC	DHV
<5m/s	36%	54%	39%
<10m/s	74%	90%	88%
<15m/s	92%	99%	99%

Figure 3 represents the ground level wind roses daytime and nocturnal wind roses at different locations at the ORM. Topographical components appear at ground level determining a marked change in the day wind-vector pattern. The effect is obviously related to heat from the Sun and the results in a up draught causing diurnal winds normal to the topographical contours and differences of about a 10% of the mean wind speed in nocturnal and daily wind speed.

Note that at the GTC site the nocturnal wind-speed frequency peaks in the ENE direction, whereas at the NOT site there is a western dominant component (between WSW and WNW) and other ESE component.

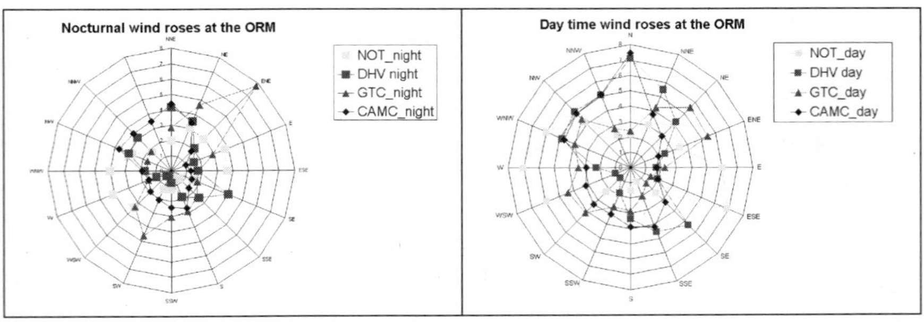

Figure 3. Nocturnal (left) and daily (right) wind roses at different AWS locations in the ORM (Varela et al., 2006).

4. Troposphere Wind Regimes from 200mbar to Ground Level

The propagation of the wind flux in height and its correlation with the ground level winds are crucial to understand the influence of the trade winds over the image quality. There exists a linear correlation between the wind value at 200mbar and the atmospheric turbulence, therefore the wind at 200mbar pressure level has been proposed as a relevant parameter for astronomical site selection (Sarazin & Tokovinin, 2002; García-Lorenzo et al., 2005). Nevertheless, at ground level can appear local wind components associated to the local topography and to the climatology of the site that can disturb the optical conditions. The useless of the 200mbar wind speed is conditioned to the continuity between high and low pressure level wind regimes.

Troposphere winds at the ORM for the period 1980-2002 have been selected from the climate diagnostic model and the balloon NCEP/NCAR databases. Ground level winds are provided by local automatic weather stations. The correlation at the ORM is supported by the Figure 4 of the monthly wind speed profiles from 200mbar to ground level (NOT is at 775mbar) day and night time separately, although differences appear only at ground level due to the topography. We can see:

1. Seasonal wind behaviour.
2. Wind speed connection at different pressure levels.
3. The lower average 200mbar wind speed is attained at the ORM when comparing with other astronomical sites and then as the first in adaptive optics suitability ranking (García-Lorenzo et al., 2005).

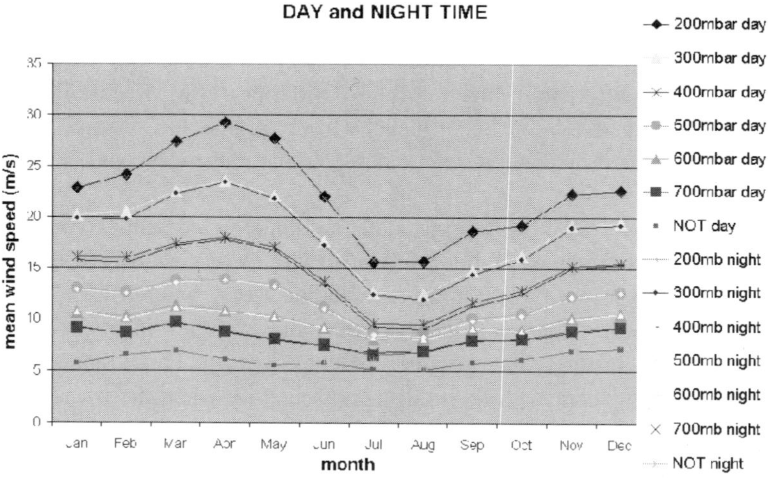

Figure 4. Monthly wind profiles at the ORM from 200mbar to ground level (NOT is at 775mbar).

This correlation is corroborated in Figure 5 where we show the troposphere wind rose at the ORM. Wind vector-pattern shows a dynamical spiral structure according with trade wind and the inversion thermal layer scenarios. High altitude winds can be an useful parameter for astronomical site evaluation because correlate with the ground surface layer winds.

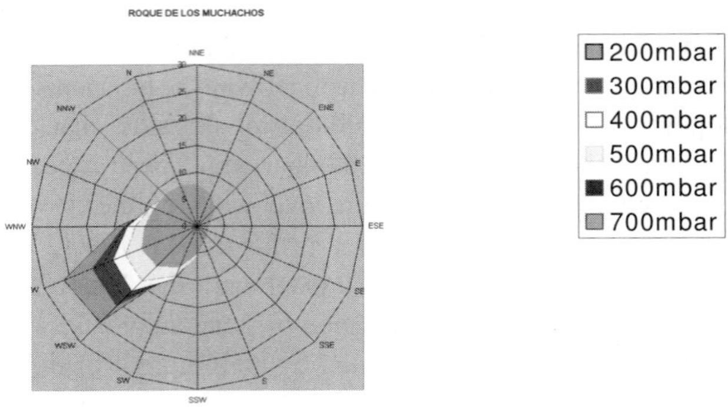

Figure 5. Troposheric wind roses at the ORM. Different colours represent the pressure levels indicated in the right cell.

5. Ground Wind and Seeing Dependency

Our experience of more than 15 years of meteorological and seeing monitoring at the ORM, sudden short–term episodes of seeing worsening have been detected and associated to abrupt changes in wind directions and gusts. This effect caused a double peaked seeing distribution function.

In a first approach to estimate the importance of ground level wind speed on the optical seeing conditions we have analysed the seeing values obtained by a Differential Image Motion Monitor (DIMM) installed at the ORM and operated by the Sky Quality Group of the IAC from 1997-1999 (http://www.iac.es/project/sitesting/site.html) separating low and high wind speed conditions.

In table 3 we show the seeing statistics under wind values smaller than 8.3m/s (the most frequent regime) and otherwise, when winds are faster than 8.3m/s. As we expected typical mean and median seeing values of 0.6” and 0.7” are obtained in the first case, nevertheless, these values increase till about 20-30% under faster wind regimes. Better seeing conditions occur in summertime, coincident with the trade wind scenario: when the inversion layer drops and wind speed are lower.

Ground level winds depend on the climatology and on the local topography and strongly determine the optical seeing conditions of the site.

Table 3. Seeing statistics under different wind speed regimes for the period 1997-1999 at the ORM.

Wind speed < 8.3m/s	1997	1998	1999
Ndata	41102	40512	36759
Mean seeing (")	0.65	0.68	0.71
Standard deviation(")	0.28	0.31	0.30
Median seeing (")	0.57	0.61	0.65
Wind speed > 8.3m/s			
Ndata	7712	6190	9886
Mean seeing (")	0.77	0.89	0.97
Standard deviation(")	0.41	0.44	0.45
Median seeing (")	0.66	0.75	0.86

Ground level winds depend on the climatology and on the local topography and can determine the optical seeing conditions of the site.

References

1. A.M. Varela, C. Bertolin, Muñoz-Tuñón, J.J. Fuensalida and S. Ortolani, MNRAS, in press (2008).
2. C. Muñóz-Tuñón., J. Vernin and A.M. Varela, Astron.Astrophys.Suppl.Ser. 125, 183 (1997).
3. C. Muñóz-Tuñón., A.M. Varela and T. Mahoney, New Astronomy Reviews, **42**, 409, Special Issue "Site Properties of the Canarian Observatories", C. Muñoz-Tuñón Ed., Elsevier, 1998.
4. T. Mahoney T., C. Muñoz-Tuñón C. and A.M: Varela., New Astronomy Reviews, **42**, 417, Special Issue "Site Properties of the Canarian Observatories", C. Muñoz-Tuñón, Ed., Elsevier, 1998.
5. A.M. Varela and Muñoz-Tuñón, to be send to MNRAS, in preparation (2009).
6. A.M. Varela, C. Muñoz-Tuñón, B. García-Lorenzo and J.J. Fuensalida, Proc. Of SPIE, Vol.6267, 62671X (2006).
7. G. Lombardi, V. Zitelli, S. Ortolani and M. Pedani, PASP, 119, 292 (2007).
8. M. Sarazin, ESPAS Site Summary Series : Mauna Kea, Issue 1.2 (2003).
9. S.A. Ehgamberdiev, A.K., Baijumanov., S.P. Ilyasov., M. Sarazin., Y.A. Tillayev Y.A., A. Tokovinin and A. Ziad., Astron.Astrophys.Suppl.Ser. 145, 293 (2000).
10. M. Sarazin and A. Tokovinin, in Vernet et al. Eds., Proc. 58th ESOConference Wokshop, Beyond Conventional Adaptive Optics, ESO Publication, Garching, 321 (2002).
11. B. García-Lorenzo, J.J Fuensalida, C. Muñoz-Tuñón & E. Mendizabal,. MNRAS, 356, 849 (2005).

SNODAR II: PROBING THE ATMOSPHERIC BOUNDARY LAYER ON THE ANTARCTIC PLATEAU

C.S. BONNER*, M.C.B. ASHLEY, J.S. LAWRENCE, D.M. LUONG-VAN, J.W.V. STOREY

Department of Astrophysics, School of Physics, University of New South Wales Sydney, 2052, Australia

The height of the atmospheric boundary layer on the Antarctic plateau is of particular interest to meteorologists and designers of optical telescopes for Antarctica. "Snodar" was developed at the University of New South Wales to measure the height of the boundary layer at Dome A and Dome C on the Antarctic plateau. Snodar operates between 3 kHz and 14 kHz and has a vertical resolution of 1 m or better. Snodar was deployed to Dome A in 2008 and was found to have sufficient sensitivity to be able to measure the height of the atmospheric boundary layer with a single pulse. We present here the technical details of a second-generation Snodar, which will be deployed to Dome A in 2009.

1. Introduction

Acoustic radars have been used by meteorologists to probe the lower atmosphere for decades. Acoustic radars transmit an intense pulse of sound into the atmosphere and listen for the very small portion of energy scattered by temperature and velocity inhomogenities[1]. The intensity of the scattered pulse is proportional to the intensity of the atmospheric turbulence and the Doppler shift of the scattered pulse provides information about wind velocity. The remote sensing nature of acoustic radars have made them an invaluable tool in exploring the nature of the atmospheric boundary layer (ABL)[2,3,4,5]. The ABL over the Antarctic plateau is of interest to meteorologists, atmospheric chemists, and astronomers alike. In recent years acoustic radars have been used by astronomers for site testing and characterisation[6,7]. Travouillon[7] has shown that data from acoustic radars can be used to directly estimate the ground-layer contribution to astronomical "seeing" (i.e., image degradation due to the earth's atmosphere).

* Email: cbonner@phys.unsw.edu.au

1.1. *Instrument Background*

At Dome C, the turbulence in the free atmosphere above the ABL has been found to be exceptionally calm and stable[8,9]. However, the winter-time ABL, which is typically tens of metres thick, was found to be very turbulent, contributing up to 87% of the winter-time astronomical "seeing" at ground level[9]. The PLATeau Observatory (PLATO) was developed at the University of New South Wales and deployed to Dome A[10] in January 2008 as part of the Polar Research Institute of China PANDA traverse. At an altitude of 4093 m, Dome A is the highest point on the Antarctic plateau as shown in Figure 1. One of the motivations for PLATO was to investigate the ABL at Dome A. Unfortunately, no commercially available acoustic radars could probe the first few tens of meters of the atmosphere with the required resolution of 1 m or better. "Snodar" is an acoustic radar designed specifically to autonomously measure the height, strength, and variability of the ABL on the Antarctic plateau. The original design of Snodar is presented by Bonner *et al.*[11]. We present here technical details of the second-generation Snodar.

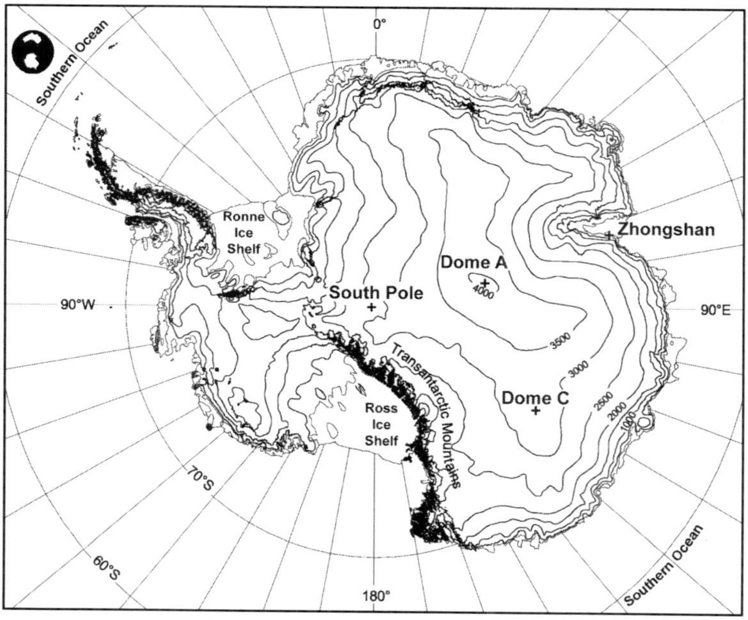

Figure 1 Map of Antarctica showing the Chinese coastal station Zhongshan and the high plateau stations South Pole, Dome C and Dome A. Basic map courtesy of the Australian Antarctic Data Centre.

2. Instrument Design

Snodar was the first acoustic radar deployed to Dome A in February 2008. Snodar successfully operated fully autonomously and was able to determine the height of the atmospheric boundary-layer (ABL) from a single pulse. Data were collected for a short period before the failure of an electromechanical relay crippled the instrument. Parts of the instrument were redesigned in 2008 to expand its capabilities and increase its reliability.

The original Snodar used a fixed-gain low-noise preamplifier and a 60 W Class-AB power amplifier. A USB sound card provided high-speed analog IO. All signal processing was performed in real time on a PC/104 running GNU/Linux (Debian etch, with a read-only filesystem). On-site data reduction reduces the 2.7 MB of raw data by a factor of approximately 600 to 4.5 KB. Data files are then compressed using bzip2 before being written to disk. The final turbulence profile is reconstructed by post processing the data to correct for temperature gradients, noise and atmospheric attenuation.

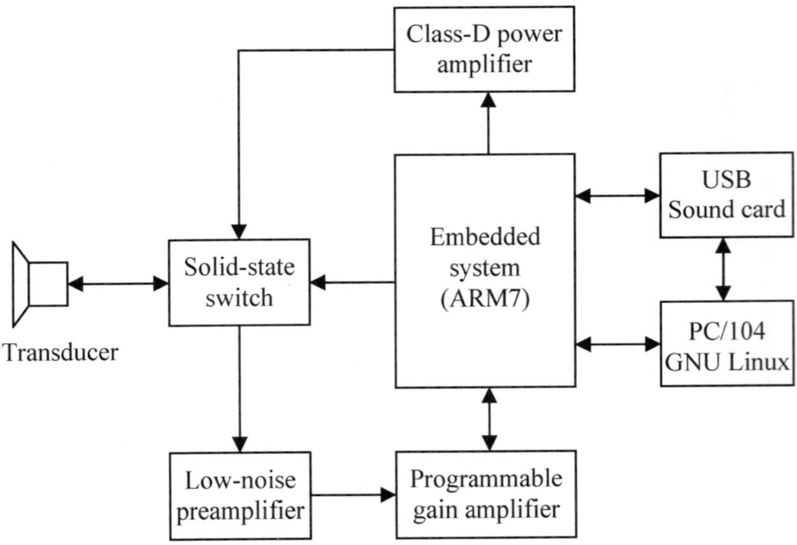

Figure 2 System overview of the second-generation Snodar.

The second-generation Snodar incorporates several enhancements including a 12-bit programmable gain amplifier following the low-noise preamplifier, a 150 W Class-D power amplifier and an embedded system as shown in Figure 2. The programmable gain amplifier has a software selectable gain between 1 and

4095 which increases the dynamic range of the instrument. The electrical efficiency of the instrument is significantly increased by using a Class-D power amplifier. The embedded system is based on a 40 MHz ARM7TDMI microprocessor and has the ability to perform all IO and digital signal processing. A USB sound card and a PC/104 running GNU/Linux (Debian etch, with a read-only root filesystem) will be used at Dome A to expand the system's capabilities and increase maintainability through the PLATO facility.

The PC/104 communicates with the embedded system via RS-232C while the USB sound card is connected to the embedded system via two 3.5 mm stereo sockets. Snodar can be turned into a low-cost, low-power instrument for remote installation by removing the PC/104 and solely relying on the embedded system. In such a case a radio modem, Iridium modem or Ethernet-based RS-232 device server can be used for communication with the instrument. A secondary RS-232C interface has been added to the embedded system to facilitate the connection of auxiliary devices such as weather stations or a secondary Snodar.

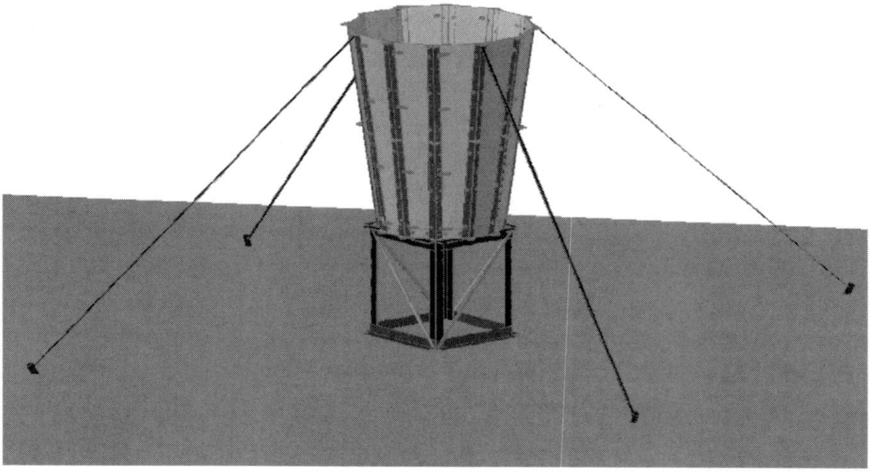

Figure 3 A CAD rendering of Snodar installed in Antarctica. The 1.6 m tall sound cone sits on a 0.8 m open air frame. Four guide wires are used to secure the structure. A 0.95 m off-axis parabolic reflector is located at the bottom of the sound cone.

2.1. *Acoustic Design*

The second generation of Snodar uses a 0.95 m f/0.6 off-axis parabolic reflector housed in a 1.6 m tall aluminium sound cone and operates between 3 kHz and 14 kHz. A rock-wool based sound absorber is used to reduce the reverberation time and provide acoustic isolation. Heaters have been installed on the back of the parabolic reflector and covered with several inches of insulation. The heaters have enough power to sublime small portions of snow and prevent frost formation. A low emissivity coating has been applied to the top surface of the parabolic reflector to passively hinder frost formation and increase heating efficiency. As shown in Figure 3, Snodar has been placed on a 0.8 m tall open-air frame to reduce snow drift. Guide wires are used to secure the structure.

2.2. *Calibration*

Although it is straightforward to accurately derive the height of the turbulence layer for an acoustic radar, deriving accurate estimates of the turbulence intensity is more difficult[12]. As the height of the ABL can be defined through relative changes in the temperature structure function constant (C_T^2), absolute calibration of the scattering coefficient is seldom required. However, if the temperature, pressure and C_T^2 profiles are known then the refractive index structure function constant (C_N^2) can be derived and its contribution to the astronomical "seeing" calculated[7]. For this reason it is desirable, but not essential, to calibrate the intensity data from Snodar. The fact that Snodar must operate over a 40° C temperature range, unmanned for a year poses an issue with the stability of the instrumental sensitivity. To address this, we suspended a sphere of 50 mm diameter made from solid phenolic resin over the aperture of the sound cone as shown in Figure 4. The sphere is referred to as the "calibration sphere" and has two purposes. Firstly, as the height of the calibration sphere above the ice surface is accurately known the echo can be used to remove any systematic offset errors in the scattering heights. Secondly, at any given wavelength the magnitude of the echo from the calibration sphere should be constant throughout the year. This will allow the efficiency of the system to be monitored and the transducer sensitivity to be corrected during post processing.

2.3. *Diagnostics*

The accumulation of snow and ice on exposed surfaces poses the largest issue for autonomous operation of Snodar in Antarctica. If ice forms on the transducer or parabolic reflector the gain of the system will be reduced. This can have a significant impact on the signal to noise ratio of data. During 2008 the echo from the calibration sphere allowed the gain of the system to be monitored and any changes corrected for. A WiFi based camera will be installed inside Snodar's antenna for 2009. The camera will transmit images through the PLATO facility allowing snow and ice accumulation to be monitored and the heating of the dish optimised.

An array of analog and digital IO has been added to the embedded system to provide a simple system control and data acquisition (SCADA) system. Several of the digital outputs drive high-current high-side switches that are used to control heaters, fans and power to the WiFi camera. The analog inputs allow bus voltages and currents to be monitored.

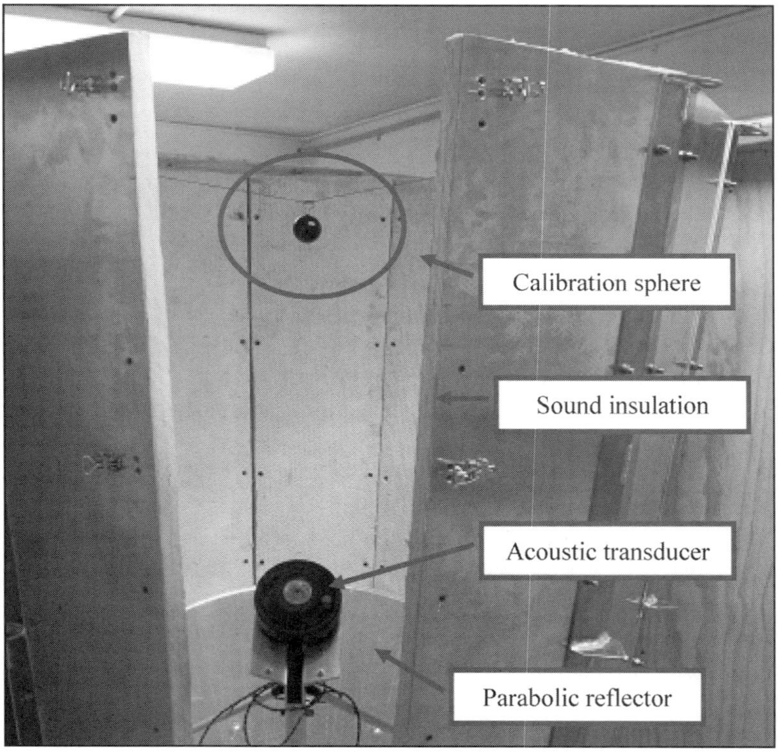

Figure 4 View of the aluminium sound cone, parabolic reflector and calibration sphere. The calibration sphere is suspended by thin wires. Image credit Xuefei Gong.

3. Future Work

The second-generation Snodar will be installed at Dome A in January 2009 by the Chinese PANDA traverse. The electronics of the original Snodar will also be upgraded at the same time. Operating two Snodars side-by-side will allow the different antenna designs to be compared. This will be used to optimise the antenna design for a third generation of Snodar in terms of acoustic performance and sensitivity to snow and ice. The design goal for the third-generation Snodar is that it can be installed in remote locations with a nominal power consumption of only a few hundred milliwatts. A large bank of batteries and solar panels will be used to provide power to the instrument.

Acknowledgments

The authors thank the Polar Research Institute of China PANDA traverse team for transporting and installing Snodar at Dome A. This research is financially supported by the University of New South Wales and the Australian Research Council.

References

1. C.G. Little, *Proc. IEEE*, **57**, 571 (1969).
2. M.A. Kallistratova, *IOP Conference Series: Earth and Environmental Science* **1**, 012001 (2008).
3. A.D. Culf, *Antarctic Science*, **1**, 363 (1989).
4. S. Argentini, G. Mastrantonio, A. Viola1, P. Pettre and G. Dargaud, *Boundary-Layer Meteorology* **81**, 75 (1996).
5. W.D. Neff, *PhD Thesis*, Wave Propagation Laboratory, Boulder Colorado, USA (1981).
6. T. Travouillon, M.C.B. Ashley, M.G. Burton, J.W.V. Storey, and R.F. Loewenstein, *Astron. & Astrophys.* **400**, 1163 (2003).
7. T. Travouillon, *Proc. SPIE*, **6267**, 626720 (2006).
8. J.S. Lawrence, M.C.B. Ashley, A. Tokovinin, T. Travouillon, *Nature* **431**, 278 (2004).
9. A. Agabi, E. Aristidi, M. Azouit, E. Fossat, F. Martin, T. Sadibekova, J. Vernin, A. Ziad, *The Publications of the Astronomical Society of the Pacific* **118**, 344 (2006).
10. J.S. Lawrence, G.R. Allen, M.C.B. Ashley, C. Bonner, S. Bradley, X. Cui, J.R. Everett, X. Feng, S. Hengst, J. Hu, Z. Jian, C.A. Kulesa, Y. Li, D. Luong-Van, A.M. Moore, C. Pennypacker, W. Qin, R. Riddle, Z. Shang, J.W.V. Storey, B. Sun, N. Suntzeff, N.F.H Tothill, T. Travouillon, C.K. Walker, L. Wang, J. Yan, J. Yang, H. Yang, D. York, X. Yuan, X. Zhang, Z. Zhang, X. Zhou, Z. Zhu, *Proc SPIE* **7012**, 701227 (2008).
11. C.S. Bonner, M.C.B. Ashley, J.S. Lawrence, J.W.V. Storey, D.M. Luong-Van and S.G. Bradley, *Proc. SPIE* **7014**, 70146I (2008).
12. S.D. Danilov, A.E. Gur'yanov, M.A. Kallistratova, I.V. Petenko, S.P. Singal, D.R. Pahwa, B.S. Gera, *Meas. Sci. Technol* **3**, 1001 (1992).

ADAPTIVE OPTICS AND INTERFEROMETRY: PRESENT AND FUTURE SYSTEMS AND THE IMPACT OF TURBULENCE

PETER WIZINOWICH

W. M. Keck Observatory, 65-1120 Mamalahoa Highway
Kamuela, HI 96743, USA

This paper provides an overview of an invited presentation on the astronomical capabilities provided by existing and future adaptive optics (AO) and interferometer facilities, and the limitations and design constraints imposed by atmospheric turbulence.

1. Why do Adaptive Optics and Interferometry?

Astronomers use AO and interferometers to improve angular resolution. These systems can also provide improved signal and reduced sky background.

1.1. *Angular Resolution*

Turbulence in the earth's atmosphere blurs the images obtained by ground-based telescopes. The resultant "seeing-limited" angular resolution of a ground-based telescope is given by $\theta \sim \lambda/r_0$, where the atmospheric coherence length parameter, r_0, is typically 10 to 20 cm at a wavelength, λ, of 0.5 μm, and grows with wavelength as $r0 \sim \lambda^{6/5}$. At a wavelength of 2 μm the image size, θ, is typically 0.8 to 0.4 arcseconds.

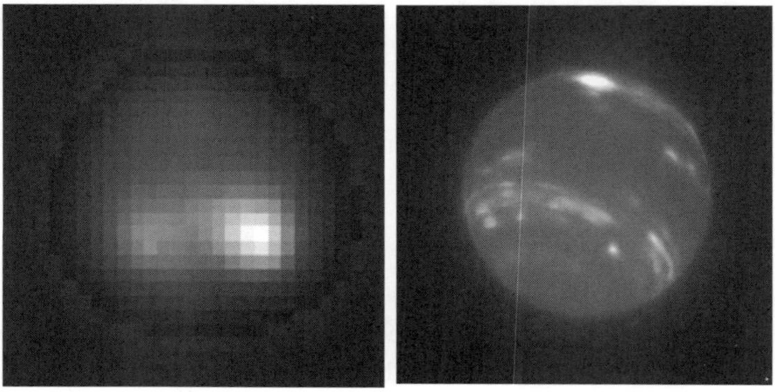

Figure 1. Keck seeing-limited image (left) and AO-corrected image (right) of Neptune.

AO can improve the resolution of ground-based telescopes by sensing and correcting for the distortions introduced by atmospheric turbulence. With good AO correction the angular resolution then becomes limited by diffraction from the telescope's aperture. The resultant "diffraction-limited" angular resolution is given by $\theta \sim \lambda/D$ where D is the diameter of the telescope aperture. For a 10 m diameter telescope the diffraction-limit at 2 μm wavelength is $\theta \sim 0.04$ arcseconds; an improvement in angular resolution by a factor of 10 to 20 versus the seeing-limit. Figure 1 illustrates the angular resolution improvement for observations of Neptune using the Keck 10 m telescope.

Interferometry offers a way to further improve angular resolution by combining the light from two or more telescopes. If the piston and tip-tilt between the telescopes is corrected then the angular resolution in the direction between two telescopes is given by $\theta \sim \lambda/B$, where B is the baseline distance between the two telescopes. For the twin Keck telescopes with an 85 m baseline the angular resolution is improved to 0.005 arcseconds at 2 μm wavelength. The core of the diffraction-limited Keck image of a star is crossed by interference fringes with a spacing given by the 85 m spacing between the two Kecks as shown in Figure 2.

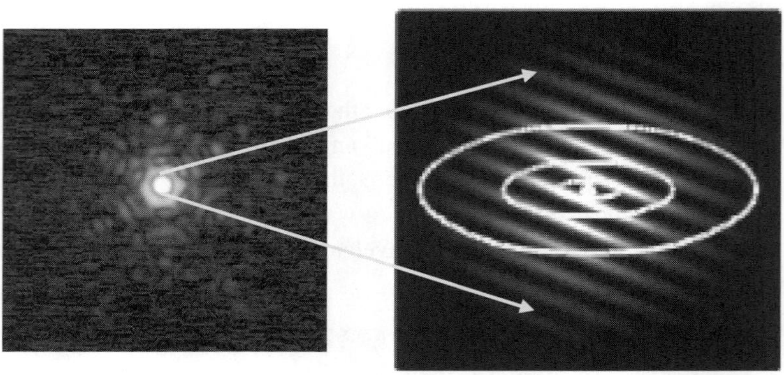

Figure 2. The central core diffraction-limited AO-corrected image of a star from a single Keck telescope (left) is crossed by simulated fringes produced by the interference between light from the two Keck telescopes (right). The orbits of three planets around a star are shown for comparison to the angular resolution of the interferometer.

1.2. *Signal and Background*

The signal obtained with a telescope is proportional to $(D/\theta)^2$ T, so the signal increases quadratically with a decrease in image size at the expense of some throughput, T, losses due to the additional optics required for AO and interferometry.

Smaller detector pixels can be, and need to be, used for higher angular resolution systems. This results in a quadratic reduction in the sky background per pixel with the pixel size. Unfortunately the additional optics required in an AO system can significantly increase the background at thermal wavelengths.

1.3. *AO Concept*

The basic AO concept is shown schematically in Figure 3. The turbulent wavefront transmitted through the telescope is reflected off a deformable mirror. Part of the light, typically the near-infrared light, is transmitted to a science camera, while part of the light, typically the visible light, is reflected to a wavefront sensor. A Shack-Hartmann type wavefront sensor is shown in Figure 3. In this wavefront sensor the telescope aperture is divided into a number of sub-apertures by an array of lenses. Each lens produces an image on the detector and the tip and tilt of the wavefront over that sub-aperture is measured by the change in image position on the detector. The tip and tilt of all the sub-apertures are used to calculate the wavefront. Commands are then sent to the actuators behind the deformable mirror to essentially flatten the wavefront reflected from the deformable mirror. When the AO control loop is closed the image on the science camera goes from being seeing-limited to diffraction-limited.

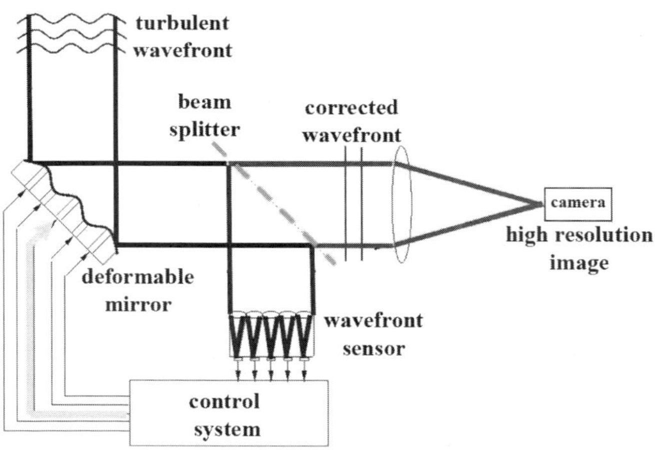

Figure 3. A simple AO system schematic.

A bright enough star is needed to measure the wavefront aberrations in each wavefront sensor sub-aperture in a fast enough time that the atmosphere has not changed significantly. Similarly, the number of sub-apertures and deformable mirror actuators must be chosen so that a tip-tilt correction over that sub-aperture sufficiently corrects for the turbulence. Since both the atmospheric coherence length and coherence time are proportional to $\lambda^{6/5}$ the problem gets more challenging at shorter wavelengths.

The problem of a bright enough star close enough to the science object has limited natural guide star (NGS) AO to a few percent of the sky. Astronomers have created their own laser guide stars (LGS) to allow AO to be used over 50% or more of the sky. The scientifically operational LGS AO systems in use today are generally based on exciting the sodium atoms in the earth's mesosphere, although LGS based on lower altitude Rayleigh scatter from dust and aerosols have also been used.

1.4. *Interferometer Concept*

The interferometer concept is shown schematically in Figure 4 and Figure 5. In addition to distorting the wavefront incident on each telescope, turbulence also introduces an atmospheric optical path difference, OPD, or piston between the wavefront incident on the two telescopes, as well as a relative tip and tilt. If the telescopes are pointed at an elevation other than straight up there is also an elevation OPD introduced between the two telescopes. These OPDs are shown in Figure 4.

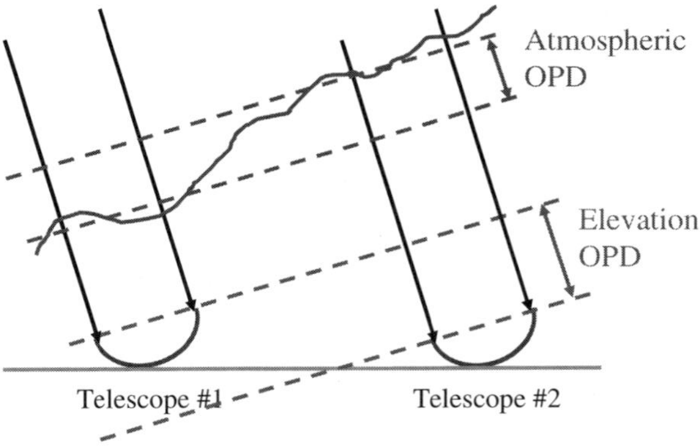

Figure 4. Optical path differences (OPD) between two telescopes of an interferometer introduced by changes in the elevation of the telescopes and by atmospheric turbulence.

In the case of the Keck Interferometer, AO systems on the Nasmyth platforms correct the wavefront incident on each telescope. A series of mirrors is then used to relay the AO-corrected light to the basement where two levels of OPD correction are provided by long delay lines used to statically match the average elevation OPD during an observation and fast delay lines used to dynamically match the elevation and atmospheric OPD.

The wavefronts from the two telescopes are subsequently combined at a beamsplitter as shown in Figure 5 before being input to a detector. The signal measured on a single detector pixel will be high when the OPD to that pixel is zero or a small multiple of one wavelength and will be low when the OPD is equal to half a wavelength (or 3/2ths or 5/2ths, etc.). As the OPD is varied the measured intensity on a single pixel varies as shown by the interference patterns at the bottom of Figure 5. The interference pattern would just be sinusoidal if a single wavelength were used. As the range of wavelengths increases the sinusoidal pattern gets narrower versus OPD until only a single sinusoid would remain. The position of the delay lines to zero the OPD can be determined by scanning the OPD through this interference pattern.

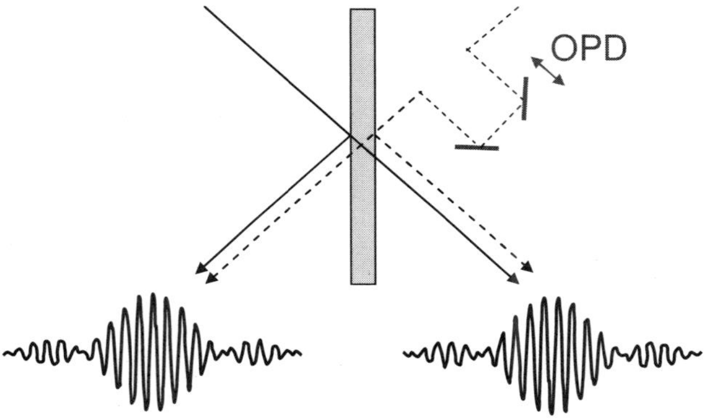

Figure 5. Light from two telescopes (shown as solid and dashed lines) is combined at a beamsplitter. The interference patterns shown at the bottom of the figure are produced as the OPD is varied. The peak intensity is obtained when the OPD is zero.

2. Current and Future Facilities and Performance

2.1. *Adaptive Optics*

Table 1 lists the operational and planned AO facilities on existing large telescopes (D > 6 m). The operating AO facilities are all referred to as single conjugate AO (SCAO) systems in the table since a single conjugate plane is reimaged onto a deformable mirror. This single conjugate plane is normally, but not always, the primary mirror of the telescope. In the MMT case the telescope's secondary mirror is the deformable mirror. The Gemini-North, Keck, MMT and one of the VLT facilities are all based on Shack-Hartman wavefront sensors, while the Subaru and the remainder of the VLT systems are based on curvature sensors. The Gemini-North, Keck II and VLT LGS AO systems all use sodium-wavelength lasers. The MMT uses a Rayleigh laser.

Table 1. Operational (O), planned (PL), and proposed (PR) AO facilities on existing large telescopes.

	diameter	SCAO		GLAO	MCAO	MOAO	ExAO
Telescope	(m)	NGS	LGS	LGS	LGS	LGS	NGS
Gemini - N	8	O	O	PR			
Gemini - S	8	PL			PL		PL
Grantecan	10	PL					
Keck I	10	O	PL				
Keck II	10	O	O			PL	
LBT	2x 8	PL	PL	PL	PR		PL
Magellan (x2)	2x 6.5	PR					
MMT	6.5	O	PL				
Subaru	8	O	PL				
VLT	1x 8	O	O	PL	PL	PL	PL
VLTI	4x 8	O					

All of the future extremely large telescopes (D > 20 m) are planning to have AO facilities as an integral part of their first light capabilities.

Several categories of future systems are referenced in Table 1 including ground layer AO (GLAO), multi-conjugate AO (MCAO), multi-object AO (MOAO) and extreme AO (ExAO) systems. The purpose of GLAO, MCAO and MOAO is to increase the size of the AO-corrected field. The purpose of ExAO is to be able to make high contrast measurements such as detecting a planet next to a star.

GLAO, MCAO and MOAO all require some level of atmospheric tomography to determine turbulence versus altitude. When lasers are used, the wavefront sensing is referred to as laser tomography AO (LTAO). Each laser is used to probe the atmosphere in a slightly different direction in order to distinguish the turbulence contributions at multiple altitudes. In the GLAO case only the turbulence near the ground is corrected; this provides modest correction (since not all of the turbulence is corrected) over a wide field. In the MCAO case two or more deformable mirrors are located conjugate to different altitudes to provide better correction over a modest field. In the MOAO case the turbulence is well corrected over multiple narrow fields using a single deformable mirror for each narrow field.

2.2. *Interferometers*

The VLTI (four 8 m telescopes and four 1.8 m telescopes) and Keck (two 10 m telescopes with an 85 m baseline) Interferometers are the only operational optical/infrared interferometers using large telescopes. The LBTI will combine the light from the two co-mounted LBT 8.4 m mirrors as an interferometer with a 14.4 m baseline. In the US other interferometer facilities include the CHARA array (six 1 m telescopes with baselines from 31 to 331 m) and the NPOI array (seven 0.5 m siderostats with baselines from 19 to 100 m). Baselines as long as 800 m have been proposed by connecting the existing six (or more) AO-corrected telescopes on Mauna Kea, Hawaii and demonstration experiments are underway.

3. Performance

In the ideal world, AO would provide a high Strehl ratio, diffraction-limited and stable point spread function (PSF) across the science field anywhere in the sky with a high signal-to-noise ratio. A perfect simulated Keck PSF (i.e., the image of a point source) is shown on the left-center of Figure 6; the diffraction-limited core is surrounding by a halo structure due to diffraction from the telescope aperture. The Strehl is the ratio of the energy in the core of a diffraction-limited image versus that in the perfect image; the Strehl ratio is 1 for an image limited only by diffraction.

Due to atmospheric turbulence the energy in the PSF core is reduced and re-distributed non-uniformly into the halo. This energy distribution varies with time and field location due to changes in atmospheric turbulence.

3.1. *Atmospheric Parameters Effecting Performance*

The design, performance and optimization of AO and interferometer systems depend on site-dependent atmospheric conditions, all of which vary with time. The profile of atmospheric turbulence and wind versus altitude impact both the phase and amplitude of the wavefront from an astronomical source. The spatial scale of the turbulence impacts the subaperture size to be used in the wavefront sensor and deformable mirror design and the resultant fitting error. The temporal scale of the turbulence impacts the correction bandwidth. The height of the turbulence impacts the size of the corrected field and the acceptable off-axis distance of an NGS used for wavefront sensing, as well as the effectiveness of LGS AO. The outer scale of the turbulence impacts the magnitude of low order terms such as tip-tilt and how much actuator stroke is needed to correct these terms. For ExAO the impact of amplitude variations in the atmosphere on scintillation in the image must be considered.

The refractive index of air is a function of both temperature and water vapor density. The temperature dependence is nearly achromatic whereas the water vapor sensitivity increases with wavelength. The water vapor seeing is significantly less than the dry air turbulence and is not an issue for AO. However, water vapor seeing and dispersion can be an issue for interferometry (e.g., between fringe tracking at 2 μm and science observations at 10 μm).

3.2. *Adaptive Optics*

Figure 6 illustrates the performance of a NGS system (van Dam et al.). The plot shows how Strehl ratio degrades as a function of guide star magnitude (each magnitude represents a 2.5 times fainter guide star) and as the seeing degrades. The star images used to produce a few points on this plot are also shown. The bottom three images show how the angular resolution ($\theta \sim \lambda/D$) improves at shorter wavelengths (J = 1.25 μm, H = 1.65 μm and K = 2.2 μm) while the image quality degrades due to the poorer seeing ($r_0 \sim \lambda^{6/5}$).

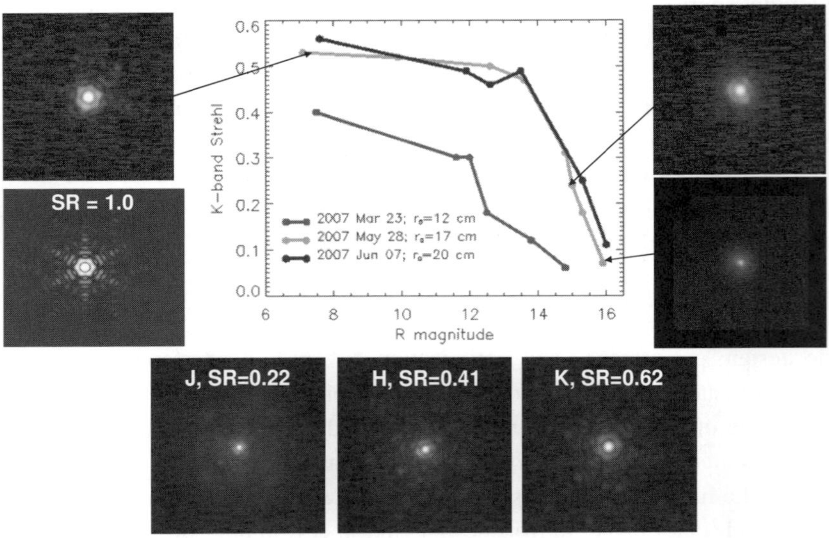

Figure 6. Keck NGS AO performance versus NGS magnitude, seeing conditions (r0) and wavelength. The images are of the stars used for the wavefront sensing and to calculate the Strehl ratio. The SR = 1.0 image is a simulated image of the perfect diffraction-limited image at K-band (2.2 μm).

The primary, and huge, performance advantage offered by LGS over NGS is that the LGS can be pointed at a science target anywhere in the sky. An NGS is still needed to provide tip and tilt information at high bandwidth but the NGS can be much fainter, as well as further from the science object. The LGS performance is poorer than for a bright NGS primarily for two reasons: 1) the atmosphere sampled by the LGS is a cone instead of a cylinder, because the

LGS is at a finite distance from the telescope instead of being at infinity like a star and 2) the LGS is not a point source like a star because the sodium layer (or Rayleigh scatter) has some thickness which produces an elongation of the LGS as seen from an off-axis wavefront sensor subaperture. For the case of a sodium wavelength LGS the altitude, density and structure of the sodium layer vary with time which further complicates the problem.

Figure 7 shows the performance results from a survey of field brown dwarfs between 2005 and 2007 with the Keck LGS AO system (Liu et al.). No data was censored. This data set represents a range of seeing and sodium conditions, magnitude and separation of the NGS star used for tip-tilt, and technical performance of the system. The system allows NGS to be used up to 60" off-axis. When combined with Strehls of 0.2 or more this allows LGS observations to be performed over roughly 2/3rds of the sky.

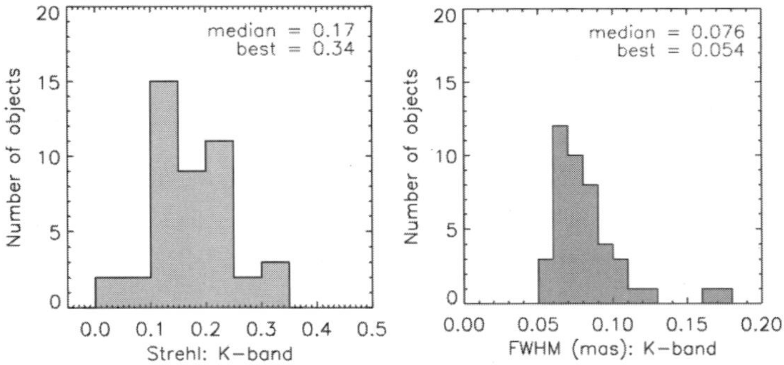

Figure 7. Variability of Strehl and FWHM from a 2005 to 2007 survey of field brown dwarfs with the Keck LGS AO system.

3.3. *Interferometry*

Compared to other current techniques, interferometers offer higher spatial resolution at optical/infrared wavelengths by factors of up to 30, higher sensitivity to the dust around stars by at least a factor of 10, higher astrometric accuracy by a factor of 100 at narrow angles to factors of 3-5 over the full sky and higher dynamic range and photometric precision.

The big limitation of interferometers is sky coverage. Bright targets are needed for AO correction (at least tip-tilt) and fringe tracking, and interferometers using separated telescopes have essentially no field of view. Approaches are being taken to improve sky coverage including using LGS to remove the AO limitation and using bright nearby targets for fringe tracking while doing science on faint targets (similar to NGS AO).

4. Astronomical Science

Astronomers are interested in using a range of techniques to study astronomical sources including imaging, spectroscopy, photometry, astrometry, coronagraphy or nulling, and polarimetry. All of these techniques can be used and enhanced with AO or interferometry.

4.1. *Adaptive Optics*

The scientific productivity of the two most productive large telescope AO facilities can be seen in Figure 8 which shows the refereed science papers published through 2007. The VLT has four AO-fed science instruments each fed by different AO systems. The AO systems that feed CRIRES, SINFONI and the Interferometer are all curvature sensor based systems while NACO is a Shack-Hartmann system. An LGS upgrade has recently been implemented on one telescope in order to be used with NACO and SINFONI. All of the VLT AO results through 2007 have been with NGS AO.

The Keck II telescope has a single combined NGS and LGS AO system that feeds multiple science instruments (NIRC2, OSIRIS and NIRSPEC) as well as the Interferometer. The Keck I NGS AO system only feeds the interferometer. The NGS and LGS results are shown separately in Figure 8. A total of 44 AO science papers were published in 2007 versus a total of 303 papers for all Keck instruments. The demand is high as illustrated by 141 time allocation committee assigned Keck II AO science nights in 2007, including 93 LGS, 39 NGS and nine (NGS) Interferometer nights.

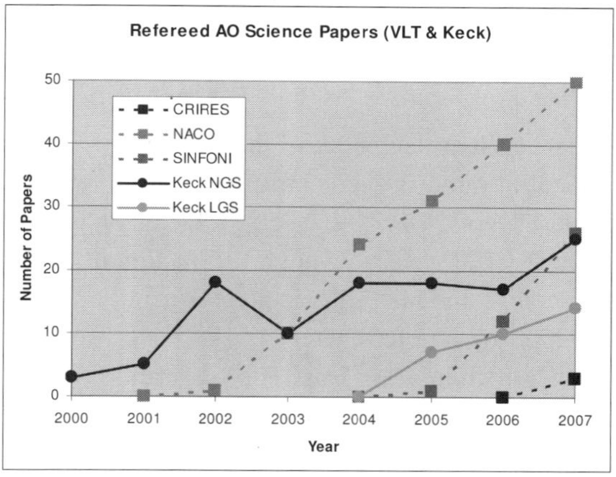

Figure 8. Number of refereed AO science papers versus year for three VLT AO systems (dashed lines) and the Keck NGS and LGS AO systems.

Prior to 2007 the Keck AO science results were dominated by galactic and solar system science, as seen in Figure 9. Extragalactic targets were in general too faint to be observed with AO. The jump in extragalactic science in 2007 is directly attributed to the improvement in sky coverage provided by LGS AO. The LGS AO assisted extragalactic science through 2007 has primarily included observations of high redshift galaxies, but has also included observations of supernovae (the supernova are bright enough to be used as tip-tilt guide stars), galaxies lensed by galaxies and stellar populations in other galaxies.

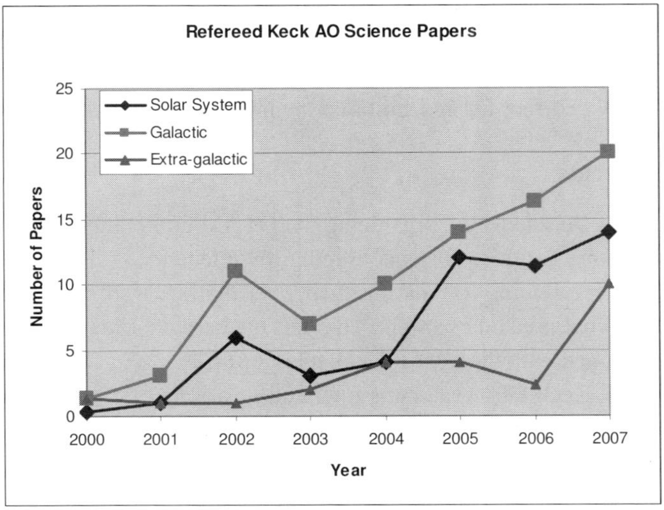

Figure 9. Refereed Keck AO science papers by science area.

4.2. *Interferometry*

Science productivity is growing on the world's interferometers. The VLTI had 32 science papers, between three different science instruments, in 2007.

The simplest science mode of an interferometer is to measure the fringe visibility of a science target compared to a point source calibrator. This approach can be used to measure stellar and disk diameters and binary orbits. Visibility measurements can also be made at various spectral resolutions.

Both the Keck and LBT interferometers are designed to be able to null the light from a star (destructive interference) in order to measure the quantity of zodiacal dust around the star. Interferometers can also be used to make high precision astrometric measurements such as those being made at NPOI and planned for VLTI and Keck.

Finally interferometers with sufficient telescopes and baselines can be used to produce images as the result of measurements at different spatial frequencies and orientations.

5. Conclusion

The ground-based astronomical community has developed AO and interferometer facilities that correct for atmospheric turbulence in order to perform high angular resolution science. Despite these technological achievements, turbulence continues to strongly limit ground-based astronomy. Meteorologists and optical turbulence experts could significantly improve the scientific productivity of ground-based telescopes by helping to determine how we can optimally correct for and calibrate the impacts of optical turbulence.

Acknowledgments

Grazie to Elena Masciadri for organizing the OTAM conference and for inviting me and to Simone Esposito for encouraging my attendance. Thanks to Jacques Beckers and Rene Racine for sharing their presentations with me in advance to inform me of what areas they might have left uncharted. Data presented in this paper was obtained with the W. M. Keck telescopes.

The W. M. Keck Observatory is operated as a scientific partnership among the California Institute of Technology, the University of California, and the National Aeronautics and Space Administration. The Observatory was made possible by the generous financial support of the W. M. Keck Foundation.

References

1. M. van Dam et al., *Keck Adaptive Optics Note* **489** (2007).
2. M. Liu, SPIE 7015-7 (2008).

A NEW SENSOR FOR LASER TOMOGRAPHY ON ELTs

E. GENDRON*, M. BRANGIER, G. CHENEGROS, F. VIDAL, Z. HUBERT, G. ROUSSET and F. POUPLARD

LESIA, Observatoire de Paris, UPMC, Université Paris Diderot, 5, place Jules Janssen, 92195 Meudon Cedex, France
**E-mail: eric.gendron @ obspm.fr*
www.obspm.fr

We present a new concept for an optical differentiation wavefront sensor, featuring a high number of phase measurements across the pupil, with a linear response versus the phase gradient. We present measurements obtained in the lab with this sensor though turbulence on a longitudinally-elongated laser spot.

Keywords: Adaptive optics; wavefront sensor.

1. Introduction

We present in this article a new type of wavefront sensor. It belongs to the class of differentiation wavefront sensors (see review papers from Hénault F.,[3] and papers from Oti *et al.*[12]). It allows to measure the gradient of the phase. This sensor has the advantage of beeing linear on a wide range, allows a high spatial sampling of the pupil, and uses only 4 pixels per measurement. Those characteristics makes it a promising candidate for open-loop operation with laser guide stars (LGS) on future extremely large telescopes.

Moreover, it is known that the noise behavior of wavefront sensors depends on the spatial extension of the source : the noise variance grows together with the spot surface. This effect does not exist here, where the noise is independent from the extension of the object.

Another property is that the sensitivity does not depend on the object extension neither. The response of a quad-cell to a spot movement depends on the illuminated surface of the spot (this is the effect called 'centroid gain') . This effect do not exist with our sensor.

2. Description

2.1. *Principle*

Optical differentiation wavefront sensors work by optically Fourier- filtering the pupil image, placing a varying amplitude transmission filter in the focal plane that will transform the phase gradient into intensity variations in the pupil plane. Instead of using a spatialy varying transmissive filter, we will manufacture a spatialy varying polarization-rotater. A first Wollaston prism splits the incoming beam into 2 linearly polarized beams, with orthogonal directions of polarization. One of the beam will be used to get the X-gradient of the phase, the other for the Y-gradient. On both beams, we place the central element of our system : a thin plate that makes the polarization to rotate by an angle α, that varies continuously across the focal plane from -45 to +45 degrees, as described on the left of figure 1 (α is a linear function of the position). We will explain how this "plate" works in the following section. Then, a second Wollaston prism is used, as an analyzer. The axis of this second Wollaston prism is rotated by 45 degrees from the first one, so that 4 pupil images are formed at the output (a field lens is inserted somewhere, to form pupil images).

On a non-aberrated beam, all the rays cross the same point (the centre) of our variable rotatory plate, and the polarization is rotated by 0 degrees for any point of the pupil. Hence, the amplitude spreads equally on the axes of the second Wollaston prism, and the intensity is the same for the four pupil images. Instead, an aberrated ray will cross the focal plane at a particular location, characterized by a certain amount of polarization rotation. The second Wollaston prism will translate this rotation into a difference of intensity in the pupil planes. As α is a linear function of the position, getting α is like getting the impact of the ray in the focal plance, i.e. the phase gradient.

Adopting a simple ray-tracing approach to describe the behavior of this sensor, calling $I_A(x, y)$ and $I_B(x, y)$ the intensity at a pupil location (x, y) in a couple of pupil images formed by the second Wollaston prism, we can retrieve the rotation angle α by :

$$\alpha = \frac{1}{2} \text{ Arcsin} \left(\frac{I_A - I_B}{I_A + I_B} \right) \tag{1}$$

As the quantity $(I_A + I_B)$ is the total incoming intensity, we see that the sensor is insensitive to scintillation.

2.2. *The Spatialy Variable Rotatory Plate*

Some optical materials are said to be *optically active*. In these materials, the plane of a linearly polarized light will rotate as the beam propagates into the medium, proportionally to the travelled thickness. Any optically active material exists in two flavours, called dextrorotatory (clockwise rotation) or levorotatory (counter-clockwise). Monocrystals of quartz are optically active, but the optical activity is usually masked by the linear birefringence. Optical activity of quartz is only fully revealed when the light propagates parallel to its optical axis, where its linear birefringence vanishes. At $\lambda =$ 589 nm, the rotatory power of quartz is $21.68°/mm$ (2.08 mm make 45° rotation).

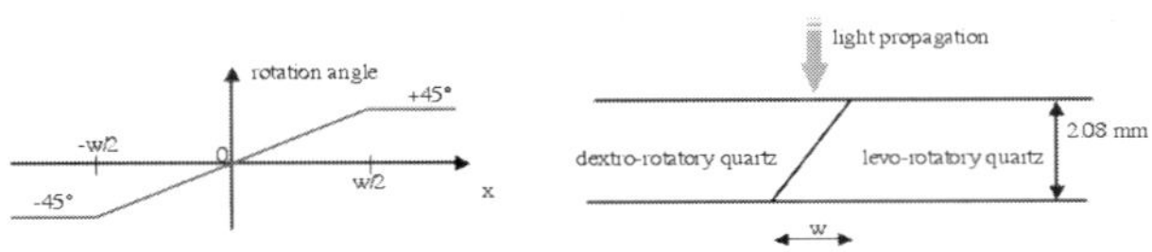

Fig. 1. Left : Variation of the rotation angle of the polarization versus position. Right : Optical design of the spatialy variable rotatory plate. Two plates of a monocrystal of quartz are assembled together. One is dextrogyre, the other laevogyre. The light propagates perpendicular to the plate. The quartz optical axis is parallel to the light propagation. The thickness is 2.08 mm, so that the polarization turns by +45° in one crystal, by −45° in the other, and of course by any angle in between (as described by the left figure in the zone where the crystals overlap).

We assemble two plates made in a monocrystal of quartz, as described on figure 1. The width of the wedge (called w on the figure 1) corresponds to the zone into which the sensor will behave linearly. The analyzed beam is to be focused right in the middle of the plate thickness, right in the middle of the wedge. This particular point defines the center of the wavefront sensor. This is the point where any length travelled in the dextro-plate equals the length in the levo-plate : any rotation will vanish (contrarily to what happens for any other ray). It should be noticed that no particular edge of any material marks the 'zero' of the wavefront sensor (contrarily to to the edge of the knife for Foucault test, or the peak or edges between the faces of a pyramid sensor). Thus, diffractive effects are minimized.

We emphasize that

- the sensor is not sensitive to the polarization of the incoming light at all, since the first Wollaston prism acts as a polarizer, and thus erases whatsoever previous polarization property.

- the rotatory power of a material is highly dependent on the wavelength λ. The dependence is usually something close to λ^{-2}. When used in polychromatic light, one can consider that the sensor will have a sensitivity that varies with λ^{-2}.

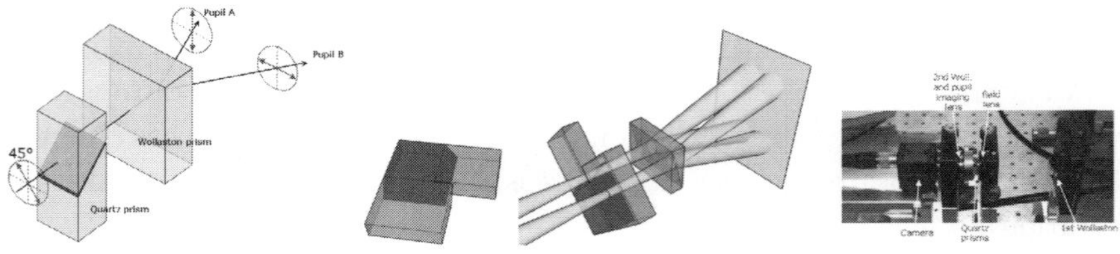

Fig. 2. Left: The polarization rotates by an amount proportional to the height the ray went through the wedge, from -45° to +45°. A Wollaston prism analyzes and splits the beam to form 2 pupil images. Middle left: Picture showing the technical implementation : two orthogonal sides of a quartz plate have been polished with a wedge, and two others plates from the other enantiomer have been cemented to it. Middle right: 3D representation of the setup : light is coming from the left. The two beams have already been separated by a first Wollaston prism. They are both linearly polarized. They are both focused at the center of the quartz wedges. The field lenses have not been represented here. Right: the setup in our lab. Light is coming from the right. The total length from the first Wollaston to the CCD plane is about 100 mm.

3. Image Processing

Several steps are required in order to get the gradient signal out of the images :

- From equation 1, it is clear that the linearity of the CCD response versus the received flux is of key importance. We have calibrated it, and we have to compensate from a slight non-linearity in the processing.
- The fine subtraction of background is even more important than with a Shack-Hartmann, and the background should be finely calibrated.
- Flat-field also matters a lot, since it impacts directly on the detected photometric signal.
- Due to the non-zero thickness of the quartz plate, the amount of rotation of the polarization has to be corrected from a geometrical effect. One has to apply to the gradient image a kind of "flat-field" correction, which expression (developed at first order in a Taylor series) is : $(1+\frac{e}{w}\theta)(1+\psi^2/2)$, with θ and ψ the angles corresponding

to pupil coordinates, respectively parallel and orthogonal to the direction of differentiation.

- Due to aberrations of the optics, the pupil image may reveal some optical distortion (1.5% in our present design). A special operation -that we call pupil morphing- needs to be performed so that the 4 pupil images accurately match together, pixel-to-pixel. The calibration of this morphing is done beforehand only once, by inserting a square-grid mask in the pupil, and measuring its relative deformation with sub-pixel accuracy in the 4 pupils (correlation).

4. Experimental Results

4.1. *Optical Setup*

We built a prototype sensor, currently tested in our lab within an optical setup that reproduces a f/17 focus of a 4.2m telescope. We have chosen a wedge width of w=1.07 mm (see fig. 1), equivalent to 2.9 arcsec on-sky. In order to get 4 pupil images arranged on a square, we put the Wollastons' axes orthogonal to each other, and added a quartz plate in between to rotate the polarizations by 45°. The output pupil diameter is 2.2 mm, leading to 228 pixels into the pupil diameter (45000 points in total). The sensor is fed by a fiber-source at $\lambda = 620$ nm, of various diameters. Pupil is at infinity, and may include Kolmogorov-type phase screen.The source can be moved in, x, y, z with precision translation stages.

4.2. *Operation*

The first remarquable thing is that the pupil images are of a very high quality, with no diffracted light outside it (see fig. 3). Rings on the edge of the pupil (derivative of a discontinuous function) do exist, but they are just 2 pixels wide, and not brighter than $\pm 15\%$ of the average pupil intensity.

We have studied the interaction between optical activity and linear birefringence. We found that a bias is introduced, that is about 0.2% on the pupil edge for a f/15 beam, and behaves as $(f/D)^{-4}$. Although negligible, a solution can be to use a purely optically active crystal, such as $NaClO_3$.

Our wavefront sensor has manufacturing flaws on both sides of the wedge. It is now aging, and the wedge is not cemented any more at the edge, letting a tiny air gap to appear at both sides, that make a total reflection of lignt. This flaws prevents us from using the very whole area of the wedge. Results are fine when the light passes through the central 0.6 or 0.7 mm, but problems occur on the edges.

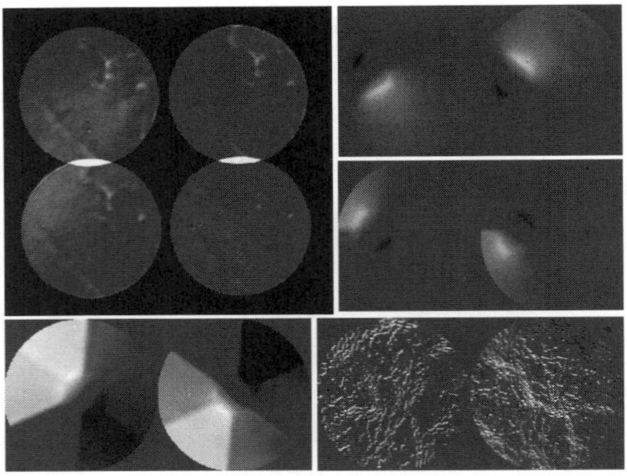

Fig. 3. Upper left : Raw sensor image, background subtracted only. The patterns inside the pupil are just due to inhomogeneity of the incoming flux. Pupil overlap here because the incoming beam is at f/15 (the sensor is designed for f/17). Right and bottom : after image processing (flat-field, sub-pixel morphing, combination of images according to equation 1) one gets the X and Y gradients of the phase. The differentiation directions are at 45°. Right : phase gradients when applying a voltage on an electrode of a 31-elements bimorph mirror from CILAS. Bottom left : central electrode of the same bimorph mirror. Bottom right : phase screens used on our AO test bench, manufactured by etching successive binary masks into silica. The very high resolution of the sensor reveals the stair-like structure of the screens (although steps are only 10 nm high).

5. Linearity and Sensitivity Properties

We particularly focused on the interesting linearity properties of the sensor. Compared to the Shack-Hartmann, the main interest of our sensor is to reduce the required number of pixels, while keeping nice linearity properties and a constant gain.

5.1. *Linearity on Unresolved, Extended, and 3D Sources*

We have tested the sensor with an unresolved and extended point-sources of 50, 100, 200, and 600 μm. The test consists in translating the source by a known amount across the field, and measure the consequent tilt with the sensor in terms of this translation. The experiment was conducted with and without a fixed Kolmogorov-type aberrated plate in the pupil.

In any case, we could verify

- an extremely linear response in terms of the source position
- a slope of this response (the sensor gain) which is independent of the source diameter : no "centroid gain" effect
- no centroid gain effect with the presence (or absence) of seeing

Each time, we could experimentally check a extremely good linearity with the source position.

We could simulate an LGS beacon by defocusing a spot during the exposure time of the camera. We performed the same tests as before with this source, trying different diameters, with or without phase screens. Results are unchanged : the sensor gain is fixed and equal to what it was for 2D sources.

6. Noise Behavior

When the angular extension of the source impacts on the measurement noise of any wavefront sensor, here the noise variance is a fixed value that only depends on the flux level. This is because the measurement is made right in the pupil plane, so that there is no blurring of the signal due to the spatial extension of the object.

6.1. *Theoretical Considerations*

We can compute the noise variance on the phase difference between edges of a pixel, assuming the flux of photons N, per frame and per "pupil sample", is divided equally between the two gradient channels, and then equally divided between the two pupils A and B. We assume photon noise, and a detector read-out noise of σ_e electrons rms per pixel. The noise variance is given by:

$$\sigma_\phi^2 = \left(\frac{2w_{sky}}{\lambda/d_{subap}}\right)^2 \left(\frac{2}{N} + \frac{8\sigma_e^2}{N^2}\right) \tag{2}$$

where w_{sky} is the full size of the wedge expressed as an angle as seen on the sky.

6.2. *Noise Behavior in Terms of Source Extension*

We have measured the noise variance, by acquiring on a stabilized light source a set of 100 gradients on a fixed wavefront and computing the time-variance for each pixel (shown vs flux on Fig. 4). The results

- really show a perfect match with theoretical expectations
- show strictly no dependence of the noise versus source diameter.

One should insist on this point : measurements actually fit theory remarquably, probably much better than with any other sensor where many assumptions are required. The same calculus in the case of a Shack-Hartmann

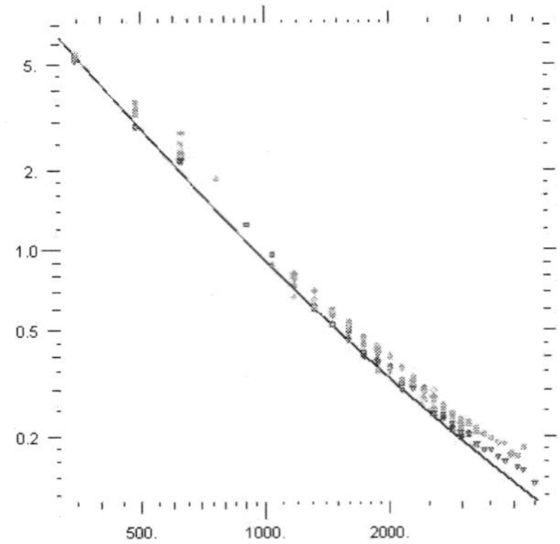

Fig. 4. The graph shows the variance of the polarization rotation (in degrees^2) in terms of the flux. The solid line is the theoretical expectation. The different points are for various sizes of object, ranging from point-source to 200 microns wide (equiv. to 0.6 arcsec on-sky). This shows that the experimental noise measurements meet the theoretical expectations, and that the noise level is independent from the source extension.

implies assumptions on threshold value, pixel sampling, subaperture field of view, width of the spot and adequation of a matched filter to the spot size. We emphasize that this independence to the spot profile really allows to relax the constraints on the optical quality of the laser beacon, and this may have a nice financial impact on the laser optical quality, projector, etc.

References

1. Oti J.E., Canales V.F., Cagigal M.P., "Improvements on the optical differentiation wavefront sensor", MNRAS **360**, 1448-1454 (2005)
2. Oti J.E., Canales V.F., Cagigal M.P., "Analysis of the signal-to-noise ratio in the optical differentiation wavefront sensor", Optics Express, vol. 11, Issue 21, p.2783 (2003)
3. Hénault, F. "Wavefront sensor based on varying transmission filters: theory and expected performance", Journal of Modern Optics, vol. 52, Issue 14, p.1917-1931 (2005)

IMAKA: IMAGING FROM MAUNA KEA

OLIVIER LAI*, JEAN-CHARLES CUILLANDRE

Canada France Hawaii Telescope,
65-1238 Mamalahoa Hwy, Kamuela, HI 96743, USA
**E-mail: lai@cfht.hawaii.edu*

MARK R. CHUN

Institute for Astronomy, University of Hawaii,
640 North Aohoku Place, Hilo, Hawaii 96720, USA

RAY CARLBERG

Department of Astronomy & Astrophysics, University of Toronto,
50 St. George Street, Toronto, Ontario M5S 3H4, Canada

HARVEY B. RICHER

Department of Physics & Astronomy, University of British Columbia
6224 Agricultural Road, Vancouver, B.C. V6T 1Z1, Canada

The CFH Telescope has a long history of wide-field imaging, and this is reflected in its users community. Many projects are competing in the $A \cdot \Omega$ race but so far, all are limited by atmospheric turbulence. Wisely choosing a site with a very thin ground layer and excellent free-atmosphere seeing would allow very wide fields to be corrected by a GLAO system.[1,2] The characteristics of the turbulence of Mauna Kea are perfectly matched to such an instrument making the site an integral part of its design. The goal of this project is to achieve exquisite image quality over the largest possible field of view, with a goal of a FWHM of not more than 0.3" over a square degree field in the optical domain. This goal depends crucially on the thickness of the ground layer and its prevalence. But over such large fields the probability of finding sufficiently numerous and bright natural guide sources is high, although a constellation of laser beacons could be considered to ensure homogeneous and uniform image quality. The image is then limited by the free atmosphere seeing (0.2" to 0.4"). This can be further improved by an Orthogonal Transfer CCD camera (Tonry et al.[3]) which can correct local image motion on isokinetic scales from residual high altitude tip-tilt. Knowing the vertical turbulence profile, the outer scale and the isokinetic patch plays a critical role at the design level.

Keywords: Atmospheric turbulence; Adaptive optics; ground layer.

1. Introduction

The Canada France Hawaii Telescope, with its 3.6 meter diameter primary mirror, has a long and successful history of wide field and high angular resolution imaging. Its current instrument suite is state-of-the-art and pressure on observing time is expected to remain high until 2012. In this pre-ELT period, can CFHT still remain viable? To answer such a question, a call for ideas was launched in 2007. There are already many wide field and survey projects planned for this period (HyperSuPrimeCam, Pan-Starrs, LSST), many with larger $A \cdot \Omega$ than could be achieved with the CFHT. However, the site characteristics of Mauna Kea, namely a very thin ground layer, could mean that the image quality could be improved over very large fields using the technique of Ground Layer Adaptive Optics (GLAO[1,2]). Improving the image quality improves the sensitivity and the overall astrometric accuracy. 8 meter class telescopes are also considering GLAO instruments but the field of view of these telescopes is limited by their optical design to a few to tens arcminutes, providing a natural niche for our older telescope. The hawaiian word ʻīmaka means a lookout, a scenic viewpoint, and CFHT equipped with this instrument would be an ʻīmaka as we would be using Mauna Kea and its characteristics to look at the sky in a special way.

1.1. *Ground Layer Adaptive Optics*

GLAO is a technique developed by Rigaut et al.[1] and developped by Tokovinin[2] whereby only the fraction of the turbulence common to the entire field, namely the pupil and ground layer, is corrected. This produces an "improved-seeing" image, not a diffraction limited one. Depending on the relative strength of the ground layer and its thickness, the amount of correction and the field over which it can be achieved are considerable. Ultimately, the FWHM cannot be smaller than the free atmosphere seeing. Two independent studies reveal that the site of Mauna Kea is extremely well suited to this technique: Gemini commissioned a site-testing study using a combination of SLODAR and LOLAS for a possible Ground Layer Adaptive Optics module. This study, carried out by M. Chun, R. Avila, T. Butterly and R. Wilson, reveals that the ground layer is very thin (between 50m and 100m), and the median free atmosphere seeing (uncorrelated to ground layer strength) is around 0.4", Figure 1. This result is confirmed by the TMT site study,[4] which finds that Mauna Kea is a superior site in terms of free atmosphere seeing compared to the best chilean sites, e.g. Cerro Armazones, even though the total seeing value favors the latter. Tokovinin[2] provides expressions for the volume of atmosphere that is partially corrected, the "gray" zone. Geometrically, the decorrelation of the correction varies linearly with field and altitude; given a ground layer thickness, it is thus possible to compute the field which can effectively corrected $H_{max} = \lambda/\beta\theta_0$. λ is the wavelength, β is the goal image quality, and θ_0 is the radius of the corrected field. So if H_{max}=70m, λ=700nm, and $\beta \simeq$ 0.3", we find that θ_0 is approximately 30', so the corrected field diameter can be up to 1°.

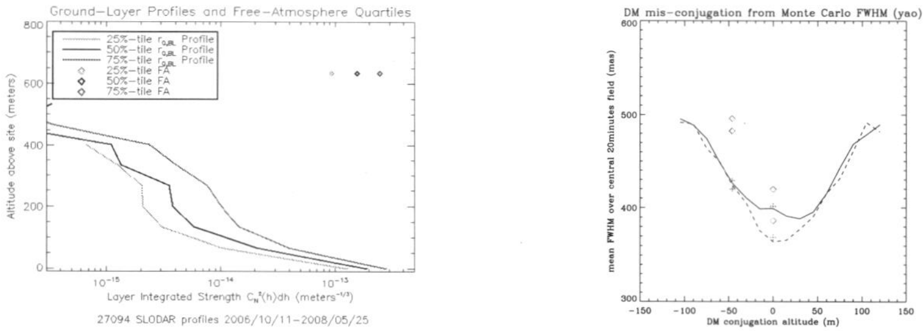

Fig. 1. Left: C_n^2 as a function of altitude for 25%-ile, median and 75%-ile. Also shown is the integrated free atmosphere seeing (diamonds). Most of the turbulence is confined to with 50 meters to the ground. Results obtained by Chun et al, Gemini GLAO site study. Right: effect of Deformable mirror conjugation on FWHM, with a 10×10 SH (full line) and 20×20 (dashed).

2. Astrophysics

GLAO will provide images about 1 magnitude deeper per unit observing time than natural seeing, reduce confusion noise in crowded fields, and open up new science opportunities for "barely resolved" objects.

Solar System: Kuiper Belt 30km size objects, formation and collisional evolution. Also Jupiter's moons.

Extra-solar planets: by transit, in dense star fields. However, will need millimagnitude accuracy.

Stellar physics: High proper motion accuracy. Search for exotic stellar objects in the field.

Galaxy formation & Evolution: Deep red surveys are good tool to discriminate z∼ 6 objects that will be routinely probed by JWST and ELTs. Morphological galaxy survey up to z∼1. At low redshift (up to Virgo cluster), star formation history from multi-band photometry on resolved stellar population.

Cosmology: Equation of State parameter, w at redshift 1.2 from supernovae. Strong lensing of distant supernovae.

3. Instrument

The question of whether this instrument can actually be built is not trivial: in a classical AO relay at the Cassegrain focus, the beam has to be reflected off the deformable mirror at an angle to the plane of the pupil to clear itself. This introduces a variable altitude DM conjugation across the pupil, as well as a pupil shear on the DM as a function of field. This effect is in fact prohibitive for deformable mirrors < 200 mm in diameter. However, if larger deformable mirrors are available, this solution should be revisited; as we shall see in section 4, the only apparently viable solution, an Adaptive Secondary Mirror (ASM) may be limited by the mis-conjugation of the secondary mirror with respect to the pupil, figure 1. In terms of design, an ASM is quite elegant as it reduces the number of optics to mainly a wide field corrector. The CFHT F/8 secondary mirror is not easily directly replaced with an adaptive component: it is 8.6m above the primary and 1.5m in diameter. As cost is generally proportional to D^2, it is important to keep the ASM small. This requires the secondary mirror to be close to focus. Unfortunately, the closer to focus it gets, the further away from the pupil its (virtual) image is formed (the conjugation is at negative altitudes). Since the thickness of the correctable gray zone is proportional to the corrected field, the conjugation constraints for ‘ĪMAKA are difficult to meet.

3.1. *Optical Designs*

Optical layouts were designed by John Pazder (HIA) and Harvey Richardson to assess the optical feasibility. Harvey Richardson developed a design which provides diffraction limited performance over the entire field. It requires a concave secondary ahead of focus, a concave tertiary conjugated to the (curved) pupil plane, an aspheric convex quaternary, followed by a 7 element wide field corrector. The deformable mirror would thus have to be $\sim$ 600mm in diameter, concave and with a central hole. A simpler design with image quality within tolerances for GLAO was developed by John Pazder and requires a 1 meter diameter convex aaptive secondary mirror, conjugated to -46m altitude, followed by a 3 piece corrector.

3.2. *Wavefront Sensing*

GLAO is especially well suited to laser guide stars, especially Rayleigh beacons, as they are insensitive to high altitude turbulence and mostly measure the ground layer. Also the constellation can be fixed on the sky, which means the wavefront sensors can be fixed with respect to the field. Rayleigh beacons are cheaper than sodium beacons, and can be easily spectrally separated. However, the complexity and cost associated with lasers, as well the fact that there may be sufficiently many bright stars for wavefront sensing on such wide fields may make them unnecessary. The probability of finding more n stars with magnitude m (or brighter) is given by:

$$P(m, > n) = 1 - \sum_{i=1}^{n} \mu(m)^i \frac{e^{-\mu(m)}}{i!} \tag{1}$$

where $\mu(m)$ is the star density count. When using the Besançon star count model,[5] we find that the probability of finding 6 stars with m< 11 is 50% at the NGP; the sky coverage increases to 95% if wavefront sensing can be achieved on 6 stars of magnitude 12, figure 2. Small pick-up mirrors on articulated arms might be necessary to pick up the beams from the bright stars and send the to wavefront sensors, although interesting alternative options exist and should be further studied. For example, new low noise high frame rate chips such as those developed by Teledyne, would allow for near-IR wavefront sensing; the main advantage is that a dichroic beam-splitter would eliminate the footprint of the mirrors and articulated arms. Conversely, pyramids could be used instead of mirrors on the articlated arms, and on-chip pyramid sensing could be performed, figure 2.

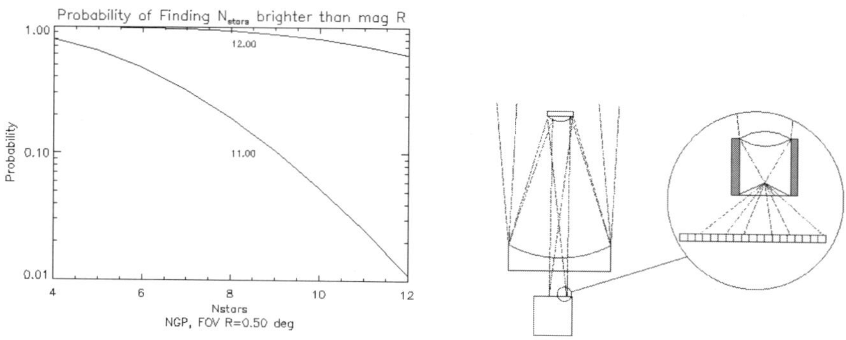

Fig. 2. Left: Probability of finding $> n$ guide stars of magnitude m or brighter at the NGP. If the limiting magnitude for wavefront sensing is 12, 95% of the sky is accessible. Right: Wavefront sensing could be achieved on-chip with deployable pyramids in front of the OT-CCD chips.

4. Simulations

Analytical and Monte Carlo simulations of the performance of the GLAO system were performed. PAOLA, an analytical PSF computation tool based on power spectra based on Rigaut et al,[6] and developed by Jolissaint et al,[7] suffered from aliasing problems due to the large size of the field and the fine sampling required on the MTF to properly reproduce the low frequency, large residuals of GLAO. However, these results confirm it is possible to achieve FWHM of 0.44", 0.38" and 0.35" at V, R and I respectively.

The difficulty of Monte Carlo simulations is that large phase screens required: at 10km, 1° is 175m, so >2000 pixels are required if a few pixels sample r_0. Nonetheless, such simulations were carried out using yorick (with the yao package, F. Rigaut, private communication) and in IDL (simul.pro, F. Rigaut, private communication). They were compared and found to agree in a comparative study of performance as a function of guide star magnitude. Extensive yao simulations were used to evaluate performance as a function of number of guide stars, order of AO system, sampling frequency, atmospheric conditions and DM conjugation altitude. The yao simulations were also used to compute the PSF uniformity across the field (figure 3), but were limited to temporal 9000 steps (with 1msec time sample = 9 second exposure) which was not enough to average atmospheric residual variations.

All the previous simulations assume the 0.65" (50th percentile) seeing/profile. But seeing, and therefore performance, is a probability distribution. David Andersen (HIA) computed the free-atmosphere seeing as a limit of GLAO performance, by analyzing 7000 Cn^2 profiles and assuming that all the turbulence below 600 or 70m can be corrected.This shows that 0.3" performance can only be achieved 20% of the time in optimistic scenario of correcting all the ground layer. A solution to improve the image quality is to correct the next dominant term in the residual phase error budget, namely the high atmosphere tip tilt. This will be definition have smaller isokinetic angles, so the tip-tilt correction has to done locally. This is well adapted to the Orthogonal Transfer CCD (OTCCD), developed by Tonry et al.[3] A collaboration between CFHT and UH could be envisaged to develop a clone of the PanStarrs camera: a 1.3 Gpixel CCD mosaic with orthogonal charge transfer. With a plate scale of 0.1", the 36 000 × 36 000 detector covers a 1°^2} field. The orthogonal charge transfer allows to measure and correct the local (i.e. de-correlated) tip-tilt by shuffling charges in real time on the detector. It has been demonstrated to improve on sky image quality from 0.75 to 0.5" over a few arcminutes field of view.

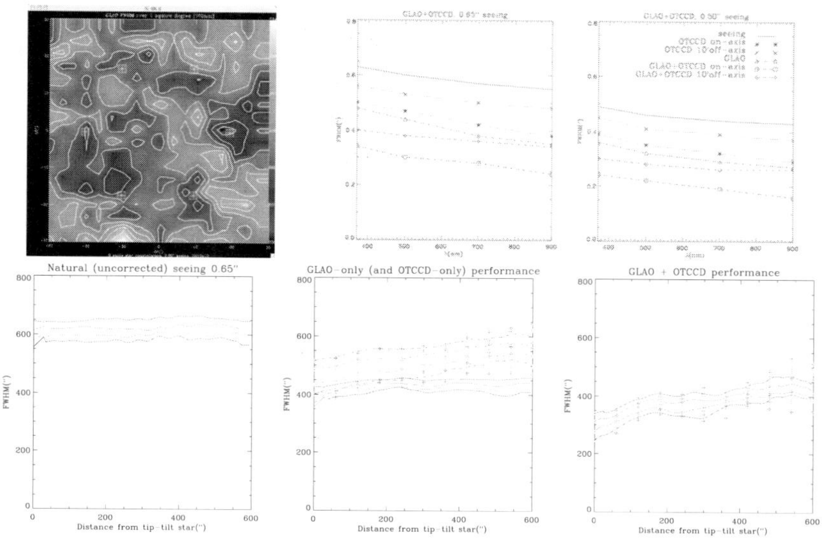

Fig. 3. Top Left: FWHM overe 1° degree field, 6 guide stars on hexagon with 20' radius, as simulated by yao. Top Right: GLAO, OTCCD and GLAO + OTCCD FWHM as a function of wavelength.Bottom: FWHM as a function of distance to guide star, natural seeing (left), GLAO or OTCCD (middle) and GLAO+OTCCD (right).

The simulated performance of GLAO+OTCCD is encouraging, although it was performed in two steps (first a GLAO to remove the ground layer, and then a second iteration correcting only tip-tilt for the free atmosphere) and assuming an (almost) perfectly corrected ground layer (figure 3). It would be important to simulate the performance end–to–end.

5. Conclusion

A feasibility study has been completed outlining possible instrument designs and an overall cost. Such capability does not come cheap, but it is a unique scientific and performance niche. The science cases & interest are well established. The instrument design needs to be refined and a Cassegrain AO relay with a large deformable mirror should be investigated. The expected performance is very strongly dependent on the relative strength of ground layer. To improve the reliability of the simulations, a LOLAS experiment on telescope itself is expected to provide in-situ measurements of the telescope, dome, ground layer and free-atmosphere seeing. This will allow to improve enable a trade study on the design of the AO system, its optical design and architecture.

At the frontier between astronomy and meteorology, a better understanding of our site's characteristics and including them as integral part of instrument design is opening the way to the next generation of ground based imaging.

References

1. F. Rigaut, Ground Conjugate Wide Field Adaptive Optics for the ELTs, in *European Southern Observatory Astrophysics Symposia*, eds. E. Vernet, R. Ragazzoni, S. Esposito and N. Hubin, European Southern Observatory Astrophysics Symposia, Vol. 582002.
2. A. Tokovinin, *Publ. Astron. Soc. Pacific* **116**, 941(Oct. 2004).
3. J. Tonry, B. E. Burke and P. L. Schechter, *Publ. Astron. Soc. Pacific* **109**, 1154(Oct. 1997).
4. M. Schöck, Site selection survey: calibration and results, in *Optical Turbulence, Astronomy meets Meteorology*, eds. E. Masciadri and M. Sarazinin These Proceedings2008.
5. A. C. Robin, C. Reylé, S. Derrière and S. Picaud, *Astron. & Astrophys.* **409**, 523(Oct. 2003).
6. F. J. Rigaut, J.-P. Veran and O. Lai, Analytical model for Shack-Hartmann-based adaptive optics systems, in *Society of Photo-Optical Instrumentation Engineers (SPIE) Conference Series*, eds. D. Bonaccini and R. K. Tyson, Presented at the Society of Photo-Optical Instrumentation Engineers (SPIE) Conference, Vol. 3353Sep. 1998.
7. L. Jolissaint, Spatial frequency analytical modeling of adaptive optics for diffraction limited coronagraphs studies on extremely large segmented telescopes, in *EAS Publications Series*, eds. M. Carbillet, A. Ferrari and C. Aime, EAS Publications Series, Vol. 222006.

DEALING WITH TURBULENCE: MCAO EXPERIENCE AND BEYOND

R. RAGAZZONI*, Y. MOMANY, C. ARCIDIACONO, R. FALOMO, J. FARINATO, M. GULLIEUSZIK, A. MORETTI

Osservatorio Astronomico di Padova, INAF, Padova, Italy
**E-mail: roberto.ragazzoni@oapd.inaf.it*

E. DIOLAITI, M. LOMBINI

Osservatorio Astronomico di Bologna, INAF, Bologna, Italy

G. PIOTTO, R. TUROLLA

Universitá di Padova, Padova, Italy

E. MARCHETTI, R. DONALDSON

European Southern Observatory, Garching bei Muenchen, Germany

Multi Conjugated Adaptive Optics, thanks to the Multi Conjugated Adaptive optics Demonstrator, MAD, of European Southern Observatory, entered in the realm of producing astronomical science after more than a decade. I review our experience with the Layer Oriented Wavefront Sensor paying special attention to the practical influence of the encountered atmospheric turbulence on the achieved science. An outcome, maybe very personal, of the perspective in the near and not so near future are outlined.

Keywords: Multi Conjugated Adaptive Optics; Extremely Large Telescope.

1. Introduction

Multi Conjugated Adaptive Optics (MCAO hereafter) is a technique for achieving correction of astronomical images distorted by the atmosphere turbulence over an "extended" Field of View.[1] The concept of "extended" here means significantly (by a factor of a few to several...) better than the isoplanatic patch size. This is accomplished by optically conjugating more than a Deformable Mirror (DM) to different heights of conjugation along the line of sight of the telescope. With a little oversimplify this means that the correction of the turbulence does no longer happens on the flat realm of the telescope pupil, just characterized by two dimensions, but it deploys also along the line of sight of the telescope, in the volume of turbulence that lies in the path of the telescope. The turbulence occurring in the conoidal shaped region on top of a large telescope acquires, in this light, a new relevance. To make a meteorological example MCAO "changes" the way one have to look in to turbulence. The same way as different problems make a different view of the same meteorological conditions. The same judgment of the meteorological conditions can be extremely different if you are looking for a leisure walking, driving or for flying, just to make an example. A combination of low ceiling with no precipitation can make flying unsafe and prohibitive, but still acceptable for both walking and driving, while a very thin fog can make vertical visibility acceptable for flying and unsafe for driving or unpleasant for walking. Adding a third dimension can makes our judgment of a situation completely different. This is true also for turbulence. A strong turbulence almost concentrated on the ground layer can make MCAO extremely easy, in spite of the total integral of the image deterioration being worse than a situation in which the degradation of the image is caused by a large number of distributed thin layers all scattered over the whole column in the line of sight with not, or little, concentration toward the ground layer. Switching to MCAO, it can pos completely new problems, so that, for example, it is irrelevant in conventional Adaptive Optics if the ground layer thickness is 10m, 100m, or 1km, while this would makes huge differences and poses strong limitations or imposed different technical approaches if the same imaging problem is attacked through MCAO. Also, MCAO is somehow generic as statement, although it clearly indicates that we care much more how is the vertical distribution of the turbulence, on top of the bi-dimensional one on the pupil surface, making the integral of the turbulence, what is generally called "seeing" to make a step back in terms of importance. If MCAO is developed at a point that the volume of the atmosphere sampled by a certain DM has, or it can be assumed to have, the same horizontal speed, the accuracy of a wavefront prediction based upon the Taylor's hypothesis (the one of "frozen layers") can makes a huge difference in the limiting magnitude, or in the Signal to Noise Ratio attainable with a certain Laser Guide Star, in the MCAO correction. Dealing with turbulence using MCAO, in other words, open a completely new realm and it is obvious to me we are just starting to scratch the surface and interesting news are awaiting us. In the following I will try to give a short background to the problem and to express my impressions on our experience with MCAO, and my initial views on what can going to happen in the near future.

2. A Brief History of MCAO

The "father" of MCAO is, without any doubt, Jacques Beckers. In a couple of short, little, probably kept for too-futuristic for some time, conference papers[1,2] wrote around the end of the 80s encompass most of the concepts that will be revisited in the decade occurring about ten years later his work. The most relevant one, that I would like to recall here, is that, in principle, and with the proper knowledge of the turbulence, and assuming that the turbulence power is uniformly distributed along the vertical axis, the introduction of a DM can account for the same volume of turbulence individuated by a certain distance for the conjugation of it. In other words, in an MCAO with a certain Field of View, each DM can compensate turbulence up to, say, a certain distance from the DM. If this distance is, just to make an example, 5km, one DM can only compensate for this turbulence, and the residual one will deteriorate the whole image. This is the concept of Ground Layer Adaptive Optics. If you place two DMs and one is conjugated to the ground, while the other is conjugated to 10km, under the same assumptions of the previous example, you can achieve compensation till 15km. Also, this makes clear that the larger the number of DMs, the larger Field of View you can provide is (almost linearly, if you are forced to place one DM conjugated to the ground). It is interesting to see a theory of a complicated thing like MCAO not using one of that odious power like 6/5, or 11/3 so much common in this field. The rest of the development of MCAO involves so much people and it is impossible for me to make a fair review of all the persons that made a development into this realm, also because of mine and mine co-workers involvement. So, I will stitch to the concept of Layer Oriented (LO) for a while. The reader interested to the technical details of course should read the related[3–5] but the important thing I would like to make clear here is that, as in MCAO the new thing is to make more than one DMs coupled optically with different "heights" in the atmosphere, Layer Oriented is the one that physically (optically) makes this also for the WaveFront Sensing part. Conventional WaveFront Sensing of several stars and "fusion" of these signals into a computer makes, virtually, the same approach, but this is achieved inside a WaveFront computer, with the advantages of flexibility and the disadvantages of Signal to Noise Ratios (just to mention a couple of things, and not the less controversial, or at least the ones for which, depending upon the type of sensing, coupling, or algorithm, things can dramatically change). In my poor and maybe isolated opinion, moreover, simulations, in the realm of MCAO, becomes more and more difficult or, better, the results from simulations are increasingly more difficult to interpret, just because of the massively augmented degree of freedom of the problem. This means, just to be a little more explicit, that the number of assumptions one have to make are much more, and that, as for a journalist writing a piece on politics, choosing this or that assumption, model, distribution, number and/or position of stars, explicit or implicit assumptions, the results can be made to vary in a rather large span of figures. It can happen that, based on a (detailed) couples of simulations based on a long chain of (realistic) assumptions, a result is obtained and that, also for some sociological behavior of the Astronomical community, a certain result is given for granted or for "typical" while, maybe this is not the case. As this paper is dealing with "my" experience on dealing with turbulence in MCAO I can say we spent a lot of time and efforts into developing a simulation code using real stars distribution as actually observed in the sky using input catalogues as we were interested into assessing the sky coverage of this and that technique. Then all of these figures can be varied by factors of "several" just because of assumptions on turbulence, or on the spectral sensitivity of detectors, and a lot of other factors, making the efforts to refine the "second digit" almost useless. Anyway, given I was outlining my personal view on the development of MCAO I think a good series of event happens at the turn of the century (wow... I always wanted to write a phrase like that...) with basically two projects taking place: namely MAD at European Southern Observatory (ESO) and the MCAO at GEMINI. It is remarkable, in that time, the first open-loop measurements on the sky,[6] made with the TNG telescope from the Canary Island sky, using an asterism suggested as a nice gem to observe in an eyepiece of an amateur astronomer's telescope in the magazine "Sky and Telescope". This piece, have shown for the first time with direct hand-on measurement in the sky, that MCAO was possible. Not that really anyone working into it doubted, but still it has been an useful thing to push toward the situation of today. The two projects I mention developed under completely different paths. ESO developed a MCAO Demonstrator (that explain the MAD acronym) that turned to reality through the efforts of Enrico Marchetti and co-workers,[7] while the GEMINI system[8] run under the visions of Francois

Rigaut and, at least for most of its initial phase, Brent Ellerbroek. MAD was supposed to be without any specific dedication to achieving this or that limiting magnitude or sky coverage, and was supposed to be used for purely technical purpose with no science in mind at all. I vividly recall a discussion in which I was the only one claiming we should focus, at least partially on making science with it. As usually happens in life, then it turned out that science becomes the main focus of ESO on MAD and in fact most of its impact on the Astronomical community is surely depending upon this. So, in fact, I made an implicit step in my narration and we arrived to nowadays where MAD has been successfully used on the sky several times.[9] The Layer-Oriented wavefront sensing has been operated on the sky as well, although in a completely independent run.[10] I still think Layer Oriented is a more efficient and "cheaper" (for example in terms of number of pixels to be used in the wavefront sensor) that the conventional sensing of individual stars, but I do not want to take this opportunity to stress this even more here.

3. MCAO (LO), Finally on the Sky

And so, after almost twenty years after the "vision" of Jacques Beckers and seven years after having got proof that MCAO was doable on a night telescope we had a run of nine contiguous night with the Layer oriented (LO) wavefront sensor onboard MAD. The first three nights were devoted to technical assessment of the performance on the sky. Efforts to make this useful to science as well as by product failed as it was required to use narrow-band filters only in order to avoid troubles in the determination of the Strehl in the observed field. So, the total number of useful night for science purposes is actually six. We observed targets ranging from pulsars around isolated stars,[11] to Globular Clusters,[12] to extragalactic objects like nearby dwarf galaxies,[13] BL-Lac, high-z galaxies, QSOs and their jets. In most of these cases we get data useful enough to produce a peer-reviewed scientific paper. No one wants to compare these efforts, produced by a handful of volunteer, with respect to the European-wide efforts made in three runs by ESO to furthermore exploit the capabilities of MAD. The important point here, is just one, and common: MCAO works, and it can produce science at the forefront of knowledge. It offers an unprecedent quality over a relatively large Field of View and, given that at NIR wavelength the diffraction limited size matches the one of HST, I have finally see combined images HST (visible) with MAD (NIR) in which the comparison between the two is not clearly in favor of the space telescope. It is remarkable that the ratio in size of HST with the VLT is so close of the one of JWST with next generation ELTs. Instead, I would like to focus on different aspects. For example, again, as in the beginning of this paper I just mention that adding a further variable the judgment and assessment of an MCAO system becomes more depending upon different assumptions, the same can be said for science capability. There has been endless discussions and an uncountable number of documents in which it is stressed that one imperative for MCAO is uniformity of the quality. In spite of this, I have seen exceptional, outstanding published results obtained with PSFs whose Strehl varies of a large factor over the Field of View of the camera. Also, the quality of the results are still very depending upon the initial turbulence conditions. And, finally, and maybe the most important thing, I still believe it is questionable if the results of the run, or of any run, of MAD in the sky can be used to make a definitive assessment of the sky coverage attainable with solely Natural Guide Stars. Any answer, in fact, will be plagues upon the assumption of the condition of the turbulence in the nights when the run actually took place. Why such a crude statement??? If the result of an Adaptive Optics system would solely depends upon, say, the seeing, you could "place" the result of the MCAO-MAD run in the context of the (well) known statistic of achievable seeing. So that if you get tantalizing results into night of exceptionally good seeing you could question its usefulness, while if the same results are obtained into night of poor or average, in statistical sense, seeing, you can place a big crown on such a technique. However, as I mentioned earlier, this is not so simple for MCAO as the number of variables in the game are much larger. Some of them are simply not known, and some are not used in that particular MCAO experiment. I mention sometime the Taylor hypothesis and clearly if one would be in the position to use it to "integrate" for a much longer exposure time the wavefront sensing measurement (please note that there is no mention of the word "prediction" in such a statement) the limiting magnitude can be conveniently augmented. In principle, the only limit to this is given, eventually, by the ratio of the size of the telescope diameter with the Frieds parameter of the focused layer, a number that could easily exceeds one order of magnitude. MAD uses solely Natural Guide Stars (NGS), so that all the technical and fundamental troubles affecting Laser Guide Stars (LGS) are ignored implicitly. However the other MCAO instrument that is going to see the light in the sky soon, the MCAO system at GEMINI, heavily rely on these artificial references. In this moment I can only states that it will be extremely interesting to see the classes of results and of performances that will be obtained. For the moment, as I have been generally optimist about the usability of NGSs instead of LGSs, it can only be seen as suspicious the statement I am doing, that among the three 8m class telescopes where LGSs are routinely usable, the results and the efficiency have been widely different. A subtle indication that, furthermore, the realm of variables playing in this difficult game is even more large, in spite of the common thought that LGSs will make the use of MCAO more uniform and seamless. It is worthless to say that we are all looking forward to see the outcome of the GEMINI system in the very near future.

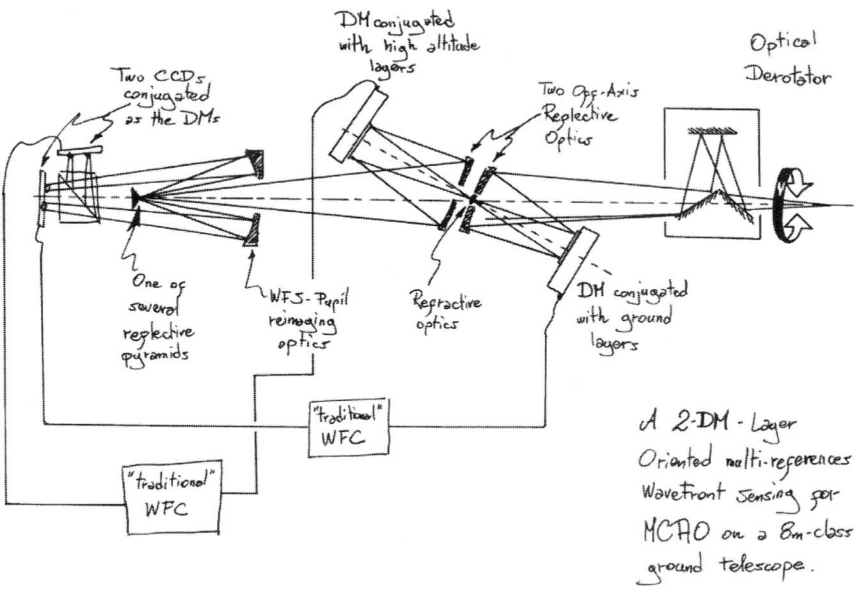

Fig. 1. A sketch of the MAD-LO "optical design", about 1999.

References

1. J. M. Beckers, Detailed compensation of atmospheric seeing using multiconjugate adaptive optics, in *Active Telescope Systems*, 1989.
2. J. M. Beckers, Increasing the size of the isoplanatic patch with multiconjugate adaptive optics, in *ESO Conference on Very Large Telescopes and their Instrumentation*, Octtober 1988.
3. R. Ragazzoni, Adaptive optics for giant telescopes: NGS vs. LGS, in *Proceedings of the Backaskog workshop on extremely large telescopes*, 2000.

4. R. Ragazzoni, J. Farinato and E. Marchetti, Adaptive optics for 100-m-class telescopes: new challenges require new solutions, in *Adaptive Optical Systems Technology*, ed. P. L. Wizinowich, Proc. SPIE, Vol. 4007July 2000.
5. R. Ragazzoni, E. Diolaiti, J. Farinato, E. Fedrigo, E. Marchetti, M. Tordi and D. Kirkman, *A&A* (2002).
6. R. Ragazzoni, E. Marchetti and G. Valente, *Nature* **403**, 54(January 2000).
7. E. Marchetti, N. N. Hubin, E. Fedrigo, J. Brynnel, B. Delabre, R. Donaldson, F. Franza, R. Conan, M. Le Louarn, C. Cavadore, A. Balestra, D. Baade, J.-L. Lizon, R. Gilmozzi, G. J. Monnet, R. Ragazzoni, C. Arcidiacono, A. Baruffolo, E. Diolaiti, J. Farinato, E. Vernet-Viard, D. J. Butler, S. Hippler and A. Amorin, MAD the ESO multi-conjugate adaptive optics demonstrator, in *Adaptive Optical System Technologies II*, eds. P. L. Wizinowich and D. Bonaccini, Proc. SPIE, Vol. 4839February 2003.
8. B. L. Ellerbroek, F. J. Rigaut, B. J. Bauman, C. Boyer, S. L. Browne, R. A. Buchroeder, J. W. Catone, P. Clark, C. d'Orgeville, D. T. Gavel, G. Herriot, M. R. Hunten, E. James, E. J. Kibblewhite, I. T. McKinnie, J. T. Murray, D. Rabaud, L. K. Saddlemyer, J. Sebag, J. Stillburn, J. M. Telle and J.-P. Veran, Multiconjugate adaptive optics for Gemini-South, in *Adaptive Optical System Technologies II*, eds. P. L. Wizinowich and D. Bonaccini, Proc. of SPIE, Vol. 4839February 2003.
9. E. Marchetti, R. Brast, B. Delabre, R. Donaldson, F. Fedrigo, C. Frank, H. Hubin, J. Kolb, J.-L. Lizon, S. Oberti, R. Reiss, C. Soenke, S. Tordo, A. Baruffolo, P. Bagnara, A. Amorim and J. Lima, MAD on sky results in star-oriented mode, Proc. SPIE Vol. 70152008.
10. C. Arcidiacono, M. Lombini, R. Ragazzoni, J. Farinato, E. Diolaiti, A. Baruffolo, P. Bagnara, G. Gentile, L. Schreiber, E. Marchetti, J. Kolb, S. Tordo, R. Donaldson, C. Soenke, S. Oberti, E. Fedrigo, E. Vernet and N. Hubin, Layer oriented wavefront sensor for MAD on sky operations, in *Adaptive Optics Systems*, Proc. of SPIE Vol. 7015July 2008.
11. R. P. Mignani, R. Falomo, A. Moretti, A. Treves, R. Turolla, N. Sartore, S. Zane, C. Arcidiacono, M. Lombini, J. Farinato, A. Baruffolo, R. Ragazzoni and E. Marchetti, *A&A* (2008).
12. A. Moretti, G. Piotto, C. Arcidiacono, A. P. Milone, R. Ragazzoni, R. Falomo, J. Farinato, L. R. Bedin, J. Anderson, A. Sarajedini, A. Baruffolo, E. Diolaiti, M. Lombini, R. Brast, R. Donaldson, J. Kolb, E. Marchetti and S. Tordo, *A&A* **493**, 539(January 2009).
13. M. Gullieuszik, L. Greggio, E. V. Held, A. Moretti, C. Arcidiacono, P. Bagnara, A. Baruffolo, E. Diolaiti, R. Falomo, J. Farinato, M. Lombini, R. Ragazzoni, R. Brast, R. Donaldson, J. Kolb, E. Marchetti and S. Tordo, *A&A* **483**, L5(May 2008).

PUPIL SLICING ADAPTIVE OPTICS: MAKING EXTREMELY LARGE TELESCOPES DIFFRACTION LIMITED AT SHORT WAVELENGTHS

JACQUES M. BECKERS
Unaffiliated
Scottsdale, AZ 85255, USA
jbeckers@cox.net

Extremely Large Telescopes (ELTs) will have aperture diameters up to 42 meters. Adaptive Optics (AO) at short wavelengths (< 1 micron) will be very hard to implement at these wavelengths because of the limited number of actuators on state-of-the-art deformable mirrors and because of the limited brightness of Laser Guide Stars (LGSs). For 1 arcsec seeing at 500 nm wavelength deformable mirrors (DMs) with about 150000 actuators will be needed and LGSs of a brightness of V = 8 to 9. That exceeds our present capabilities by a factor of about 100 and 2 magnitudes respectively. One might expect both to improve with time. We propose to combine the techniques of "pupil slicing" and AO to sharpen the telescope images at short wavelengths to the size of Airy disk of the pupil slices. I refer to this technique as "Pupil Slicing Adaptive Optics" or PSAO. At 500 nm wavelength that would correspond to the Airy disk of an approximately 5 meter diameter aperture, or a FWHM of 0.02 arcsec. As DMs increase in their number of actuators, the size of the pupil slices increases thus improving the angular resolution. Ultimately the full angular resolution of, for example, a 42 meter aperture would be reached (0.0024 arcsec at 500 nm). Of course, this does not resolve the issue of the limited brightness of LGSs. For it one has to wait for more powerful lasers and the development of perspective elongation correction techniques. Alternatively one would accept limited sky coverage (0.1%) when using natural guide stars (NGSs). Particularly interesting is the PSAO technique for high resolution spectroscopy where the smaller image sizes even for many slices results in a significant decrease in spectrograph dimensions.

I refer to the paper titled "Very high-resolution spectroscopy for extremely large telescopes using pupil slicing and adaptive optics" by myself and Torben E. Andersen & Mette Owner-Petersen for a detailed description of the PSAO concept and its application to very high resolution spectroscopy. It has been published in *Optics Express* **15**, No.5, 2007.

WHY IS THE VLT VERY EFFICIENT?

F. COMERÓN

European Southern Observatory
Karl-Schwarzschild-Str.2
D-85748 Garching bei München, Germany
E-mail: fcomeron@eso.org

The operations model of the ESO Very Large Telescope (VLT) heavily relies on a full-scale implementation of Service Mode observing. In this contribution we review the main features of ESO's approach to Service Mode at the VLT, we outline the advantages offered by this mode, and the challenges faced when implementing it given the wide diversity of instrumentation and instrument modes currently available at the VLT and the VLT Interferometer (VLTI). We give special emphasis to the part of this challenge directly derived from the evolution of the atmospheric conditions, which drive the short-term scheduling of the different scientific programmes competing for the available time.

Keywords: Very Large Telescope; observatory operations; service mode; flexible scheduling; queue scheduling.

1. The Very Large Telescope: Overview

Almost ten years after the beginning of operations of its first 8.2m unit, the ESO Very Large Telescope (VLT) is a great science machine in full production [1]. The VLT complex is located on Cerro Paranal, a mountain of 2635 m altitude in the Atacama desert in northern Chile. It includes four identical Unit Telescopes (UTs) that can operate either independently from each other, or combining their light beams coherently to work as an interferometer. In the latter mode, known as the VLT Interferometer (VLTI), the UTs can be complemented with four movable 1.8m Auxiliary Telescopes (ATs) in order to provide additional baselines (see Fig. 1). At the time of writing this, the UTs host 11 permanently mounted instruments that cover a broad range of observing techniques (imaging; single, multi-object, and integral field spectroscopy; imaging polarimetry and spectropolarimetry; time-resolved imaging and spectroscopy; adaptive optics-assisted imaging, spectroscopy and polarimetry...) over a broad range of wavelengths (0.3 μm to 20 μm), offering spectral resolutions of up to $\sim 10^5$ between 0.3 μm and 5μm in the visible and infrared. Two additional VLTI instruments operate in the near and thermal infrared, also offering spectroscopic capabilities, and a third interferometric instrument facility offering phase-referenced astrometry is currently undergoing commissioning. A second generation of instruments has already begun installation at the VLT and will be completed over the next few years, offering enhanced observing possibilities. Second generation VLTI instruments are also being planned.

Fig. 1. A view of the VLT facility showing the enclosures of the four 8.2m Unit Telescopes. The four Auxiliary Telescopes appear in the foreground.

Science operations of the VLT observatory directly involve approximately 80 staff members, including operations astronomers, user support astronomers, telescope and instrument operators, support scientists, data handling operators, and quality control scientists. Depending on their functions, the different operation groups are based either at the observatory site on Cerro Paranal or at the ESO headquarters in Garching bei München (Germany).

The VLT has a large number of users worldwide, most of which are affiliated with European institutes in the fourteen ESO member countries. For the semester extending between April and September 2008, 702 VLT or VLTI observing runs corresponding to programmes submitted by 507 different principal investigators have been scheduled for execution. The community of VLT users is by no means a closed one: of the 507 principal investigators quoted above, 169 are in this category for the first time, implying a healthy rate of new users every semester. The scientific productivity leads to a remarkable average rate of nearly two refereed papers based on VLT observations being published every day of the year. All scientific observations obtained with the VLT are stored in the ESO Science Archive Facility. Over 2,500 hours of science observations are performed each semester, with a high level of efficiency: the weather downtime on Paranal is below 7%, and the technical downtime is even lower being currently below 3%.

2. Service Mode and the VLT Operations Model

The high scientific efficiency of the VLT is largely due to the excellent quality of its facilities and instrumentation and to the high competence of the staff devoted to its operation and maintenance, but also to an innovative end-to-end operations model that heavily relies on devoting a major fraction of the available time (over 60%) to operations in Service Mode, also known in other facilities as flexible or queue scheduling; see [2] for further details and references to its implementation at other facilities.

In Service Mode, astronomers are requested to specify not only their list of targets, instrumental setups to be used and exposure times, but also the environmental conditions (sky transparency, seeing, lunar illumination...) that allow the science goals to be reached. The requested time is then allocated but without reference to a fixed set of dates defined long in advance. Instead, the time in which the observations composing a given programme are executed is normally spread over a whole semester, as the execution takes place only when the conditions are as good as specified by the astronomer or better.

The VLT operations model was designed from the outset to include a large fraction (at least 50%) of Service Mode observing, as this has always been recognized as a key element in its scientific productivity and efficiency. Its main advantages have been described in some detail elsewhere [2], and can be summarized as follows:

- Adaptation to the external conditions, allowing the operations astronomers in charge of deciding the nightly sequence of observations to ensure a good match between the actual conditions and those requested by the different programmes that may be executed.
- The best observing conditions go to projects that really need them, rather than to programmes that were scheduled long in advance on a time slot that offered much better conditions than those needed.

- Top rated projects get the highest execution priority, because the scientific merit as judged by the time allocation committee is used as a criterion to decide which observations to execute when the prevailing conditions allow the execution of more than one programme.
- Some science cases require execution exclusively under periods of excellent conditions so that their goals can be fulfilled. Typical examples are deep fields, whose detection limits depend on the average image quality during execution, and where excellent image quality is frequently required to yield the intended scientific results. Such programmes would be either impossible or extremely inefficient to execute in the classical mode of preallocated dates, as the suitable conditions would be seldom encountered.
- Long monitoring series (months, years) become easily manageable by being executed by observatory staff, instead of requiring the presence of the astronomer who requested the observations.
- Target of Opportunity programmes, requiring the observation at short notice of unpredictable events, are less disruptive on operations since their impact can often be minimized and ultimately charged to the lowest priority programmes.
- Time is made available to programmes that can make use of worse-than-average conditions. This is particularly useful for projects of rather risky or experimental nature, or those that do not have a very high scientific priority but still can produce useful results.
- It is possible to continue executing scientific programmes in case of instrument malfunctions, since 2-3 instruments are available at the VLT at short notice and there is usually a pool of programmes to execute at each of them at any given time.
- The time spent on calibration observations is optimized, since calibrations can be shared among several programmes and the short-term schedule can be prepared so as to maximize the number of observations executed on a given night that use a restricted number of calibrations.
- Finally, the science archive is populated with data obtained under well defined procedures, calibrated through a uniform calibration plan.

As can be seen from this list, several of the advantages of observing in Service Mode are directly linked to the properties of the atmosphere and to the ability of the observatory to adapt to them and optimally exploit the existing conditions at any given time.

The VLT end-to-end operations model in Service Mode starts with a sophisticated long-term scheduling that takes into account scientific priority, programme needs, statistical expectation on the conditions to be encountered, and many other factors. Next, fully pre-defined observations to be executed for the programmes selected for execution during the semester are collected, optimized, and validated. At execution time, a number of science operations policies are enforced with the general goal of enhancing the overall efficiency of the observatory, even if they may occasionally impact the efficiency with which specific programmes are executed. Many heterogeneous factors are taken into account in the short-term schedule, such as scientific priorities, execution conditions constraints, time criticality, need to minimize the number of instrument changes, proximity of targets in the sky, etc. At the same time, a well established calibration plan that ensures the accuracy of physical quantities derived from the observations down to a certain quantitative level is executed. Finally, as already mentioned, all observations are stored in the ESO archive, not only to provide other scientists the possibility of reusing the data for new research, but also as a valuable resource for operations to track the performance of the facility when needed.

The acceptance of Service Mode among the ESO community is high, as demonstrated both by the steadily increasing demand with respect to Visitor Mode since the beginning of VLT operations (Fig. 2) and by the feedback directly received from users via a dedicated questionnaire that probes in detail multiple aspects of the end-to-end model from a users' perspective [3].

3. Dealing with a Changing Atmosphere

Some of the complexity in the management of a full scale implementation of Service Mode observing arises from the multiple constraints that define the suitable conditions for the execution of an observing programme, and also from the fact that those constraints vary from one instrument or instrument mode to another. As noted earlier, current VLT instrumentation covers a wavelength range from 0.3 mum to 20 μm. The seeing is crucial for the quality of the observations at the shorter wavelength end, whereas it is irrelevant at the long wavelength end where the telescope always operates at the diffraction limit. Conversely, the precipitable water vapor column has rahter marginal importance for most applications at visible wavelengths, but it is essential in determining the feasibility of thermal infrared observations.

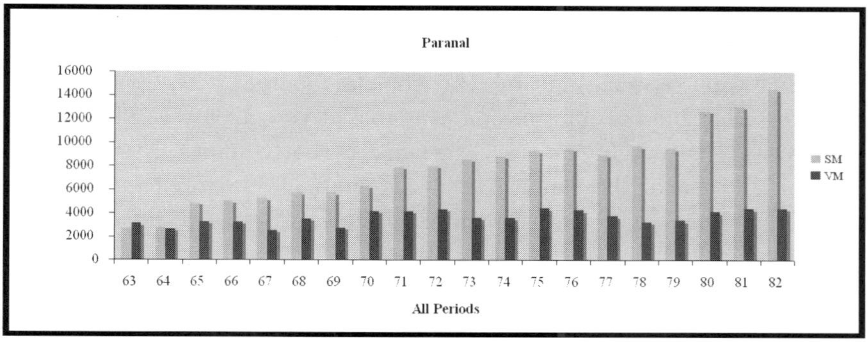

Fig. 2. Evolution of the demand of observing time in Service (light bars) and Visitor (dark bars) modes since the beginning of VLT operations. Period 63 was the first in which the VLT was offered, starting in April 1999, whereas Period 82 started in October 2008. Despite the increase from one to four telescopes, the time requested in Visitor Mode has kept nearly constant for ten years, whereas the time requested in Service Mode has steadily increased. From an initial ratio slightly in favor of Visitor Mode, the demand of time in Service Mode currently exceeds that in Visitor Mode by a factor of almost 3.

Most of the instruments currently available at the VLT operate under natural seeing conditions, but an increasing number (NACO, SINFONI, CRIRES, in addition to the VLTI) are assisted by Adaptive Optics (AO), using either natural or laser guide stars. The balance will shift even more towards AO-assisted instruments in the future with the second generation instruments, which will include the extreme AO instrument SPHERE designed to directly detect extrasolar planets.

The availability of AO instrumentation at the VLT, and particularly of a laser guide star facility, has stressed the increased operational complexity that such instrumentation implies. Not only is the image quality delivered by AO instrumentation more sensitive to the pattern of atmospheric turbulence than for seeing-limited observations; in addition, the longer system setup times involved by the tuning of the AO system introduce additional operational overheads and a potential leak of time useful for science observations if the conditions turn out not to be suitable for the planned observations once they have been executed. Besides this, successful operation of the laser guide facility at the VLT has turned out to be more sensitive to the atmosphere transparency than anticipated. Furthermore, the highly successful recent experimentation with Multi-Conjugate Adaptive Optics (MCAO) recently carried out at the VLT [4] has also demonstrated further operational constraints specific to this mode. All these factors show that advanced, real-time equipment for the characterization of atmospheric turbulence and its forecast represents an important benefit for the operation of current, and especially future astronomical facilities.

The rather limited information available at present at the observatory in this respect leads to some well recognized shortcomings. No prediction of the evolution of the conditions is systematically factored in the short term scheduling decisions, which are mainly based on current conditions and some ad-hoc forecasting based on historical knowledge and the experience of the operations astronomer. The impact of observations that turn out to be executed outside constraints is contained by limiting the length of individual observations to approximately one hour, in order to allow the operations astronomer to retain the necessary flexibility to adapt to changing conditions. However, this often leads to fragmentation of observations and to an increase of the operational overhead that could be reduced if a reliable forecast of the conditions on a timescale of a few hours were available. We note however the recent improvements in understanding of the seeing on Paranal made possible by combining sets of measurements obtained with a variety of instrumentation, including the image quality measured at the instrument focal plane [5].

The uncertainty in the short-term evolution of the atmosphere leads to two short-term scheduling strategies, which are combined in different proportions by the science operations staff in charge of their execution. A 'risky' strategy leads to the start of the execution of observations when the user-specified constraints are very close to the current conditions. The advantage is that this strategy makes use of the best possible conditions and gives chances of execution to the most demanding programmes, but at the cost of normally decreasing efficiency by requiring frequent repetition of observations that happened to be outside constraints. Conversely, a 'conservative' strategy consists of starting the execution of observations whose constraints are significantly less demanding than the current conditions, thus leading to the observation being completed within constraints even if the conditions deteriorate substantially during the execution. While this second strategy results in a higher likelihood that the observation are completed within constraints and thus do not need repetition, it decreases the likelihood of execution of the most demanding programmes, and may withhold the achievement of the best scientific performance that the facility can give.

The trade-off between both strategies leads to a sizeable fraction of observations, estimated to reach $\sim 20\%$ of the available science time on

Paranal, needing repetition because conditions went outside constraints during the execution. This is thus a significant source of inefficiency that can only be improved through a better ability to forecast the evolution of conditions in the short term.

4. Conclusions

The end-to-end operations model of the VLT, evolved through nearly ten years of operations, has established a paradigm in the operation of major ground-based facilities and constitutes an essential reference for the planning of the operations of future facilities, including those operating in very different wavelength domains like ALMA. It is nevertheless important to wonder where the next steps in further increasing the efficiency of the facility may lie.

Can atmosphere prediction tools further help VLT operations? In the classical Visitor Mode observing, weather prediction tools are a useful resource for the observer but hardly a decisive factor in the scientific outcome of an observing run. Nevertheless, in Service Mode, which fundamentally relies on the ability to adapt to the external conditions, such tools can provide a boost of the efficiency by enabling a more robust short-term scheduling and a reduction of the overhead caused by observations needing repetition, even for those programmes having the most stringent demands on the conditions. The implementation of atmosphere characterization and prediction tools, and their integration in the observation scheduling process, probably represents the next major step in increasing the efficiency of the VLT and of other forefront ground-based astronomical facilities.

Acknowledgements

I am pleased to thank my colleagues Francesca Primas and Stéphane Marteau for providing material for this paper.

References

1. de Zeeuw, T., 2008, The Messenger, 132, 2.
2. Comerón, F., 2004, in Organizations and Strategies in Astronomy, vol. 5, ed. A. Heck, Kluwer Acad. Publ.
3. Primas, F., Marteau, S., Hainaut, O., Mathys, G., Romaniello, M., Sterzik, M., 2008, The Messenger, 131, 36.
4. Diolaiti, E., et al., 2008, SPIE, 7015, 21.
5. Sarazin, M., Melnick, J., Navarrete, J., Lombardi, G., 2008, The Messenger, 132, 11.

FUTURE-LOOK SCIENCE OPERATIONS FOR THE LBT

RICHARD F. GREEN
LBT Observatory, 933 N. Cherry, Tucson, AZ 85721-0065, USA
Email: rgreen@as.arizona.edu

The reconfigurable optics of LBT provides considerable versatility for observers to choose instruments and modes in near real time. Current values and predictions about both cloud cover and turbulence will become a critical factor in those choices, as the suite of available instruments expands with time. Two modes of operation are envisioned, both of which involve changes of focal station in response to conditions. One is partner-observer queue mode for facility instruments, which includes natural guide star adaptive optics. In that case, queue planning will use predictions and actual seeing to accommodate those programs requiring the best conditions. The other is a block-scheduling mode for new or complex capabilities, such as interferometry and early days of laser guide star operations. In that mode, site information will be essential for shifting to backup programs on nights that prove to be unsuitable for stable fringe tracking or reliable laser projection.

1. Description of the Large Binocular Telescope

The Large Binocular Telescope (LBT) contains two 8.4-meter diameter borosilicate honeycomb primary mirrors on a stiff common mount. The two mirrors provide equivalent light-gathering power of a single 11.8-meter diameter primary. Since they are spaced at 14.4 meters center-to-center, the maximum baseline is 22.65m, yielding angular resolution of 10 milliarcseconds in the J band. The telescope has a pair of prime foci at f/1.14 and a number of Gregorian focal stations at f/15 [1].

The unique aspect of the telescope is that compensation for atmospheric turbulence will be achieved by a pair of adaptive secondary mirrors [2]. The active surface on each is a thin shell of Zerodur glass, 91 cm in diameter and 1.6 mm thick. The shape and orientation is controlled with a bandwidth of 1 KHz by 672 voice coil magnets glued to the back surface. The advantage of this configuration is that it avoids the introduction of warm re-imaging optics into the beam, which add thermal noise in the near-IR where the biggest benefit accrues from adaptive correction of turbulence.

The LBT is situated at an elevation of 3200m on Mt. Graham in southeastern Arizona. The site is favorable orographically, because it drops off steeply in the direction of the prevailing wind during the most stable atmospheric conditions. The median site-delivered seeing is <0.7", and the fraction of usable hours is >70%. The telescope achieved first light with one prime focus camera in October 2005. Partner astronomers are currently using the binocular prime focus two weeks per month for scientific programs, while the first Gregorian instrument is in commissioning.

2. LBT Instrument Complement

The LBT will support three categories of instruments. There are three pairs of facility instruments that co-point to obtain complementary information of a single field: the Large Binocular Cameras (LBC) at the prime foci; the Multi-Object Double Spectrographs (MODS) for optical work at the direct Gregorian foci; and two LUCIFERs, with near-IR cameras, long-slit and multi-object spectrographs. There are two strategic instruments that require a more substantial effort for optimization and development of routine operation. They are two coherent beam combining Fizeau interferometers: the LBT Interferometer working from 3-5 and 10-20 microns, and LINC-NIRVANA, working from 1-2.5 microns. Finally, there is a PI instrument, PEPSI, the Potsdam Echelle Polarimetric and Spectrocopic Instrument, that is a high-stability, high-resolution, fiber-fed bench spectrograph.

2.1. *LBC*

The Large Binocular Cameras are a pair of 36 Mpixel CCD mosaic imagers, with 27'x27' fields of view [3]. One camera is optimized for blue and the other for red, achieving higher throughput than single instruments panchromatic over the full optical range. The design provides for offset guiding and continuous measurement of low-order optical distortions for sub-1-Hz active optics adjustment of the primary mirror focus, collimation, and shape. The pixel sampling is optimized for a range around median seeing.

2.2. *LUCIFER*

The instrument currently in commissioning at the bent Gregorian focus after its first light is LUCIFER 1, the near-IR camera and spectrograph [4]. It covers 0.9-2.5 μm with five observing modes: seeing-limited imaging, long-slit and multi-object spectroscopy, and diffraction-limited imaging and spectroscopy. Its

4'x4' field of view feeds a 2Kx2K Hawaii-2 HgCdTe array, with 27 filters and 3 gratings. It has 10 fixed slit masks, and cold exchangeable cassettes that hold 23 custom-cut multi-slit masks. An internal jukebox mechanism exchanges focal plane masks, and the user has a choice of 3 cameras for appropriate image pixel sampling. The diffraction-limited image scale is set for the performance of the natural guide star AO system under good conditions (with high delivered Strehl ratios).

2.3. *MODS*

This pair of double spectrographs will enable three observing modes: long-slit, multi-slit, and direct imaging. The spectral range is 0.31-1 μm over a 6'x6' field of view. The blue and red channels are independently optimized for high throughput. Custom slit masks can be loaded into a 24-slot cassette for mechanized exchange. To maintain the high throughput, the number of folds was minimized; the long structure contains a closed-loop flexure compensation system to maintain resolution during long exposures. Delivery from Ohio State is expected in mid-2009 [5]. The image scale is set for very good natural seeing conditions; performance would benefit from the potential upgrade to rapid tip/tilt correction for far-red observations.

2.4. *PEPSI*

Although classified as a PI instrument, PEPSI is being built to facility standard for long-term high-resolution spectroscopic investigations [6]. It is a fiber-fed echelle spectrograph on a 6m x 2m optical table in a chamber controlled for absolute temperature and pressure in the base of the telescope pier. It contains a 20x80cm R4 echelle working in double-pass Littrow with resolutions of 120,000 and 300,000. It has separate visual and red arms, each with 3 VPH cross-dispersers, and covers 0.39-1.05 microns in 3 exposures. Polarizers will yield 4 spectra per echelle order, projected onto 10.6Kx10.6K CCDs. The goal is S/N=10 in 1 hour at R=120,000 for a star with V=20.1 mag. Large fiber diameters yield high throughput in good conditions and make this the go-to instrument in marginal conditions.

2.5. *LBTI*

The LBT Interferometer is a thermal IR beam combiner and nulling interferometer. It is optimized for 8-13 μm and 3-5 μm with two channels integrated into the camera system [7.8]. It has a cryogenic beam combiner with two additional Fizeau ports for future expansion. The anticipated sensitivity is 100 μJy, sufficient to image hot Jupiters and zodiacal disks, and the ultimate field of view with uniform diffraction-limited resolution is 40". The nulling capability is anticipated to provide suppression of starlight of $\sim 10^4$.

2.6. *LINC-NIRVANA*

This Fizeau beam combining imager covers the range of 1-2.4 μm with a pixel scale of 5.1 milliarcseconds and field of view of 10" feeding a 2Kx2K Hawaii-2 detector. The paradigm is a warm optical bench, with expected sensitivity of K~26 mag at 5σ in one hour. It will sharpen the image of the phase reference star and target through a multi-conjugate adaptive optics system, to heights of 100m and 8-14 km [9]. This instrument is likely to perform best under the most stable atmospheric conditions for solid control of differential atmospheric piston.

3. Partner Queue Mode Observing and Prediction of Conditions

The versatility of configuring the optics for the LBT allows the use of more than one facility instrument pair on a given night. In principle, all 3 pairs will be available for use at any time, with a reconfiguration time of about 20 minutes [10]. In practice, calibration and other overheads are likely to constrain the use to only two per night.

The partner institutions have chosen a relatively lean operations mode compared to some other large-aperture telescope observatories. Therefore, the partner astronomers share the responsibility for the success of their scientific programs with the LBTO support staff. Each major partner institution selects their own complement of observing programs for each semester with no subsequent merging time allocation process. LBTO for now schedules their observing time in partner blocks of six nights per lunation. A partner observing team comes to the site for their scheduled time with a queue of programs from their entire cadre of proposers. The observers' challenge is to match their choice of instrument and configuration to conditions: moonlight, clouds, seeing, and stability. The more information they can have in advance, the more advantage they can take of their choices for each night.

The most reliable set of predictions available – rise, set, and phase of the moon – will dictate the shift from a darktime instrument, such as the optical imager or faint object spectrograph, to the near-IR, far red, or optical high-dispersion spectroscopy. Although some primary observations can only be acquired when the moon is down, the distinction of such programs is becoming less relevant. Part of the issue is demand for the most competitive capabilities.

Another is the impact of moonlight on the S/N ratio for secondary sources, such as natural guide stars for optical wavefront correction or interferometric phase reference. In the end, the highest ranked programs will command the best conditions, regardless of instrument choice.

The next level of accuracy in prediction is cloud cover. Truly cloudless, photometric conditions may obtain 40-50% of the time. Much of the marginal time includes thin cirrus, which can accompany stable, low turbulence conditions. Cirrus will dim both the science target and accompanying reference stars, and will make laser guide star projection difficult. Thicker cloud conditions rule out all but the brightest natural guide stars for AO (or even telescope tip/tilt correction). Such clouds can be accompanied by higher turbulence in any case, making the result of image quality correction less dramatic. These conditions would drive observers to work on bright stars, probably with high-dispersion spectroscopy. An anecdotal oddity of sites in the American Southwest (and Chile) is a brief interval of extremely low turbulence just prior to the arrival of a frontal system. The site visual seeing can be <0.25" FWHM for one to a few hours. Configuration and calibration require a choice by mid-afternoon for the coming evening. Current weather forecasts are adequate to give the overall picture of low and high pressure systems with reasonable accuracy for the coming twelve to fifteen hours. It is obviously more difficult to predict conditions when the mountain summit is on the edge of an instability for clouds to form. In that case, however, it is an issue of opening or closing, with the underlying choice of instrumentation still likely to be valid.

The most challenging prediction is that for the best conditions, in terms of the lowest turbulence and greatest stability. Such conditions will give the best limiting performance of the adaptive optics system and the interferometers. At the present time, there is no formal modeling effort or measured prediction of the turbulence profile above Mt. Graham on any given night. A Differential Image Motion Monitor is now being commissioned; time has been scheduled for cross correlating its measurements with those of the Generalized SCIDAR campaign ongoing at the Vatican Telescope [11]. We also expect delivery of a MASS from Andrei Tokovinin at CTIO within the next year. When they are both in routine service, they will provide a real-time measure of the turbulence profile and the resulting seeing. The AO system will yield an estimate of the end result as well. It will be commissioned with a sufficient amount of programmed AI that it will optimize its gain and other settings for each observation. Based on an initial one- to two-minute sampling to determine r_0 and τ_0, it will select the number of modes to correct and the update rate given the

S/N ratio of the natural guide star. To achieve the highest Strehl ratios from the AO system, we will need to determine the conditions in near real-time. Those nights will be most promising that have the best initial measures from the MASS and DIMM, the most stable thermal condition of the telescope, and on which the AO system measures of r_0 and τ_0 are consistently good.

4. Laser Guide Star Operations

As seen in these proceedings, much of the power in the C_n^2 profile above Mt. Graham is in the ground layer within 200m of the summit [11]. The LBT consortium is planning to exploit that fact by producing a laser guide star system to correct for the ground layer turbulence [12]. It will consist of a pair of 3-beam Rayleigh laser constellations, which is modeled to produce routinely a factor of 2-3 improvement in image concentration over the full 4'x4' field of view of LUCIFER. The resulting median image quality is likely to be <0.25" FWHM in the K-band at 2 μm. The goal is implementation on the telescope in 2011. The addition of a low-power sodium laser to each constellation would provide an upgrade path to achieving high Strehl ratio on axis, and reconfiguration of the wavefront sensing would enable tomographic AO. Operation of the system will require reasonable transparency, but will be effective for ground layer correction over a wide range of conditions of atmospheric turbulence. That is fortunate, because it will need to be block scheduled initially until operations can be made routine.

5. Interferometry

As discussed above, the two interferometers will exploit good conditions to combine the beams from the two primaries coherently. LBTI will work in thermal infrared. Single aperture imaging at 10 μm achieves relatively high Strehl ratios with tip/tilt correction alone. Coherent beam combination is, of course, much more challenging. The effectiveness of the Fizeau combination with LBTI will depend on low mechanical vibrations and local thermal stability of the telescope, along with atmospheric piston variations that are well within the control bandwidth. A challenge for the developers of the telescope control system is to maintain co-pointing and co-collimation of the two channels with (near) zero differential piston in the light paths. Then the question becomes identifying conditions in which the atmospheric differential piston is low and relatively stable. Research will be required to correlate those stable conditions with other externally measurable quantities, such as the turbulence profile derived from the MASS.

LINC-NIRVANA must solve the same problem of coherent beam combination, but it's ten times more demanding in the wavelength range of 1-2 μμm. The Fizeau combination gives a 10" field of view with nearly uniform (elliptical) diffraction-limited response, making the LBT interferometers unique. The raw data signal from a point source is an Airy diffraction pattern with superposed Young's fringes. The central slice contains the highest spatial frequency information. The key will be to get a good signal from the faintest possible phase reference star in order to maximize sky coverage. For that purpose, LINC-NIRVANA will employ 2-3 layer multi-conjugate adaptive correction with multiple natural guide stars in a large patrol field.

Both interferometers provide contradictory demands on the schedule. These instruments are likely to achieve greatest success under the most stable conditions. At the same time, they are classified as "strategic instruments", because they will require expert operation by members of their teams. The best scheduling to optimize the match to conditions is in a queue; to optimize the availability of key personnel requires a pre-assigned block. For extended commissioning and team-supported scientific observing, block scheduling will certainly be the initial mode. In that case, a backup program will be an obvious requirement for conditions of insufficient stability. Rapid progress in early commissioning could be enhanced by starting in the statistically best times of year, which are roughly the months of May and October. Should investigation reveal reliable correlations between atmospheric piston stability and other site measures, such as the MASS turbulence profile plus prevailing wind direction, those measures will provide the start-of-night indication of suitability to proceed with interferometric observations.

6. Aspirations

LBTO is working to optimize the data return from a facility that relies on compensation of atmospheric distortions to approach the diffraction limit of a 23-m diameter telescope. Fortunately, the design allows versatility of optical configuration, so that instrument choice can be made on relatively short notice. At the same time, the availability of reliable predictions by mid-afternoon would add efficiency and observing hours for the coveted best conditions. In general, it is known that the long direction of the ground wind rose (SSW) is correlated with the most stable conditions, typically established by a high-pressure dome. Can additional inputs from temperature (and temperature profiles), the jet stream and particular location of the center of high pressure be correlated with turbulence profiles from continuing GS and MASS measurements to aid prediction? A combination of empirical correlations and microclimate modeling would be particularly powerful in generating statistical predictions. If such correlations were successful, predictors could be developed to identify which family of turbulence profiles and associated piston variations would obtain for the coming night. A telescope opening at sunset configured to take advantage of the conditions that are then actually found to prevail would be a very powerful tool for achieving the limiting science goals to which the partner astronomers aspire.

Acknowledgements

The LBT is an international collaboration among institutions in the United States, Italy and Germany. The LBT Corporation partners are:
* The University of Arizona on behalf of the Arizona university system
* Istituto Nazionale di Astrofisica, Italy
* LBT Beteiligungsgesellschaft, Germany, representing the Max Planck Society, the Astrophysical Institute Potsdam, and Heidelberg University
* The Ohio State University
* The Research Corporation, on behalf of The University of Notre Dame, University of Minnesota and University of Virginia

References

1. J. Hill, R. Green, J. Slagle, D. Ashby, et al., *Proc. SPIE,* **7012,** 701203 (2008).
2. A. Riccardi, M. Xompero, D. Zanotti, et al., *Proc. SPIE,* **7015,** 701512 (2008).
3. R. Speziali, A. Di Paola, E. Giallongo, et al., *Proc. SPIE,* **7014,** 70144T (2008).
4. H. Mandel, W. Seifert, R. Hofmann, et al., *Proc. SPIE,* **7014,** 70143S (2008).
5. R. Pogge, B. Atwood, S. Belville, et al., *Proc. SPIE,* **6269,** 62690I (2006).
6. K. Strassmeier, M. Woche, I. Ilyin, et al., *Proc. SPIE,* **7014,** 70140N (2008).
7. P. Hinz, T. Bippert-Plymate, et al., *Proc. SPIE,* **7013,** 701328 (2008).
8. J. Wilson, P. Hinz, M. Skrutskie, et al., *Proc. SPIE,* **7013,** 70133A (2008).
9. T. Herbst, R. Ragazzoni, et al., *Proc. SPIE,* **7013,** 701326 (2008).
10. R. Green, J. Hill, J. Slagle, J. Brynnel, et al., *Proc. SPIE,* **7016,** 70160A (2008).
11. E. Masciadri, J. Stoesz, F. Lascaux, S. Hagelin, S. Egner, these proceedings.
12. S. Rabien, N. Ageorges, R. Angel, et al., *Proc. SPIE,* **7015,** 701515 (2008).

PREPARING FOR OPERATIONS WITH THE EUROPEAN EXTREMELY LARGE TELESCOPE (E-ELT)

F. COMERÓN

European Southern Observatory
Karl-Schwarzschild-Str.2
D-85748 Garching bei München, Germany
E-mail: fcomeron@eso.org

The advent of the giant telescopes such as the 42-m European Extremely Large Telescope (E-ELT), expected before the end of the next decade, is bound to produce new paradigms in the operations of astronomical observatories. It will also produce further evolution of many novel concepts that have been successfully introduced in the operation of the current generation of 8/10m-class telescopes, mainly those related to flexible scheduling. The integration of adaptive optics capabilities in the design of the E-ELT and the high performance made possible by new technologies in astronomical instrumentation have enabled very ambitious science cases that push the capabilities of the telescope and its instruments to the limits. However, the exploitation of such capabilities will require a careful planning of science operations aiming to fully optimize the use of the available observing time and the way in which scientists will interact with the facility. The characterization and short-term forecast of optical turbulence are anticipated to play a very important role in optimizing E-ELT operations. We review the most important features of science operations foreseen at the E-ELT, some requirements linked to the variety of adaptive optics modalities that will be available at the telescope and its instruments, and how the current design of the E-ELT facility and its operations model intends to respond to these needs.

Keywords: European Extremely Large Telescope; adaptive optics; observatory operations; flexible scheduling.

1. The European Extremely Large Telescope

The next generation of giant optical telescopes is being developed through several worldwide projects that aim at building facilities of unprecedented scale and power, with the expectation of entering into operations by the end of the next decade. Together with facilities under construction like the Atacama Large Millimetre Array (ALMA) and the James Webb Space

Telescope (JWST), the Extremely Large Telescopes (ELTs) represent a quantum step in astronomical technology.

The European ELT (E-ELT) projected by ESO is Europe's response to this challenge, and is the most ambitious project currently under planning. The baseline design of the E-ELT consists of a fully steerable telescope with a segmented primary mirror of 42m diameter at the core of a 5-mirror anastigmatic configuration, correcting astigmatism, coma and spherical aberration over a large field of view several arcminutes across. The telescope is designed to allow observations from the blue to the thermal infrared, aiming for a complement of instruments covering a large parameter space in terms of wavelength and spectral resolution. The telescope design includes two Nasmyth platforms able to simultaneously host several instruments, two gravity invariant foci fed by a vertical light beam, and a coudé focus (Fig. 1). The location of the E-ELT observatory is still to be decided at the time of writing this.

The E-ELT is designed from the outset as an adaptive telescope. The fourth mirror in the five-mirror configuration is deformable, and the fifth corrects for tip-tilt. In this way, the telescope itself provides ground layer adaptive optics (GLAO), using three natural and four laser guide stars. Multi-Conjugate Adaptive Optics (MCAO) and Laser Tomography AO (LTAO) can be provided by post-focal AO modules, whereas Multi-Object Adaptive Optics (MOAO) and Extreme Adaptive Optics (XAO) are capabilities integrated in some of the proposed instruments.

In addition to the post-focal AO modules, eight prospective instruments are currently undergoing preliminary design. They cover a wide diversity of modes, including multi-integral field unit spectroscopy, high resolution spectroscopy both in the visible and near-infrared, wide field imaging, high contrast imaging, mid-infrared imaging and spectroscopy, and visible multi-object spectroscopy. Combined, these instruments propose to make use of MOAO, GLAO, MCAO, LTAO, XAO, with only one of them (CODEX, the high resolution, single object spectrograph) having seeing-limited operation as a possible baseline. It is doubtful in view of the enormous investment of resources needed to build all those proposed instruments that all of them become a reality, at least in the early years of the E-ELT. However, a wide range of wavelengths, instrument techniques and AO modes can be guaranteed even if only a few of them are built.

The E-ELT project is currently in advanced Phase B (detailed design) status, with the goal of having a proposal for construction ready by mid 2010 enabling the beginning of regular operations in 2018. The estimated cost of the facility reaches nearly 1,000 million Euro.

Fig. 1. The concept of the European Extremely Large Telescope (E-ELT) showing the location of the primary mirror and the two Nasmyth platforms, each supporting several instruments. The structure raising from the center of the primary mirror is the adaptive relay unit, holding a large deformable mirror and an also large mirror for tip-tilt correction, which sends the light to either Nasmyth platform. The light beam can also be relayed to the gravity invariant foci located under each Nasmyth platform, or to a coudé focus located below the telescope structure.

2. Using the E-ELT

The E-ELT is designed to be a versatile facility able to execute a very broad range of science programmes, using a wide variety of observing techniques. Its science case has been fed with the input of a large community [1], and is reflected in the individual cases of each instrument undergoing design studies. The E-ELT is therefore expected to be used by a large community of mostly European users, probably similar in size or even larger than that having access to the VLT or to ALMA.

The operations model of the E-ELT must carefully take into account the need to make it available to all the community able to scientifically exploit it. This requires that use of the E-ELT does not bocome confined to a few groups having an expert level of technical knowledge, but that the requirement on the latter is kept to a level similar to that of other world-class facilities. In that respect, the E-ELT should be "just one more" among the front line of ground- and space-based facilities that astronomers will be using by the end of the next decade, even if one with unique capabilities, based on similar operations principles and tools.

Due to these requirements, it is anticipated that the E-ELT will execute observations for a number of programmes most of the nights, rather than being blocked for the execution of a single major programme for a long time span. Another powerful reason for such approach is the dependency of AO performance on the atmospheric turbulence profile and transparency, which require a short-term selection of the observations that can deliver scientifically useful results under the prevailing conditions.

Although the E-ELT should be sufficiently accessible to allow any scientists having a strong science case to fully define the observations and strategy needed to carry it out, it should also permit the access to advanced functions needed to fine-tune telescope and instruments to exploit their technical capabilities to the edge. Such capabilities are expected to be required by some of the most ambitious science cases, such as those aiming at the direct detection of terrestrial planets around nearby stars. Operations planning must ensure that the requirements of each observing programme are properly captured, that appropriate observing strategies can be designed, and that the execution of observations can proceed according to such requirements and strategies.

The need to execute normally several programmes on any given night, combining different instrument, instrument modes, and AO modes, and adjusting on a short timescale (typically below one hour) to the changing external conditions, strongly advocates for an extensive use of Service Mode observing similar to that currently implemented at the VLT, which is reviewed elsewhere in this volume. In fact, the experience acquired in operating the VLT for the last ten years naturally leads ESO to take it as the starting point for the planning of E-ELT operations. There are no elements in the E-ELT baseline reference design that preclude the application of such a model for the E-ELT, particularly concerning fundamental technical specifications such as preset time, switch times between instruments in stand-by mode, or expected setup times for the AO systems.

3. Scheduling of the E-ELT

At the VLT the four Unit Telescopes make twelve foci (two Nasmyth and one Cassegrain) simultaneously available, thus making it possible to have instruments permanently mounted at their respective foci. The situation will be in principle different at the E-ELT, where not all instruments may be permanently mounted. However, the number of focal stations is relatively large. Each adapter-rotator can host up to three instruments, or two instruments plus a post-focal AO module, or it can allow the optical beam to be relayed to a gravity invariant focus or to the coudé focus. In this way, up to seven instruments plus two AO modules can be hosted at any given time. However, not all focal stations are suitable for any of the instruments. Due to their requirements on volume, weight or stiffness some instruments are being designed to be installed at one particular focal station. For this reason, and given that the number and selection of instruments that will be actually built is not yet known, it is not possible to determine at the present moment whether or not change of instruments at the focal stations will be necessary, or with which frequency. Given the size and complexity of the instruments, their exchange is expected to be a rather major activity to be carefully planned in view of its impact on operations.

Operational efficiency requires a sufficient number of instruments to be in stand-by at any given time so that the programmes to be executed by them cover as broad a range of acceptable conditions as possible, and so that they are be put in operation in a matter of minutes should the changing conditions require it. These operational requirements on the suite of available instruments can be summarized in three conditions:

- Guaranteeing suitable scientific exploitation of most atmosphere conditions, so that there are always programmes that can deliver their expected results under them. This will require the simultaneous availability at short notice of as many instrument and AO modes as possible, as well as a fast switch time among them to adapt to changing conditions.
- Guaranteeing the scientific exploitation of the best conditions, ensuring that whenever rare conditions required by the execution of some top-rated science programmes occur, the most suitable instrument to exploit them is available. Should there be a large enough number of them so that instrument changes become necessary as described above, this condition may imply that some instruments that rely on the best performance of the telescope under excellent

conditions may have to be permanently mounted. Extreme AO instruments are likely to fall within this category.
- No downtime regardless of conditions, so that whenever the telescope can operate there is an instrument able to produce useful scientific results. The large number of focal stations and multi-instrument adapters virtually ensure that this will be the case, the limitation being rather in the number and capabilities of the actual instruments that will be built.

If instruments need to be exchanged due to their number exceeding that of suitable focal stations, long-term operations planning will be needed to ensure that all instruments receive the share of time needed by the scientific programmes to be executed. A similarity of this requirement with the current situation in Paranal operations can be found in the VLT interferometer (VLTI), where VLTI time slots are carefully programmed based on the demand for VLTI time, but also for time with each of the UTs separately. Such planning will determine the schedule for transportation of instruments to/from the Nasmyth platforms.

Needless to say, the fulfillment of the conditions listed above requires not only instrument availability, but also a suitable pool of programmes whose execution requirements properly sample the range of conditions expected at the observatory site. This ultimately depends on the response of the community to the calls for proposals. However, under the very reasonable assumption that the E-ELT will be a heavily oversubscribed facility, producing the proper balance of programmes required for such sampling will need a sophisticated long-term scheduling that takes into account the statistical distribution of each parameter characterizing the external conditions. This is done already nowadays to a large extent at the ESO telescopes [2], where a complex scheduling algorithm is used to produce the long-term schedule. Some of the intricacies of the current algorithm, which simultaneously schedules the Unit Telescopes at the VLT together with the VLTI, will not apply to the single-telescope E-ELT. However, some of the parameters used to match the statistics of expected conditions to the requirements of approved programmes will need to be represented with greater detail. This applies in particular to the use of the atmospheric turbulence profiles as an input, since the distribution of seeing alone (one of the parameters used in the VLT scheduling) will not be sufficient to decide the fraction of time useable by the different modalities of AO.

Similarly, friendliness of the conditions to the different AO modalities will be an important component of the short-term scheduling of the E-ELT.

The choice of the most suitable programmes will be more complex, and will need more aids, than at the VLT. In addition, one of the essential questions to wonder at the time of scheduling the nightly sequence of observations will be along the terms, "Given the current turbulence profile and its expected evolution over the next hours, what is the right AO mode to use in the next observation?"

4. Observing Modes

As noted in Sect. 2, the combination of programmes with different environmental requirements and the proper exploitation of the unique capabilities of the E-ELT and its instrumentation, particularly the variety of AO modalities, demands large flexibility in the short-term scheduling and imposes Service Mode as the baseline observing mode, executing most (or perhaps even all) of the programmes in this mode.

At the current stage of definition of the science case for the E-ELT, a rationale can be clearly identified for a small fraction of programmes whose execution would greatly benefit from the possibility for the science user of the facility to carry out real-time decisions influencing the execution of the remainder of the observing programme. Programmes obviously in this category are target-of-opportunity ones, in which rapid reaction to a developing unpredictable event is needed in order to optimize the use of the time over which the target evolves. It is expected that the Design Reference Science Plan, to be built on the basis of the response by the community at large to a questionnaire recently circulated, will expand on the range of programmes in such category. It might also produce an estimate of the fraction of science time at the facility in which real-time decisions will be needed, with a strong caveat on the accuracy that such a projection can have a decade in advance in such rapidly evolving fields.

At the VLT and other facilities, the need for real-time decisions is handled by offering the use of the facility in Visitor Mode, in which members of the science team travel to the observatory to execute the observations in preallocated dates. In general this solution will not be applicable to the E-ELT, unless programmes to be executed in this mode have little sensitivity to AO performance. Otherwise, there would be a significant chance that a programme must be executed under conditions that prevent the achievement of its scientific goals. Given the cost of time at the E-ELT and the anticipated level of competition for it, this outcome cannot be considered acceptable.

To preserve the possibility of defining the course of execution of a programme in real time if needed, without compromising the advantages offered by queue scheduling, ESO is considering implementing flexible remote observing as an additional mode. In this scenario, users will be warned at short notice (possibly hours in advance) when the forecasted conditions are likely to be suitable for the execution of their programmes. The expectation is that users can obtain access in the meantime to equipment set up to use observation definition tools and to receive pipeline-processed data. Despite that large data volumes expected from most E-ELT instruments, network infrastructure in 10 years from now should make fast communication between a remotely located observer and the observatory feasible. Another implicit expectation in setting up this model is that the approximate forecast of the turbulence profile at the observatory will be possible, at least to the extent of predicting the degree of suitability of the expected conditions for different AO modalities, on a timescale of ~ 10 or more hours.

In view of the expected availability of Service and Remote Mode observing, which will in principle offer all the necessary advantages in terms of flexibility and interactivity, we may wonder whether astronomers will ever need to travel to the E-ELT site to use it. The Design Science Reference Plan should help answer this question by sampling a wide range of science cases and possibly identifying some where such presence is required, but the rationale for Visitor Mode at the E-ELT is not established at the moment on scientific grounds. However, the E-ELT users community can benefit from getting first hand information on how the E-ELT operates, an experience that has been recognized as useful at the VLT based on the feedback received from visiting astronomers. Furthermore, the E-ELT will be partly operated by active research astronomers for whom personal interaction with their colleagues in the community is an important component in keeping a healthy scientific activity. Taking these considerations together (see for instance [3]), maintaining a certain level of Visitor Mode observing reserved for programmes that do not depend critically on the atmospheric conditions can be considered as a strategic interest to keep the observatory and its community of users in close contact.

5. User Tools

The definition of the user toolkit is an important part of the planning for efficient operations of a major facility, as it largely determines the success of the intent stated in Sect. 2 of keeping the facility accessible with unnecessary complexity hidden from the end user. ESO's planned approach to this challenge for the E-ELT is to evolve its VLT data flow system, which includes the observation handling system tools, to a new generation able to serve the E-ELT needs as well. However, the extensive use of AO in different modalities requires the introduction of new elements and some emphasis on others that will become more critical at the time of planning observations. Among these, system (telescope plus instrument) simulators fed with atmosphere models as realistic as possible will be needed to accurately predict the data quality and to decide on the setup of different instrument elements.

Given the large choice of parameters to be tuned when designing observations in the most advanced AO modes, it seems clear that both simulation and observation preparation tools will need to handle two different levels of complexity. A first one should be addressed to common users who require standard performance of the facility, for whom a predefined set of default parameters should suffice to reach the scientific objectives. The design of tools at this level should aim at hiding complexity as much as possible from users, presenting the design and preparation of E-ELT observations as a feasible task requiring a minimum of specialization. On the other hand, an expert user level should be available giving access to a greater number of instrument and telescope parameters, addressed to users with specialized knowledge on AO observing and instrument capabilities or whose science cases require the optimal exploitation of the system performance.

The characterization of optical turbulence will play an important role in the definition of these tools, and in the identification of the key parameters to which the user will need to have access. The availability of advanced AO modes raises the question of how to specify the constraints under which observations can be executed. Scientific requirements such as on-axis Strehl ratio, energy fraction in the point spread function wings, contrast at a given distance from a bright source, or corrected field translate into constraints on the turbulence profile or system setup decisions such as the choice of guide stars or the spatial frequency of wavefront sampling, which suitable simulation tools will need to provide.

6. Concluding Remarks

The operation of a fully adaptive telescope like the E-ELT opens up new challenges in the operation of a large ground-based facility. Many of these challenges derive from the need to adapt to a changing atmosphere, and efficiency will greatly depend on being able to properly model and characterize it. There is therefore an extensive ground for collaboration between meteorology and astronomy in order to optimize observatory operations and, ultimately, to obtain the best scientific results from the facilities being planned.

On the other hand, many new operation concepts and tools can be already developed and tested at existing large facilities such as the VLT making use of the acquired experience and new technological possibilities. Among the latter, atmosphere characterization and forecasting already hold the promise of enhancing the productivity of existing observatories, as explained in the chapter on VLT operations elsewhere in this volume. The continued evolution of current observatory operation models and the incorporation of new tools and capabilities imply that the planning of operations of future facilities does not attempt to answer questions such as "how would we operate the E-ELT if we had it today?", but rather "how will we operate major astronomical facilities in a decade from now?".

References

1. Hook, I. (ed.), 2005, "The science case for the European Extremely Large Telescope", OPTICON/ESO.
2. Alves, J., 2005, The Messenger, 119, 20.
3. Comerón, F., 2004, in Organizations and Strategies in Astronomy, vol. 5, ed. A. Heck, Kluwer Acad. Publ.

EVALUATION OF THE GFS AND MM5 METEOROLOGICAL MODELS FOR PARANAL, MACON AND PACHON ASTRONOMICAL ZONES

A. CHACÓN*, O. CUEVAS and M. CURÉ

AstroMeteorology group, Physics and Astronomy Department, University of Valparaíso, Gran Bretaña 1111, Casilla 5030, Valparaíso, Chile
**E-mail: arlettedfa.uv.cl*
www.dfa.uv.cl/astromet

Paranal is located in the north zone of Chile, which is considered a privileged zones for astronomical observation for his desert conditions, not much cloudiness and low light contamination. These same condition are present in the Pachón summit where is located the Gemini South observatory. Similar features have the zone of Macón in Salta region in Argentina. This place is one of the sites pre-selected for the construction of the E-ELT. The behavior of the atmospheric conditions is important within the astronomical observation because knowing his future state it would be possible program the instruments for the night of observation. To do this we use meteorological forecasts models to simulate the behavior of the atmosphere. The period in study was year 2005, where the atmosphere conditions were simulated with GFS and MM5 models. In order to perform the analysis of the simulation in those three sites, the meteorological variables data was interpolated using geographic coordinates of each weather station and compare them with the measured meteorological variables. In this report we evaluate the temperature and in the future we will evaluate: the relative humidity, atmospheric pressure, direction and intensity of the wind. In the first part of the analysis we found that the forecast of MM5 is better than GFS model, as expected. In the second part of the analysis a Kalman filter has been implemented to minimize the systematic errors of the simulation of GFS and MM5.

1. Introduction

For the astronomy, it is very important to know the behavior of the atmosphere, because depending of his state we can select the appropriate instrument for the observation night. To forecast the future states of it, we use a numerical weather prediction model, which show the evolution of a set of meteorological variables (temperature, pressure, humidity, etc) by solving the physical equations that describe the atmosphere.

If a wrong forecast of the meteorological condition is made to astronomical sites using a not validated/verified model, it could mean a considerable loss of resources. Therefore the evaluation of the predictions of any numerical weather prediction models is very important for astronomical centers.

This research evaluated two numerical weather prediction models: Global Forecast System (GFS) model with global scale (resolution > 100 Km) and the mesoscale model MM5 (resolution<100 Km). Those models are tools to support the atmospheric forecast and the information they give is considerable, so the fields of study we selected is focused to the following (five) meteorological variables: temperature, atmospheric pressure, relative humidity, direction and intensity of the wind.

The sites were the models were evaluated are: Cerro Paranal (Fig. 1) located in II region of Chile and hold the most advanced and powerful astronomical complex of the planet, the Very Large Telescope (VLT) belonging to European Organization for Astronomical Research in the Southern Hemisphere (ESO). Cerro Pachón (Fig. 1) is located in IV region of Chile, at 10 Km of southeast of Cerro Tololo. In this place is the Gemini South telescope and Southern Astrophysical Research (SOAR) telescope [1]. Cerro Macón (Fig. 1) is located in northeastern zone of Argentina in Salta region and studies realized under the frame of selection of places for astronomical observation [2], have described this place as an optimal area for this activity. That reason this site has been pre-selected for the construction of European Extremely Large Telescope (E-ELT), by ESO.

This proceeding is organized as follows. A brief overview of GFS and MM5 models is presented in section 2. The data and methodology used are described in section 3. Detailed statistical results describing the behavior of forecast temperature by the two models presented in section 4. Finally, the concluding discussion is in section 5.

2. Models Overview

2.1. *GFS Model*

The GFS model is implemented by National Center of Environmental Prediction (NCEP), in United States.

This model represent the Earth in octants, giving to South America the octant P, which correspond 0° to -90° South latitude and -120° and 30° West longitude (Fig. 2). The model have a 1.25° of resolution between points (≅125 Km) and 13 verticals coordinates using pressure levels from 1000 hPa to 70 hPa. The output of the model is 4 times per day (00, 06, 12, 18 UTC[1]) with

[1] Coordinated Universal Time.

120 hours of forecast. The GFS model is support by the data of weather station, boats, meteorological buoys, among others.

Figure 1. Geographic map, shows the location of Cerro Paranal, Cerro Pachón and Cerro Macón.

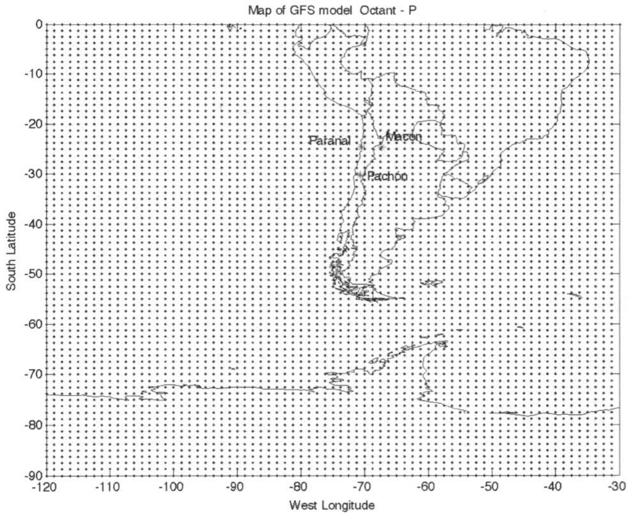

Figure 2. Octant P corresponding to South America at GFS model with the location of the three site in study.

2.2. *MM5 Model*

MM5 is a mesoscale numerical model developed by Pennsylvania State University (PSU) and the National Center of atmospheric Research (NCAR), giving technical/computing support in permanent form.

This model is a limited-area, non-hydrostatic, terrain-following sigma-coordinate model designed to simulate or predict mesoscale atmospheric circulation. The model is supported by several pre- and post-processing programs, which are referred to collectively as the MM5 modeling system.

The MM5 also have the option of nesting capabilities (Fig. 3), differing the GFS model, and works with different resolutions. This study the resolution used were:

Table 1. Resolution of the domains used in this study by MM5 model.

Domain	Cerro Paranal & Macón	Cerro Pachón
1	27 Km	81 Km
2	9 Km	27 Km
3	3 Km	9 Km
4	1 Km	3 Km

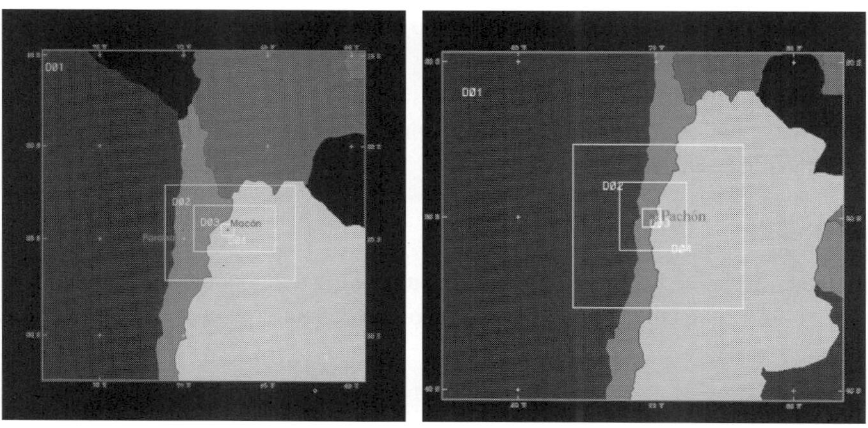

Figure 3. Nesting map used in this study by MM5 to Cerro Paranal, Cerro Macón (left) and Cerro Pachón (right).

The initial condition as well as lateral boundary condition to run this model, are given by the output of the GFS Model (Section 2.1).

3. Data and Methodology

3.1. *Data*

The period studied was the year 2005, for 2 output each night (00 y 06 UTC) for both models.

The data in-situ for the three sites was obtained using an automatic weather station (Table 2).

Table 2. Geographic coordinates and altitudes of the three sites on study.

Weather Station	Latitude	Longitude	Altitude
Cerro Paranal	24° 40' S	70° 25' W	2 635 m
Cerro Pachón	30° 14' S	70° 44' W	2738 m
Cerro Macón	24° 38' S	67 20' W	4 600 m

3.2. *Methodology*

1. Bilineal horizontal interpolation [3] using 4 near points for each site, to extract the variables from both models.
2. Spline cubic [3] for vertical interpolation using the point interpolated in item 1, in 13 pressure levels in GFS model and 30 sigma levels in MM5 model to have a meteorological variables in the altitude of sites.
3. To filter the noise for reducing any systematic errors involved in the models we use a Kalman Filter [4].
4. Statistical Analysis, lineal correlations (Cor) and cumulative distribution fuction (CDF) are calculated.

4. Results

In this proceeding we present only the results of the temperature for both models. The follow three tables shows the results of the data interpolated *without* Kalman Filter.

Table 3. These three tables are the correlation result using only the interpolation of point 1 and 2 describes in section 3.2.

Cerro Paranal/Hours	00 UTC (Cor)	06 UTC (Cor)
GFS	0.93	0.93
MM5-D1	0.91	0.91
MM5-D2	0.91	0.91

Cerro Macón/Hours	00 UTC (Cor)	06 UTC (Cor)	Cerro Pachón/Hours	00 UTC (Cor)	06 UTC (Cor)
GFS	0.90	0.92	GFS	0.93	0.92
MM5-D1	0.89	0.89	MM5-D1	0.93	0.96
MM5-D2	0.85	0.86	MM5-D2	0.94	0.96
MM5-D3	0.90	0.89	MM5-D3	0.94	0.96
MM5-D4	0.90	0.88	MM5-D4	0.94	0.96

The GFS model shows better correlation in Cerro Paranal than Cerro Macón. The highest correlation values are in domains 2 to 4 for Cerro Pachón.

The other three tables shows the results of the data interpolated *with* a Kalman Filter.

Table 4. This three tables are the correlation result using Kalman Filter.

Cerro Paranal/Hours	00 UTC (Cor-kal)	06 UTC (Cor-kal)
GFS	0.93	0.93
MM5-D1	0.93	0.93
MM5-D2	0.92	0.93

Cerro Macón/Hours	00 UTC (Cor-kal)	06 UTC (Cor-kal)	Cerro Pachón/Hours	00 UTC (Cor-kal)	06 UTC (Cor-kal)
GFS	0.92	0.93	GFS	0.93	0.95
MM5-D1	0.91	0.92	MM5-D1	0.88	0.97
MM5-D2	0.92	0.93	MM5-D2	0.89	0.97
MM5-D3	0.92	0.92	MM5-D3	0.89	0.97
MM5-D4	0.92	0.92	MM5-D4	0.88	0.97

The highest correlation values are at 06 UTC for Cerro Pachón. The correlation for Cerro Macón at 06 UTC, in average, goes up in a 4.5% respect at the value in the table 3. The results are better using Kalman filter in 06 UTC, the following graphics are the CDF at that time and with the filter.

The fig. 4 show the line of domain 2 of MM5 model is better than GFS model at Cerro Paranal. For domain 1 the 64% of the data have $\Delta T <= 1°C$, 90% of the data have $\Delta T <= 2°C$ and 96% of the data have a $\Delta T <= 3°C$.

The fig. 5 all the four domains of MM5 model exceeds the performance of the GFS model in Cerro Macón. The domain that was the best behavior was number 3 where the 22% of the data have $\Delta T <= 1°C$, 50% of the data have $\Delta T <= 2°C$ and 72% of the data have a $\Delta T <= 3°C$.

At Cerro Pachón the best performance was domain 4 from MM5 model (fig. 6) where the 63% of the data have $\Delta T <= 1°C$, 89% of the data have $\Delta T <= 2°C$ and 96% of the data have a $\Delta T <= 3°C$.

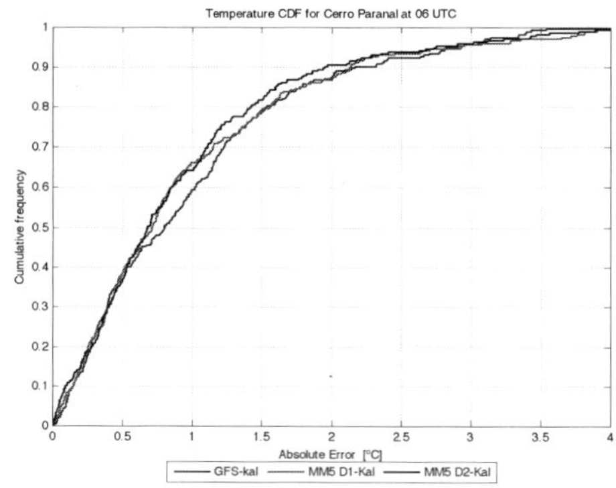

Figure 4. CDF graphic to Paranal at 06 UTC, this data are using the kalman filter.

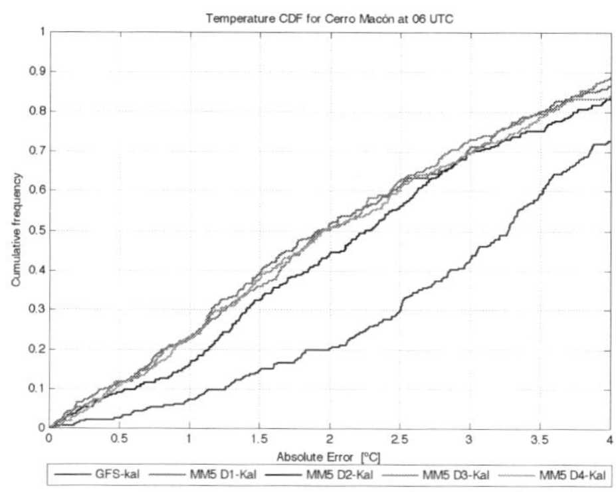

Figure 5. CDF graphic to Macón at 06 UTC, this data are using the kalman filter.

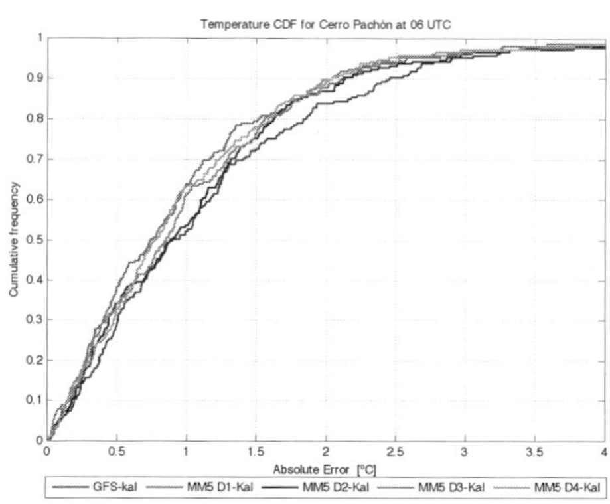

Figure 6. CDF graphic to Pachón at 06 UTC, this data are using the kalman filter.

5. Conclusions

The analysis shows that the MM5 model is better than GFS for forecasting the temperature, and the accuracy at 06 UTC is better than 00 UTC. This is because the MM5 model requires a certain period of time to stabilization of the dynamic equation used for the model in the different spatial resolutions employed.

In this study we apply a Kalman filter to minimize systematic errors that are always present in any numerical simulations. The best result we found was for Cerro Paranal and Pachón, where the 64% and 63% have a $\Delta T<=1°C$, respectively.

The complex terrain and high altitude of Cerro Macón could bring difficulties to the simulations for both models. The low performance of these models in this location does not implicate the simulations on synoptic scale behave similarly.

Acknowledgments

To Marc Sarazin and Julio Navarrete from ESO. Jean Rene Roy and Rubén Díaz from Gemini Observatories. Axel Osses and Pablo Saide from Universidad de Chile. Also to Peter Aniol from ASTELCO for his support of this work.

References

1. A. Tokovinin, T. Travovillon, 2006: Model of optical turbulence profile at Cerro Pachón. *Mon. Notice of the Royal Astron. Soc.*, **365**, 1235-1242.M.
2. Sarazin, E. Graham, M. Beniston and M. Riemer, 2002: The new tools for global servey of potencial sites for the future gigant telescopes. *SPIE conf. an astronomical telescope and instrumentation, Waikoloa, Hawaii, USA, paper SPIE-4840-54.*
3. W. Press, S. Teulkosky, W. Vettrling, B. Flannery, Numerical Recepes in Fotran, *Cambridge editorial*, **2º Edition**, 1992.
4. R. Benzi, M. Marrocu and A. Speranza 1996 in VLT-TRE-CRS-17443-0001. ESO Internal Report.

THE IMPACT OF SEISMICITY ON HIGH ANGULAR RESOLUTION ASTRONOMY: THE CASE OF THE CANARY ISLANDS

A. EFF-DARWICH

Dept. Edafología y Geología, Universidad de La Laguna, E-38205 Tenerife, Spain
E-mail: adarwich@ull.es

B. GARCÍA-LORENZO

Instituto de Astrofísica de Canarias, C/ Vía Láctea s/n, E-38205 Tenerife, Spain

L. BONATTO

Institute of Earth Sciences "Jaume Almera", CSIC, E-08028 Barcelona, Spain

L. HERNÁNDEZ-GUTIÉRREZ

Área de Laboratorios y Calidad de la Construcción, Consejería de Obras Públicas y Transportes del Gobierno de Canarias, Tenerife, Spain

R. VIÑAS

Dept. Edafología y Geología, Universidad de La Laguna, E-38205 Tenerife, Spain

J.A. RODRIGUEZ-LOSADA

Dept. Edafología y Geología, Universidad de La Laguna, E-38205 Tenerife, Spain

M.J. BLANCO

Centro Geofísico de Canarias, Instituto Geográfico Nacional, E-38001 Tenerife, Spain

C. MUÑOZ-TUÑÓN

Instituto de Astrofísica de Canarias, C/ Vía Láctea s/n, E-38205 Tenerife, Spain

Seismicity induces ground vertical and horizontal displacements that could affect the image quality obtained by telescopes in a similar fashion than atmospheric turbulence. In this work, we study the effect of local seismicity relative to atmospheric turbulence upon the image quality of astronomical observations at El Teide observatory and Roque de los

Muchachos observatory, both in the Canary Islands, Spain. Three different aspects of seismicity are studied, namely regional volcanism and seismicity (that is compared with other astronomical sites), seismic noise and possible resonances between seismic noise and the structure of telescopes.

Keywords: Site Testing – Telescopes.

1. Introduction

Some of the best astrophysical observatories in the world, namely the Canarian, Chilean and Hawaiian observatories, are located in active volcanic regions. This is not a coincidence, since topography modelled by volcanic activity is a main factor controlling the local atmospheric conditions and hence, the sky transparency that defines good astronomical sites. The Canary Islands host two astrophysical observatories in two different active volcanic islands: Observatorio del Teide in the island of Tenerife and Observatorio del Roque de los Muchachos in La Palma. Several volcanic eruptions have taken place in both islands, being the last ones in 1971 and 1909 for La Palma and Tenerife, respectively (Cabrera and Hernández-Pacheco, 1987[1]).

The characterization of astronomical sites is usually based on the analysis of the atmosphere above them (e.g. García-Lorenzo et al. 2007[2]). However, increasingly larger telescopes (10, 50 and 100 meters classes) demand stable structures and buildings, and hence geological activity will become an important parameter to take into account in astronomical site ranking. Seismic and/or volcanic hazards analysis must be considered in civil structural design. However, in the case of large telescopes, the structures have to withstand the effects associated to seismic and/or volcanic activity, but they also have to minimize the loss of operational time, recalling the extreme precision in the alignment of mechanical and optical components.

In this work, we present an analysis of seismic and volcanic risk at El Teide observatory on the island of Tenerife, and Roque de los Muchachos observatory on the island of La Palma, Canary Islands, Spain. The level of background seismic noise is also characterized and its effect on telescope operability is analyzed.

2. Seismic and Volcanic Hazard

The Canary Islands are located in a low seismic area. Largest recorded historical earthquake had a magnitude 5.2 Seismicity is mainly concentrated between the islands of Tenerife and Gran Canaria (Figure 1). The island of La Palma, hosting the Roque de los Muchachos Observatory, has very low seismic activity. Ground peak horizontal acceleration with a 10% of probability of being exceeded in 50 year corresponds to 0.06g for the eastern part of Tenerife (including the observatory) and 0.05g elsewhere (González de Vallejo et al. 2000[3]). These values are significantly lower than those expected at Paranal, 0.34g, and Mauna Kea, 0.4g (GSHAP[4]).

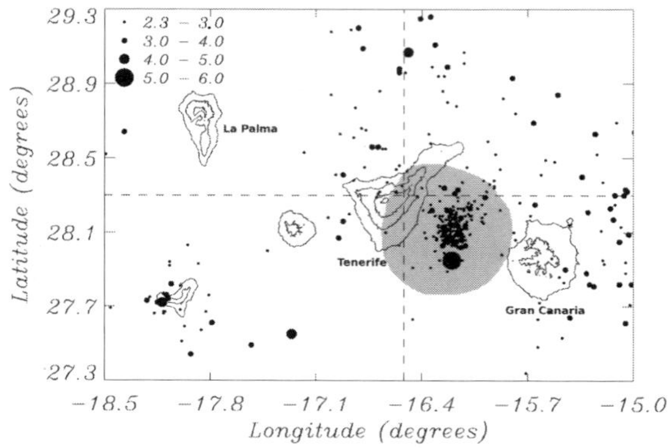

Figure 1. Seismicity in the Western Canaries for the period 1985-2008. Light grey area corresponds to the region where the ground peak horizontal acceleration with a 10% of probability of being exceeded in 50 year is 0.06g; elsewhere the 10% of probability of being exceeded in 50 year is 0.05g.

El Teide observatory has a very low probability of being affected by lava flows, since it is protected by topography from lava flows originated in the Teide-Pico Viejo volcanic complex. Volcanism closer to the Observatory is older than 30.000 years. Only two small eruptions at the beginning of the XVIII century that did not affect the area of the observatory (see Figure 2).

Some of the eruptions that have occurred in the area of Las Cañadas Caldera were accompanied by eruptive columns of ash. Ash may travel long distances depending on the wind conditions and fall on the ground. Thickness of the ash deposits could range from centimetres to several metres depending on the explosivity of the eruption and the distance travelled by the ash. A simulation of ash fall due to an eruption similar to the one that took place in Montaña Blanca approximately 2000 years ago (Araña et al. 2000[5]) is presented in Fig. 2. Hazard analysis revealed that El Teide observatory is located in a low probability area of being affected by ashfall.

Probabilities that the Roque de los Muchachos observatory could be affected by ashfall and/or lava flows are negligible, since volcanic activity ceased more than 1 Ma ago in the volcano the observatory is located on, and present day activity is located in the south of the island of La Palma (see Figure 2).

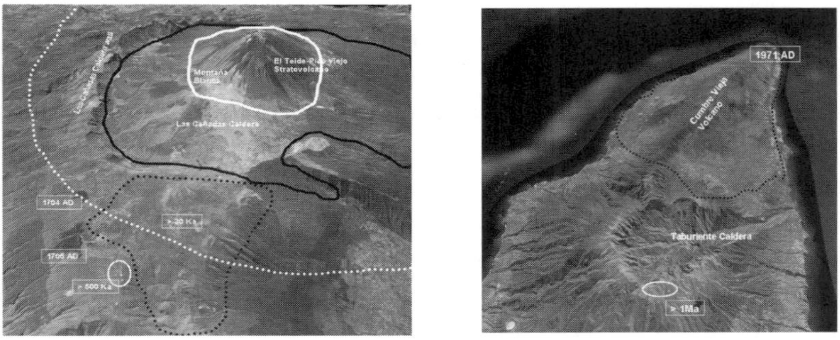

Figure 2. (left panel) Aerial view of central Tenerife taken from Google-Earth. The El Teide observatory lies within the white ellipse. The area enclosed by the black solid line corresponds to the region that has been affected by lava flows from El Teide-Pico Viejo Edifice. The area enclosed by the black dotted lines was affected by basaltic lava flows from the NE dorsal ridge. White solid and dotted lines encloses the area where there is 50% and 10%, respectively, probability that at least 10 centimetres of volcanic ash could fall after an eruption like the last one that occurred in central Tenerife. White squares show the estimated age of volcanic eruptive activity. (Right panel) Aerial view of central La Palma taken from Google-Earth. The Roque de los Muchachos observatory lies within the white ellipse. The area enclosed by the black dotted lines was affected by basaltic lava flows from the Cumbre Vieja Volcano. White squares show the estimated age of volcanic eruptive activity.

3. Seismic Noise

It has been widely observed that seismic records display the presence of continuous ground motion of variable amplitude. The source of seismic noise can be attributed to human activity, such as traffic, factories and engineering work construction; or to natural causes, namely oceanic and/or atmospheric. Human activity and wind are present in frequencies above 1Hz, whereas the ocean microseismicity lies below 1 Hz (around 0.2 Hz and 0.07 Hz), being more important in islands than in continental sites. Although seismic noise will not produce any damage on buildings, it is a continuous small vibration affecting structures. Seismic noise could be an effective tool to characterize the local soil effects to seismic loading and to analyze the coupling of telescope structures to the ground. For this purpose, a seismic station belonging to the Instituto Geográfico Nacional was installed at El Teide and Roque de los Muchachos observatories. This station is composed of a short period (1 sec), three-components Lennartz seismometer (LE-3D/1s). The acquisition unit was a Taurus (Nanometrics) with a high resolution 24-bit A/D converter, GPS time and a sampling rate of 100 mps. The station was emplaced for 30 minutes on three different locations, S1 to S3, along the two observatories (see Fig. 3). Data were analyzed in two different ways: on one hand the background seismic noise level was inferred through Fourier analyses; on the other hand the fundamental resonance frequency of the soil was computed through the Nakamura method (Nakamura, 1989[6]). Nakamura proposed a technique, which uses a horizontal-to-vertical (H/V) Fourier power spectral ratio of ambient seismic noise, measured at a single station. It has been well established that the H/V technique can effectively be applied to determine the dynamical characteristics of the underlying subsurface structure, since the H/V frequency peak (if present) represents the natural frequency of the site.

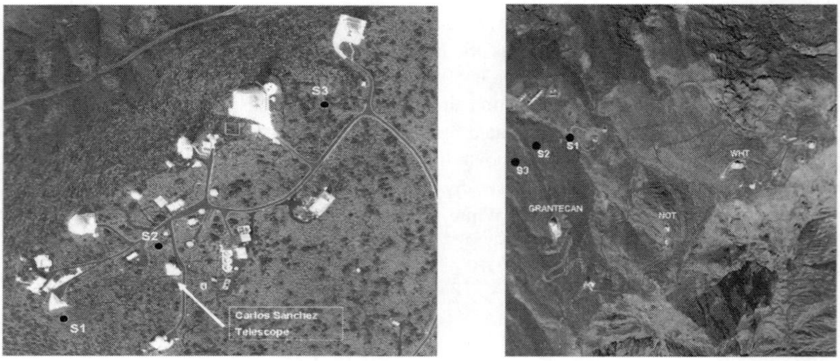

Figure 3. (Left panel) Aerial view of El Teide Observatory taken from Google-Earth. Black circles labeled as S1, S2 and S3 correspond to the locations where seismic noise measurements were carried out. (Right panel) Aerial view of Roque de los Muchachos observatory taken from Google-Earth. Black circles labeled as S1, S2 and S3 correspond to the locations where seismic noise measurements were carried out.

Ambient seismic noise power spectra calculated for sites S1, S2 and S3 are presented in Fig. 4. The distribution of power at the three sites is similar, in all cases lying between the Peterson's high and low global seismic noise power spectra (Peterson, 1993[7]). However, there is an increase in power at approximately 1Hz that is likely anthropogenic.

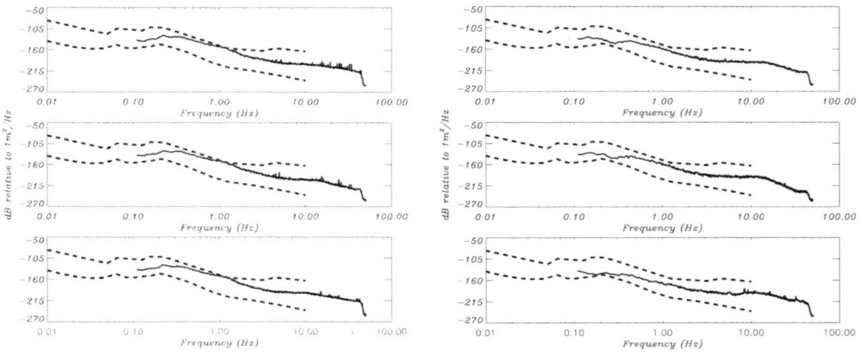

Figure 4. (Left panel) Power spectra of the displacement vertical component of the seismic signal measured at S1 (lower panel), S2 (central panel) and S3 (upper panel) at El Teide Observatory. Dashed lines correspond to the world-wide Peterson's high and low seismic noise levels. (Right panel) As in the left panel, but for Roque de los Muchachos observatory.

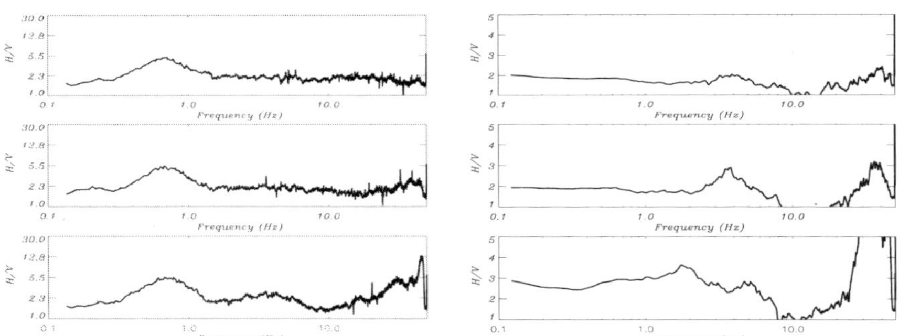

Figure 5. (Left panels) Resonance peak analysis through the Nakamura's H/V ratio of the seismic signal measured at S1 (lower panel), S2 (central panel) and S3 (upper panel) at El Teide Observatory. (Right panels) As in the left panels, but for Roque de los Muchachos observatory.

Nakamura's H/V methodology was also applied to the seismic data recorded at S1, S2 and S3, as illustrated in Fig. 5. The soil's main resonance frequency was found at 0.6 Hz for El Teide Observatory. This value is significantly different to the lower resonance mode of large telescopes (between 1 and 2 Hz). Hence we do not expect significant contributions of soil amplifications to the vibrations of the telescopes. The presence of resonance peaks at Roque de los Muchachos observatory is less clear and significant differences are found among the three monitored sites. Further measurements must be carried out.

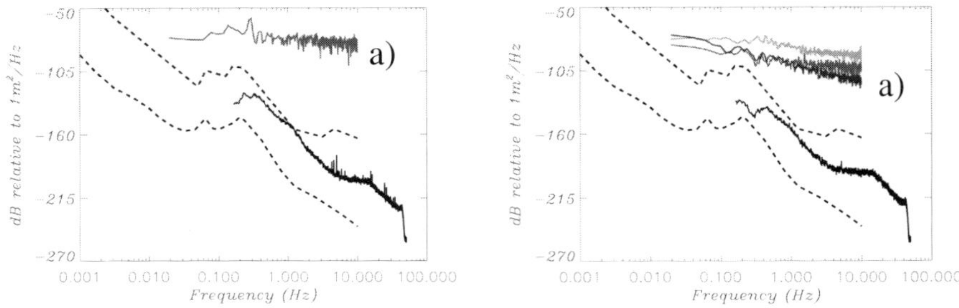

Figure 6. (Left panel) Power spectra of the seismic vertical component displacement measured at S2 at El Teide Observatory (black line). Dashed lines correspond to the Peterson's high and low seismic noise model. Solid line labelled as *a)*, corresponds to the power spectra of the variation in time of the most intense peak of the speckle patterns for a target star measured at Carlos Sanchez telescope. (Right panel) As in the left panel, but black line corresponds to S2 at Roque de los Muchachos Observatory. Solid lines labeled as *a*), correspond to the power spectra of the variation in time of the most intense peak of the speckle patterns for a target star measured at WHT, GTC and NOT telescopes, respectively.

A comparison was also carried out between the power spectra of displacement due to seismic noise and the linear displacement of a single point source, a star, along a CCD detector attached to a telescope. In this way, we wanted to compare seismic noise with the combined effect of telescope vibration and atmospheric turbulence. This could only be achieved if the CCD is capable of recording data at speeds comparable to that of a seismic station (tenths of measurements per second). In this sense, we installed FASTCAM (Oscoz et al. 2008[8]) at the 1.5 meters Carlos Sánchez Telescope at El Teide observatory and WHT, GTC and NOT telescopes at Roque de los Muchachos (see Fig. 3). For each image taken by FASTCAM, the relative position of the most intense peak of the speckle pattern is determined (see Eff-Darwich et al. 2008 for details). This position is calculated relative to an orthogonal axis system that is defined on the CCD and two time-series (one for each axis) are calculated for the sequence of images. By using the image scale and the pixel size, the time-series are calculated in meters and hence, direct comparison with seismic noise data could be carried out, as illustrated in Fig. 6. In all cases, the combined effect of atmospheric turbulence and telescope vibrations are significantly larger than the seismic noise measured in both observatories.

4. Conclusions

This work attempted to establish the actual seismic and volcanic risk at El Teide and Roque de los Muchachos observatories. Seismic risk is very low, in particular if this is compared with the much higher risk calculated for the Chilean and Hawaiian observatories. Topographic protection and large recurrence time considerably reduce risk associated to volcanic processes, namely lava flows and ashfall. Although ambient seismic noise is relatively high, most likely anthropogenic, it does not affect telescope observations. Moreover, local site effects will not amplify structural vibrations of large telescopes (10 meters or larger) that could be installed at this observatory in the future.

Acknowledgments

Authors are grateful to FASTCAM team for allowing the use of their data. AED and BGC works were partially supported by the spanish 'Ramón y Cajal' research program and the Instituto de Astrofísica de Canarias.

References

1. M.P. Cabrera and A. Hernández-Pacheco, Rev. Mat. Proc. Geol. V, 143-182 (1987)
2. B. García-Lorenzo, J.J. Fuensalida, M.A.C. Rodríguez-Hernández, Proc. SPIE, 6747 (2007)
3. Global Seismic Hazard Program, GSHAP
4. L. González de Vallejo, J. García-Mayordomo and J.M. Insua Bull. Seism. Soc. Am. 84, 2040 (2006)
5. V. Araña, A. Felpeto, M. Astiz, A. García, R. Ortiz, R. Abella, Journal of Volcanology and Geothermal Research, 103, 377 (2000)
6. Y. Nakamura, Quarterly Report of Railway Technical Research Institute (RTRI), Vol. 30 (1989)
7. J. Peterson, US Geol. Surv. Open-File Rept. 93 (1993)
8. A. Oscoz, R. Rebolo, R. López, A. Pérez-Garrido, J.A. Pérez, S. Hildebrandt, L.F. Rodríguez, J. Piqueras, and 14 coauthors, Proc. SPIE 7014, 149 (2008)
9. A. Eff-Darwich, B. García-Lorenzo, L. Bonatto, L. Hernández-Gutierrez, R. Viñas, J.A. Rodríguez-Losada, M.J. Blanco, C. Muñoz-Tuñón, Proc. SPIE 7012, 102 (2008)

TURBULENCE STRUCTURE AND METEOROLOGICAL CONDITIONS AT TEIDE AND ROQUE DE LOS MUCHACHOS OBSERVATORIES (CANARY ISLANDS)

J. A. CASTRO-ALMAZÁN*, B. GARCÍA-LORENZO and J. J. FUENSALIDA

Instituto de Astrofísica de Canarias (IAC),
Vía Láctea s/n. 38200 La Laguna, Tenerife, Spain
**E-mail: jcastro@iac.es, bgarcia@iac.es, jjf@iac.es*

The proper characterisation of the turbulence structure in an astronomical site requires an statistical study of the refractive-index structure constant $C_n^2(h)$. Our team is monitoring the $C_n^2(h)$ profiles since 2002 at the Teide and Roque de los Muchachos observatories (Canary Islands, Spain) with the g-SCIDAR technique. We have compared the turbulence profiles obtained with the radiosonde simultaneous profiles and with the NCEP I reanalysis maps at different levels. We show that turbulence measured at both observatories (being at $\sim 160km$ distant) correlate with the radiosonde data and may be explained through the atmospheric conditions in a synoptical scale. Hence, the statistical predominance of synoptical scaled phenomena may explain the similarity found in the monthly average profiles obtained at both observatories.

Keywords: Turbulence; Meteorology; SCIDAR; Radiosonde Balloons.

1. Introduction

Atmospheric turbulence is the main cause of astronomical image degradation, hence it is necessary to understand turbulence to optimise the design of compensation systems for large telescopes.

The Instituto de Astrofísica de Canarias (IAC) is carrying out an extensive program of turbulence monitoring since 2002[1,2] in the Canary Islands astronomical observatories, El Teide (OT; lon:16W5097, lat:28N3000) on the island of Tenerife and Roque de los Muchachos (ORM; lon:16W8923, lat:28N7565) on the island of La Palma, being at $\sim 160km$ distant and at $2400m$ in altitude. Observations are performed on about 4 nights per month basis, using the generalized-Scintillation Detection and Ranging (g-SCIDAR) technique. For this purpose, the IAC has developed two automated g-SCIDAR[3,4] instruments installed at the $1.5m$ Carlos Sánchez Telescope (TCS) on the OT and at the $1m$ Jacobus Kapteyn Telescope (JKT) on the ORM.

2. Technique and Data Treatment

SCIDAR has proved to be the most contrasted and efficient technique from ground level to obtain the optical vertical structure of the atmospheric turbulence. The technique is based on the analysis of the scintillation patterns produced at the telescope pupil (or at a lower conjugated plane: g-SCIDAR) by the light coming from a binary star system. Turbulence profiles as a function of height are derived through the inversion of the autocorrelation of the scintillation patterns. The result is given as the statistical refractive index structure constant $C_n^2(h)$ that is proportional to the turbulence intensity (see Ref. 5 and references therein). Dome turbulence contribution was removed from the individual profiles using a procedure based on evenness properties with Fourier analysis.[6] Turbulence profiles have been averaged from 00UT to 01UT to compare with the 00UT balloon data. Temperature and humidity balloon profiles have been plotted up to 15km to highlight the most relevant information. Thermal stability is analysed through the virtual potential temperature and the temperature profile compared with the dry adiabatic lapse rate. Wind plots have been smoothed to highlight the main trends.

3. Discussion

Local factors cause the turbulence generated in the first kilometres above the ground level (boundary layer). These factors are buoyant convection processes, like thermals rising produced by surface solar heating, and mechanical processes, like wind shears produced by the surface friction in the wind speed or by the lee waves formed by mountains or other geographic effects.[7] However, instability phenomena producing turbulence over the boundary layer may be studied in a regional and synoptical scale.[8] These phenomena, as the thermal instability or the wind shears could be affecting both observatories simultaneously, giving a similar turbulence pattern. Figures 1 and 3 show two examples of that scenario for different nights. Turbulence layers simultaneously measured at both observatories –Figs. 1, 3 (top)– correlate with the radiosonde data profiles –Figs. 1, 3 (bottom)–. Balloons are launched closer to sea level, $\sim 13km$ away from OT and more than $160km$ away from ORM.

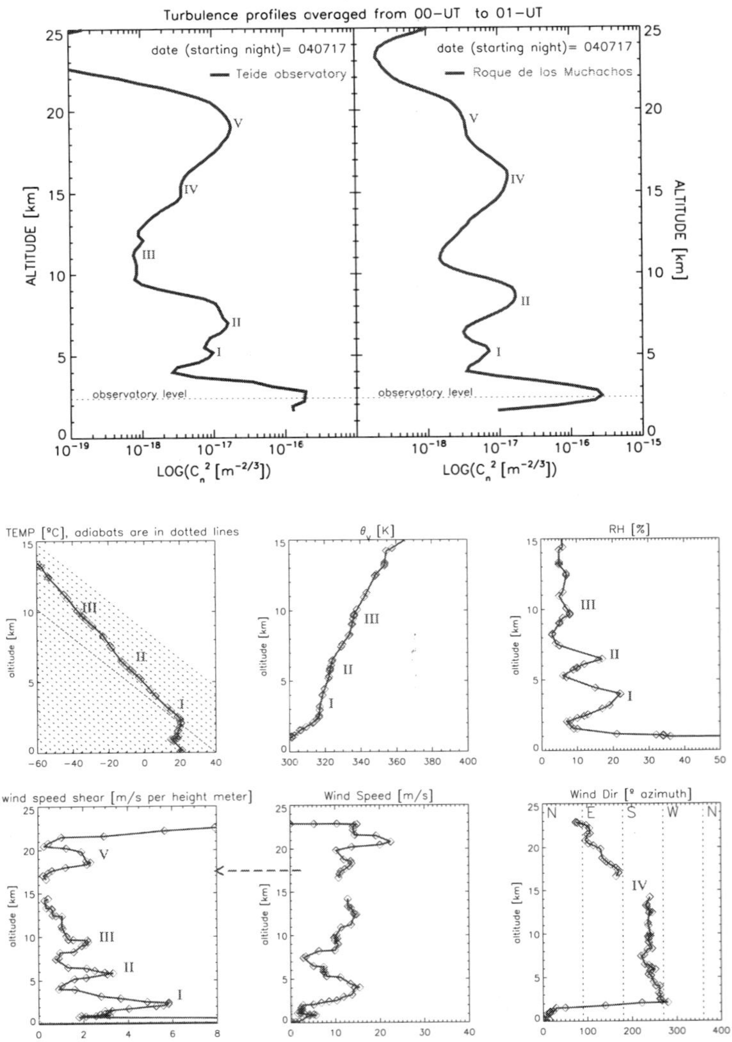

Fig. 1. Turbulence and atmospheric profiles for a selected night, on July 17th, 2004: (top) Simultaneous g-SCIDAR turbulence profiles at OT and ORM; (bottom) Balloon profiles at 00UT during the same run. Turbulence layers and their correspondences in the balloon profiles are indicated in roman figures.

Turbulence within the first $15km$ seems to be related to wet air layers incursions, implying thermal instability and wind shears. At higher altitudes, turbulence is mainly caused by wind shears due to gradients both in speed

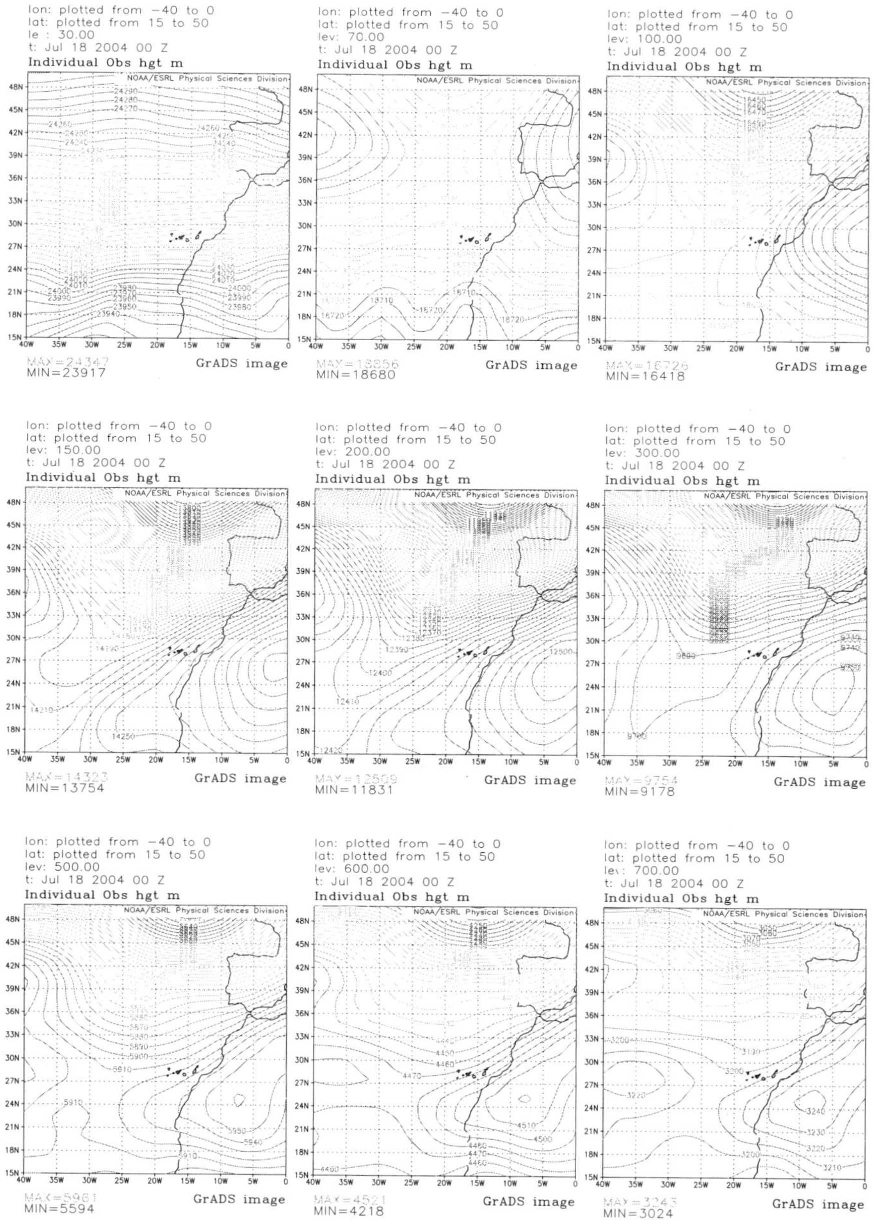

Fig. 2. NCEP I reanalysis synoptic maps for different pressure levels at 00UT during the July 17th, 2004 g-SCIDAR run (see Fig. 1).

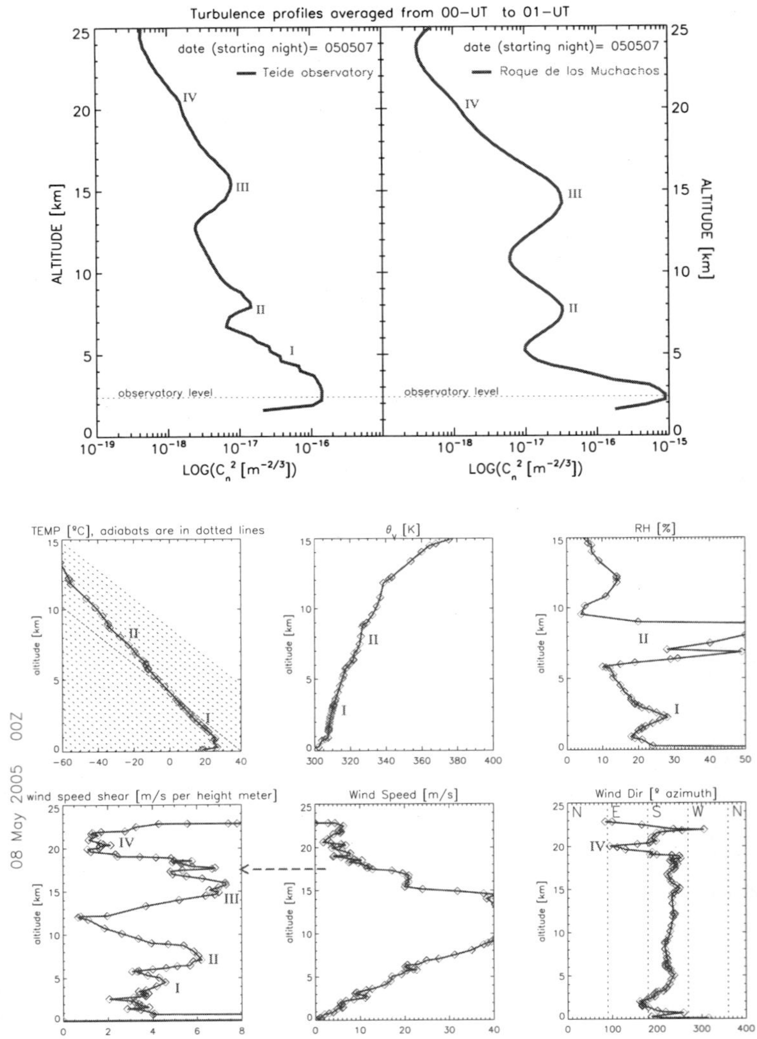

Fig. 3. Turbulence and atmospheric profiles for a selected night, on May 7th, 2005: (top) Simultaneous g-SCIDAR turbulence profiles at OT and ORM; (bottom) Balloon profiles at 00UT during the same run. Turbulence layers and their correspondences in the balloon profiles are indicated in roman figures.

and/or direction. As shown in the reanalysis pressure levels maps –Fig. 2 and 4–, the same wind pattern is covering an extensive region where the two observatories are included, so a similar turbulence profile is expected.

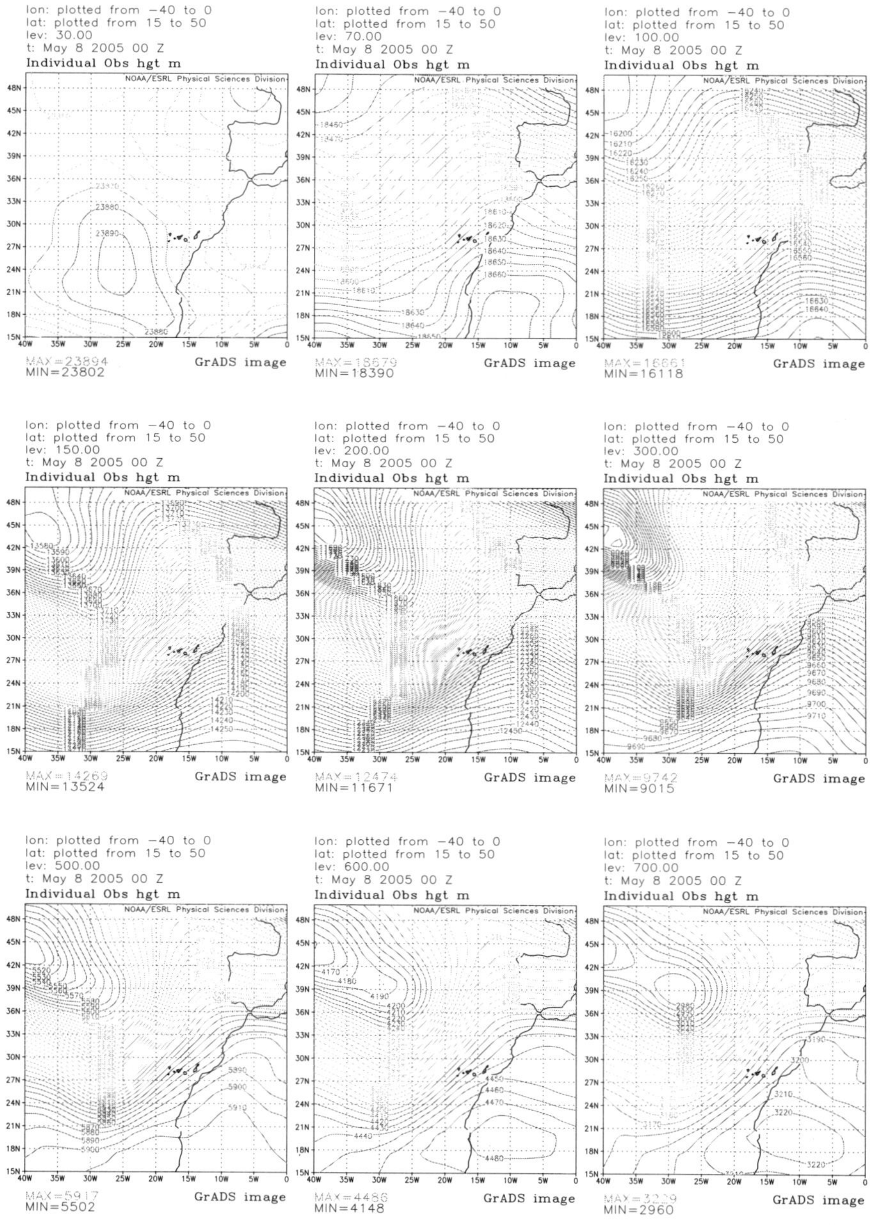

Fig. 4. NCEP I reanalysis synoptic maps for different pressure levels at 00UT during the May 7th, 2005 g-SCIDAR run (see Fig. 3).

Nevertheless, turbulent layers could not appear simultaneously above both observatories, as it is shown in Fig. 5, where the same turbulence levels at $\sim 9km$ and $\sim 14km$ are measured more than one hour later in ORM relative to OT.

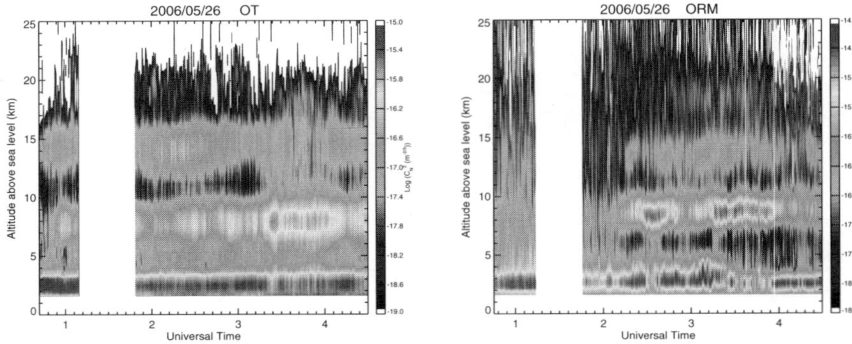

Fig. 5. Turbulence evolution for May 26th, 2006, showing a lag in the time when turbulence structures appear above both observatories: (left) OT; (right) ORM.

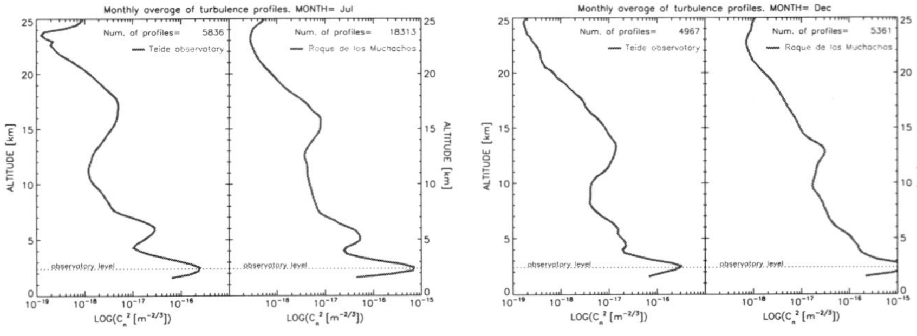

Fig. 6. Monthly average of turbulence profiles for OT and ORM: (left) July; (right) December. Complete statistical results, covering the 12 months are in Ref. 2.

The statistical predominance of turbulence generated by regional or synoptic atmospheric conditions may explain the similarity found in the monthly average profiles obtained at OT and ORM (e.g. Fig. 6, monthly mean profiles for July and December).

Acknowledgements

We have to acknowledge to José María Rodríguez González ("Chema") from the Spanish "Agencia Estatal de Meteorología" (AEM) for his useful discussions and suggestions. The Carlos Sánchez Telescope is operated at the OT by the IAC. The Jacobus Kapteyn Telescope is operated by the Isaac Newton Group at the ORM (IAC). The authors thank all the staff at the observatories for their kind support. We also appreciate the help of A. Eff-Darwich in correcting the manuscript. The NCEP I data sets and maps are provided by the "Earth System Research Laboratory" of the US NOAA (`http://www.esrl.noaa.gov/`). The balloons are launched by the AEM and the data is available through the page of the Dept. of Atmospheric Sci. of the Univ. of Wyoming (`http://weather.uwyo.edu/upperair/sounding.html`). This work has been supported by IAC and the Spanish "Plan Nacional de Astrofísica" of the Ministry of Science under grants AYA2003-07728 and AYA2006-13682. B. García-Lorenzo also thanks the support from the Ramón y Cajal program.

References

1. J. J. Fuensalida, S. Chueca, J. M. Delgado, B. García-Lorenzo, J. M. Rodríguez-González, C. K. Hoegemann, E. G. Mendizábal, M. Reyes, M. Verde and J. Vernin, Vertical structure of the turbulence above the observatories of the canary islands: parameters and statistics for adaptive optics, in *Advancements in Adaptive Optics*, eds. D. B. Calia, B. L. Ellerbroek and R. Ragazzoni, Proc. SPIE, Vol. 5490-2004.
2. B. García-Lorenzo, J. J. Fuensalida, J. A. Castro-Almazán and M. A. C. Rodríguez-Hernández, Statistical vertical turbulence profiles at the canary islands astronomical sites, in *Optical Turbulence: Astronomy meets Meteorology*, eds. E. Masciadri and M. Sarazin-2008.
3. J. J. Fuensalida, B. García-Lorenzo, J. M. Delgado, C. K. Hoegemann, M. Verde, M. Reyes and J. Vernin, *ING Newsl* **8**, 15 (2004).
4. M. A. C. Rodríguez-Hernández, J. J. Fuensalida, B. García-Lorenzo, J. M. Delgado, E. Hernádez, C. K. Hoegemann and H. Vázquez-Ramió, The hybrid shack-hartmann/g-scidar instrument, in *Optics in Atmospheric Propagation and Adaptive Systems X*, eds. K. Stein, A. Kohnle and J. D. Gonglewski, Proc. SPIE, Vol. 6747-2007.
5. R. Avila, J. Vernin and E. Masciadri, *Appl. Opt.* **36-30**, 7898 (1997).
6. J. J. Fuensalida, B. García-Lorenzo and C. Hoegemann, *MNRAS.* **389-2**, 731 (2008).
7. R. B. Stull, *An introduction to Boundary Layer Meteorology* (Kluwer Academic Publishers, Boston, 1988).
8. D. R. Lee, R. B. Stull and W. S. Irvine, *Us Air Force Global Weather Central Technical Note* **AFGWC/TN 79/001 (revised)** (1984).

STUDYING THE RELATIONSHIP BETWEEN THE AVERAGE VELOCITY OF THE TURBULENCE AND HIGH ALTITUDE WINDS AT THE TEIDE OBSERVATORY

B. GARCÍA-LORENZO*

Instituto de Astrofísica de Canarias
C/ Via Lactea S/N, 3800-La Laguna, Tenerife, Spain
**E-mail: bgarcia@iac.es*

C. DEL GIORGIO-CASTIGLIONE

Dept. Astrofísica, Universidad de La Laguna
C/ Astrofísico Francisco Sánchez, E-38205 Tenerife, Spain

A. EFF-DARWICH

Dept. Edafología y Geología, Universidad de La Laguna
C/ Astrofísico Francisco Sánchez, E-38205 Tenerife, Spain

J.J. FUENSALIDA

Instituto de Astrofísica de Canarias
C/ Via Lactea S/N, 3800-La Laguna, Tenerife, Spain

The requirements for excellent image quality of current large and future very large telescopes demand a proper knowledge of atmospheric turbulence. Thus several projects are already pursuing this aim. The precise characterization of the turbulence above a particular site requires long-term monitoring. In this sense, due to the lack of long-term information on turbulence, high-altitude winds (in particular winds at the 200-mbar pressure level, V_{200}) have been proposed (Sarazin & Tokovinin 2002,[1] S&T02 hereafter) as a parameter for estimating the total turbulence at a particular site, because records of this parameter exist from several sources. This choice is based on the idea that the greatest source for turbulence generation is related to the highest peak in the vertical wind profile, which is located at the 200-mbar pressure level globally. Moreover, S&T02 found a good correlation between the average velocity of the turbulence, V_0, and V_{200} of the form: $V_0 = 0.4^*V_{200}$ at the Cerro Pachón and Paranal Observatories in Chile. A linear relationship between V_0 and V_{200} was also found in San Pedro Mártir (Mexico) (Masciadri & Egner 2006,[2] M&E06 hereafter). We study here the possible connection between V_0 and V_{200} at the Teide Observatory (Spain).

Keywords: Atmospheric Effects – Turbulence – Site Testing – Telescopes.

1. Introduction

Astronomical observations from the Earth surface are strongly affected by atmospheric turbulence. Adaptive optics techniques try to compensate these effects on astronomical images, attempting to reach the diffraction limit of telescopes. The larger the telescope diameter is, the more difficult the proper correction of the atmospheric turbulence becomes. The image quality requeriments of current projects needs proper knowledge of the statistical behaviour of parameters describing the atmospheric turbulence, in order to design efficient adaptive optics systems (AO). In this sense, most of the important astronomical sites in the world are monitoring the turbulence structure, although decades of continuous data are needed to obtain statistically significant results. In order to solve the lack of long-term information on turbulence structure, winds at the 200-mbar pressure level (V_{200} hereafter) was proposed as a parameter for estimating the total turbulence at any particular site. This proposal was based on the relation between turbulence and velocity gradients and taking into account that the velocity profile use to have its largest peak at about the altitude of the 200-mbar pressure level. The choice of V_{200} as a parameter for site AO capabilities was supported by similar seasonal trend of the seeing and V_{200} at Mauna Kea and La Silla astronomical sites (Vernin 1986[3]), and the results founds at Cerro Pachón and Paranal (S&T02), where V0 was found to be proportional to V_{200}: $V_0 = 0.4V_{200}$ (S&T02). A good correlation between V_0 and V_{200} (of the form $V_0 = 0.56V_{200}$) was also found above San Pedro Mártir (Mexico) using simulations of C_N^2 profiles (M&E06[2]).

García-Lorenzo et al. (2005)[4] found a good agreement between high and low-altitude winds, suggesting a connection between V_0 and V_{200} at five different astronomical sites, as is suggested by definition: $V_0 = [\frac{\int_0^\infty C_N^2(h)V(h)\,dh}{\int_0^\infty C_N^2(h)\,dh}]$. The statistical wind speed at the 200-mbar pressure level was adopted as one of the parameters for ranking the suitability of different observing sites for adaptive optics (García-Lorenzo et al. 2005;[4] Carrasco et al. 2005;[5] Chueca et al. 2004;[6] Carrasco & Sarazin 2003;[7] Ilyasov, Tillayev & Ehgamberdiev 2000[8]) and V_{200} is accepted as a parameter for site evaluation.

2. The Data

The Teide Observatory (OT hereafter) (Lat: 28.30N, Lon: 16.51W, Alt: 2400m), operated by the Instituto de Astrofísica de Canarias, is located on the island of Tenerife (Spain), ~400 km distant from the coast of Morocco.

The G-SCIDAR technique (see e.g. Fuchs, Tallon, & Vernin, 1994[9]) is being used to obtain the C_N^2(h) profiles above the OT. The systematic campaigns began in November 2002, with a frequency of ∼4-6 nights every month (e.g. García-Lorenzo, Fuensalida, & Rodríguez-Hernández 2007[10]).

The wind velocity profiles were retrieved from classical meteorological radiosondes. These are launched twice daily from Güimar (Lat: 28.46N, Lon:16.37W, Alt: 105m), NOAA station 60018, on the island of Tenerife, and ∼13 km from OT. The balloons provide measurements at 00 UT and 12 UT of V_{200} above this location, although we only used the data at 00 UT.

We chose 95 nights with C_N^2(h) profiles obtained during the corresponding ballon ascent and distributed from 2003 to 2008. Figure 1 shows the

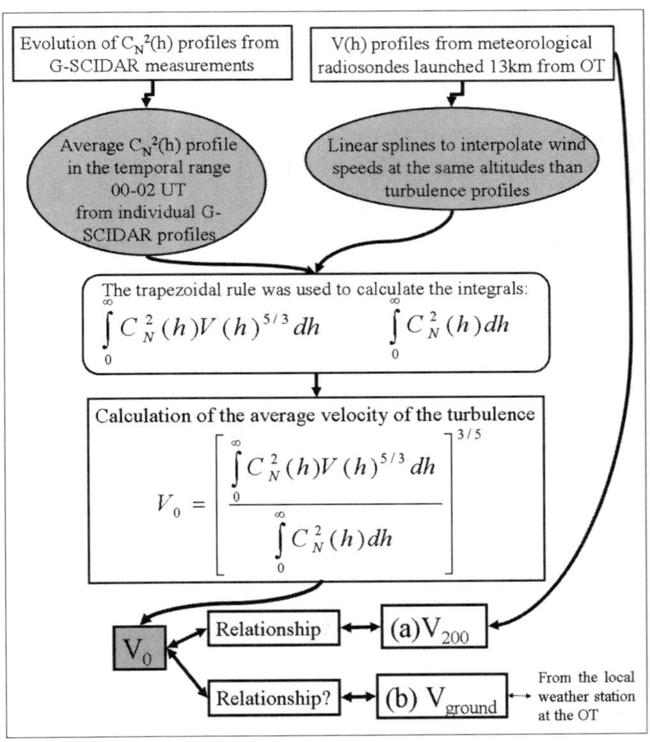

Fig. 1. Scheme of the procedure followed to study the relation between the average velocity of the turbulence and winds at: (a) 200 mbars pressure level (V_{200}; and (b) the velocity at the ground level (V_{ground}) measured by a local weather station next to the TCS where the G-SCIDAR is installed.

basic procedure we have followed to process the $C_N^2(h)$ and V(h) profiles to study the relationship V_0-V_{200} at OT. Figure 2 presents a few examples of the kind of data used in this work.

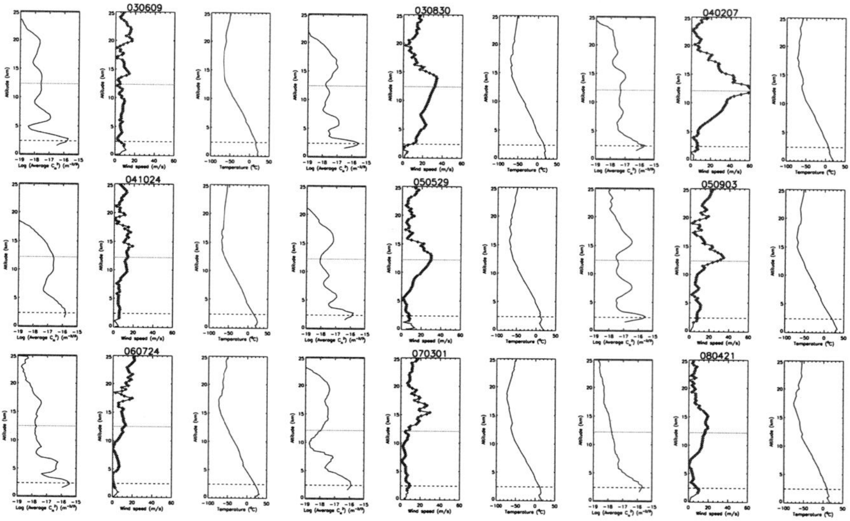

Fig. 2. Example of turbulence ($C_N^2(h)$) and wind velocity (V(h)) profiles used as input data to calculate the average velocity of the turbulence, V_0, at the Teide Observatory. Blue dots correspond to the interpolation of V(h) to the grid of $C_N^2(h)$. For comparison purposes, we have also included the temperature profile for each night.

3. Analysis of the Linear Relationship between V_0 and V_{200}

After calculation of V_0 following the procedure sketched in figure 1, we compared V_0 with their corresponding V_{200} (figure 3) which is far from the good correlation V_0-V_{200} found by S&T02[1] in Paranal and Cerro Pachón in Chile or M&E06[2] in San Pedro Mártir (Mexico). The Person's correlation coefficient and the best linear fit between V_0 and V_{200} is presented in table 1.

Following the same calculation than S&T02[1] and M&E06,[2] we have forced the best fit of the 95 data to pass through the coordinate origin, obtaining the expression: $V_0 = 0.456V_{200}$. This approah could induce relative errors ($[V_0-(0.456V_{200})/V_0]*100\%$) of 25-45 %, reaching errors larger than 100 % in some cases.

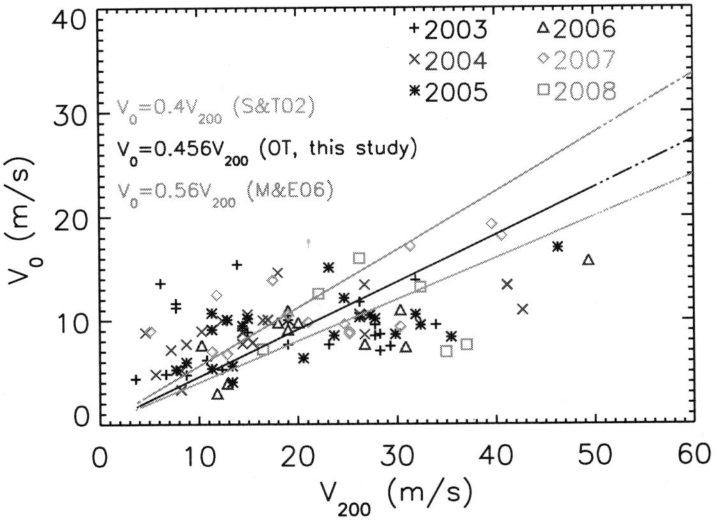

Fig. 3. V_{200} as a function of V_0. Different symbols and colours corresponds to data from different years. Black line corresponds to the best fit passing through the coordinate origin, while white and grey lines are the fits obtained for Paranal/Cerro Pachón and San Pedro Mártir, respectively.

Table 1. Pearson's correlation coefficient and best linear fit obtained from V_0 and V_{200} measurements at the Teide Observatory (Spain). The last lines (Total) include all the data in the sample (2003+2004+2005+2006+2007+2008).

Year	**Pearson's Coef.**	**Best linear fit**
2003	0.19	$V_0=0.06V_{200}+7.79$
2004	0.74	$V_0=0.16V_{200}+6.42$
2005	0.61	$V_0=0.19V_{200}+5.31$
2006	0.75	$V_0=0.23V_{200}+3.50$
2007	0.69	$V_0=0.27V_{200}+5.39$
2008	-0.15	$V_0=-0.07V_{200}+12.64$
Total	0.55	$V_0=0.16V_{200}+6.14$

4. Analysis of the Linear Relationship between V_0 and V_{ground}

As suggested by S&T02,[1] and following the procedure sketched in figure 1, we have compared V_0 with the wind velocity at the ground level (V_{ground}) measured by the weather station placed closer to the Carlos Sánchez telescope, where the G-SCIDAR measurements are carried out. In 10 of the

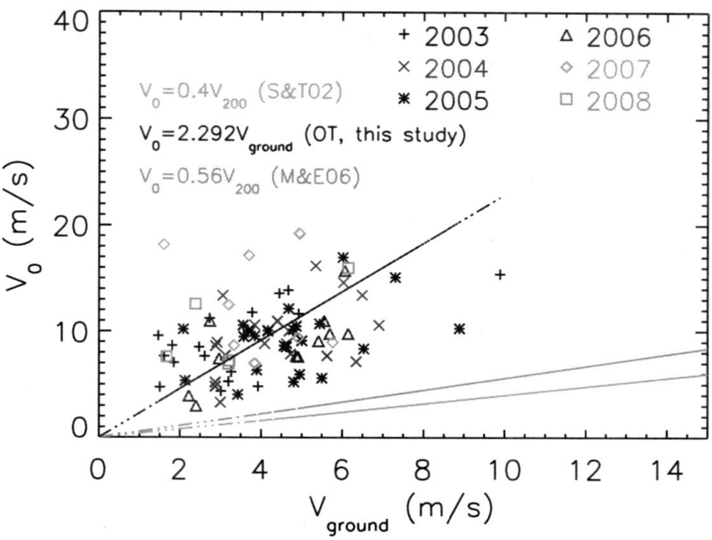

Fig. 4. V_0 as a function of V_{ground}. Different symbols and colours corresponds to data from different years. Black line corresponds to the best fit passing through the coordinate origin. White and grey lines are the best fits obtained for Paranal/Cerro Pachón and San Pedro Mártir, respectively, for V_0 as a function of V_{200}.

nights selected for this study, this station was down and we do not have local wind measurements. Figure 4 shows the average velocity of the turbulence, V_0 as a function of the wind velocity at the ground, V_{ground}. The Pearson's correlation coefficient and the best linear fit between V_0 and V_{ground} is shown in the table 2.

Table 2. Pearson's correlation coefficient and best linear fit obtained from V_0 and V_{ground} measurements at the Teide Observatory (Spain). The last lines (Total) include all the data in the sample (2003+2004+2005+2006+2007+2008).

Year	**Pearson's Coef.**	**Best linear fit**
2003	0.63	V_0=1.10V_{200}+5.26
2004	0.46	V_0=1.11V_{200}+4.61
2005	0.35	V_0=0.76V_{200}+5.70
2006	0.68	V_0=1.55V_{200}+1.84
2007	-0.31	V_0=1.20V_{200}+17.35
2008	0.70	V_0=1.65V_{200}+4.59
Total	0.38	V_0=0.83V_{200}+6.09

Following the same calculation than in the previous section and S&T02[1] and M&E06[2] for V_{200}, we have forced the best fit of the 85 data to pass through the coordinate origin, obtaining the expresion: $V_0 = 2.292 V_{ground}$. Such approach could induce relative errors ($([V_0 - (2.292 V_{ground})/V_0]*100\%)$) of 20-50 %, reaching errors larger than 100% in some cases.

5. Discussion and Conclusions

We have seen in sections §3 and §4 that the average velocity of the turbulence, V_0, shows a similar degree of correlation with V_{200} as with V_{ground} at Teide observatory. Moreover, estimations of V_0 from V_{200} or V_{ground} can induce similar errors. S&T02[1] proposed a more general formula to estimate V_0 in Paranal and Cerro Pachón sites, combining wind velocities from ground and 200 mbar level:

$$V_0 \cong [max(V_{ground}, 0.4V_{200})] \tag{1}$$

For the OT, this equation should be expressed, according to the fit forced to pass through the coordinate origin, as:

$$V_0 \cong [max(2.292V_{ground}, 0.456V_{200})] \tag{2}$$

Even using this approach (equation 2), errors larger than 100% can be reached when estimating V_0, being the mean relative errors between 20 to 40 %.

The linear relation between V_0 and V_{200} is established by the definition of V_0. This work shows up the importance of deriving atmospheric turbulence profiles, $C_N^2(h)$, to determine the coefficient of the linear fit connecting V_0 and V_{200}. According to the results of this work and in agreement with previous authors, large discrepancies may be induced when estimating V_0 from only V_{200}. Including the wind speed, other factors (e.g. buoyant convection processes, instability phenomena, etc) can be playing an important role to generate turbulence (see Castro-Almazán, García-Lorenzo, & Fuensalida 2008[11]).

Changes in wind regimes (in different seasons, for example) could have an important influence on the linear coefficient, taking into account that the approach assumed (equation 1 for Paranal and Cerro Pachón sites or equation 2 for Teide Observatory) is based only on the absolute values of the wind, and that we have not yet considered the wind directions.

Acknowledgments

This paper is based on observations obtained at the Carlos Sánchez Telescope operated by the Instituto de Astrofísica de Canarias at the Teide Observatory on the island of Tenerife (Spain). The authors thank all the staff at the observatory for their kind support. Radiosonde are launched by the Spanish "Agencia Estatal de Meteorología" and data are available at the webpage of the Department of Atmospheric Science of the University of Wyoming (http://weather.uwyo.edu/upperair/sounding.html). This work was partially funded by the Instituto de Astrofísica de Canarias and by the Spanish Ministerio de Educacion y Ciencia (AYA2006-13682 and AYA2003-07728). B. García-Lorenzo also thanks the support from the Ramón y Cajal program by the Spanish Ministerio de Educacion y Ciencia.

References

1. M. Sarazin and A. Tokovinin, in *Proc. 58th ESO Conf. Workshop, Beyond Conventional Adaptive Optics.*, eds. E. Vernet, R. Ragazzoni, S. Esposito and N. Hubin, ESO Publications, Garching, Vol. 1June 2002.
2. E. Masciadri and S. Egner, *PASP* **118**, 1604(November 2006).
3. J. Vernin, SPIE, 628, 142, 1986.
4. B. García-Lorenzo, J. J. Fuensalida, C. Muñoz-Tuñón and E. Mendizabal, *MNRAS* **356**, 849(January 2005).
5. E. Carrasco, R. Avila and A. Carramiñana, *PASP* **117**, 104(January 2005).
6. S. Chueca, B. García-Lorenzo, C. Muñoz-Tuñón and J. J. Fuensalida.
7. E. Carrasco and M. Sarazin, in *RMxAC*, eds. I. Cruz-Gonzalez, R. Avila and M. Tapia, RMxAC, vol. 27, Vol. 19September 2003.
8. S. Ilyasov, Y. Tillayev and S. Ehgamberdiev, SPIE, 4341, 181, 2000.
9. A. Fuchs, M. Tallon and J. Vernin, SPIE, 2222, 682, 1994.
10. B. García-Lorenzo, J. J. Fuensalida and M. A. C. Rodríguez-Hernández, SPIE, 6747, 11, 2007.
11. J. Castro-Almazán, B. García-Lorenzo and J. J. Fuensalida, in *Optical Turbulence: Astronomy meets Meteorology*, Conference Proceedings of the Imperial College Press World Scientific (ICP-WS)2008.

THE ANTARCTICA POLAR VORTEX: STUDY OF WINTER 2005

F. LASCAUX*, E. MASCIADRI**, S. HAGELIN and J. STOESZ

INAF/Osservatorio Astrofisico di Arcetri
Florence, Italy
E-mail: lascaux@arcetri.astro.it*
*** masciadri@arcetri.astro.it*

During winter and springtime, the flow above Antarctica at high altitude (upper troposphere and stratosphere) is dominated by the presence of a vortex centered above the continent. It lasts typically from August to November. This vortex is characterized by a strong cyclonic jet centered above the polar high. In a recent study[1] of four different sites in the Antarctic internal plateau (South Pole, Dome C, Dome A and Dome F), it was made the hypothesis that the wind speed strength in the upper atmosphere should be related to the distance of the site to the center of the Antarctic polar vortex. This high altitude wind is very important from an astronomical point of view since it might trigger the onset of the optical turbulence and strongly affect other optical turbulence parameters. What we are interested in here is to localize the position of the minimum value of the wind speed at different heights in the troposphere, particularly above 10 km. For that we studied the analyses from the ECMWF for winter 2005 at different levels. We deduced a preferential position of this minimum, tilted with altitude, in a zone between South Pole and Dome A, for the year 2005. This extensive study over one entire winter confirms the "position space" of the polar high deduced by Ref. 1.

Keywords: Polar Vortex; Atmospherical effects; Optical Turbulence.

1. Introduction

During winter, the stratospheric circulation at both poles is characterized by a large cyclonic vortex centered in a region close to the pole.[2] A strong circumpolar westerly jet isolates mid-latitude and polar stratospheres. Extended climatology of the polar vortex have been already conducted, using ECMWF (European Center of Medium-range Weather Forecasts) or NCAR-CEP data set[3,4] with methods based on potential vorticity to define the vortex. It is well known that the vortices are much more stronger in mid-winter than in summer.[5,6] More over, it appears that the Antarctica polar vortex is less variable and more persistent than the Arctic one. Indeed, above Antarctica, the polar vortex is quite persistent from August to November and centered above the polar high. Most of the meteorological studies concerning the polar vortices were aimed to link the vortex position to the stratospheric ozone depletion on the poles. However what we are interested here is a different matter: the minimum speed of the high altitude wind above the Antarctica continent. This might help us in identifying the centre of the polar vortex and we could use this information to qualify the sites for astronomical applications.

Indeed, during a previous study about site characterization for optical turbulence above the internal Antarctic Plateau using ECMWF analysis, our group investigated four different sites:[1] Dome C, Dome A, South Pole and Dome F. One of the conclusions was that in the free atmosphere, above around 10 km, the wind speed increased monotonically in winter proportionally to the distance to the center of the polar high. Thus Dome C was the site showing the highest wind speed above 10 km. This high altitude wind is very important for astronomical applications, because of its influence on some astronomical parameters. We propose in this study to confirm this hypothesis looking at a climatology of the high altitude wind speed and the corresponding vortex above the Antarctica continent for winter 2005. Due to the weak variability of the Antarctica vortex proved in other studies, we can deduce that this one-year study can provide a quantitative estimate with a good accuracy about the position of the minimum wind speed in altitude.

2. Methods

We used the ECMWF analyses from MARS catalog for every day between May 1st and September 30th, 2005, at 00 UTC. The numerical grid at this period was a TL511 spectral grid (corresponding to a N256 gaussian grid, see *http://www.ecmwf.int* for more details). The averaged distance between two horizontal grid points is of the order of $\sim$40 km. The vertical grid is composed of 60 levels, from ground level up to the 0.2 hPa level. The analysis employs the 4D-VAR assimilation scheme to assimilate a wide number of observations, including in-situ conventional data and synthetic data from satellites.

To perform our study, we focused on the wind speed at two different altitudes, 15 km and 20 km from May 1st and September 30th, every 24 hours at 00 UTC. We computed the monthly medians of the wind speed and deduced a preferential position for the minimum of wind speed in altitude for winter 2005.

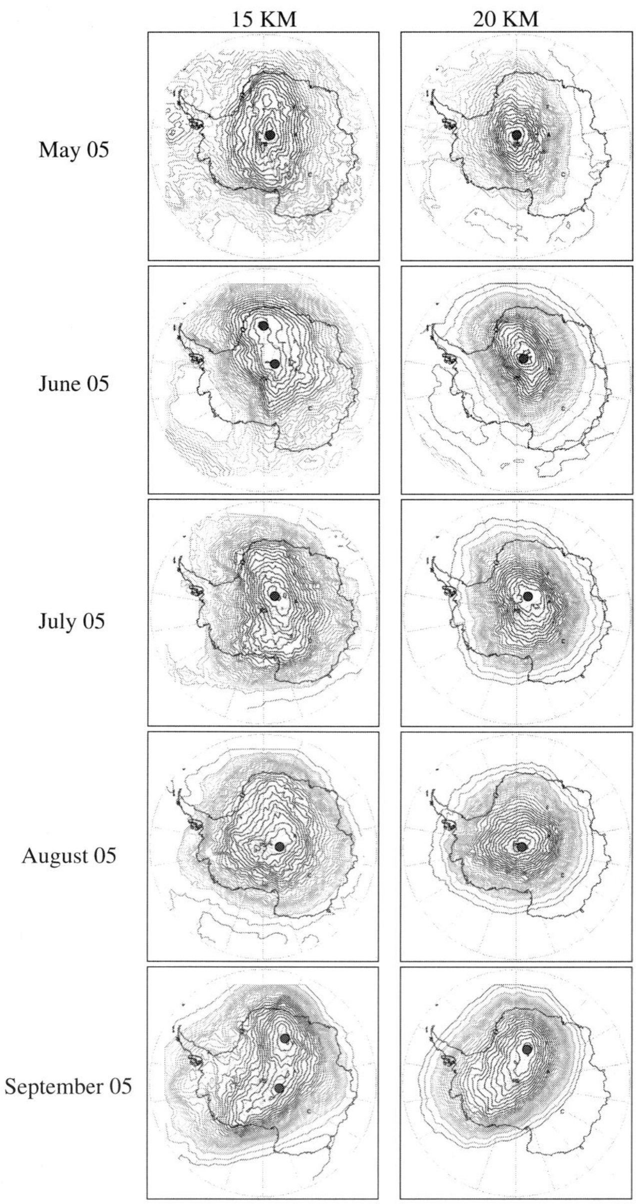

Fig. 1. Median wind speed at 15 km (column 1) and 20 km (column 2) above the Antartic continent (polar stereographic projection), for May, June, July, August and September 2005. Units in $m.s^{-1}$. Isocontours starting from 1 $m.s^{-1}$ increment of 1 $m.s^{-1}$. The red dot represents the location of the minima.

3. The Antarctic Vortex During Winter 2005: Position of the Wind Speed Minimum

Fig. 1 shows the monthly medians of the wind velocity at two different heights, 15 and 20 km, from May 2005 to September 2005.

The minimum of the median wind is clearly identified at 20 km while at 15 km it is more difficult to identify the center of the vortex that, in some months, presents double minima. The polar vortex is persistent and its center (corresponding to the minimum wind speed) does not evolve much from May to September. It remains in a area between South Pole and Dome A. Between the four sites investigated by Ref. 1, Dome C is the farthest away from the polar vortex center, and by consequence the one with the highest wind speed in altitude. This is confirmed by looking at Fig. 2, which shows the median of the wind speed at 15 km and 20 km for winter 2005 (from May to September). The reader can refer to Ref. 1 to have a comparison of the high altitude wind speed above the four sites.

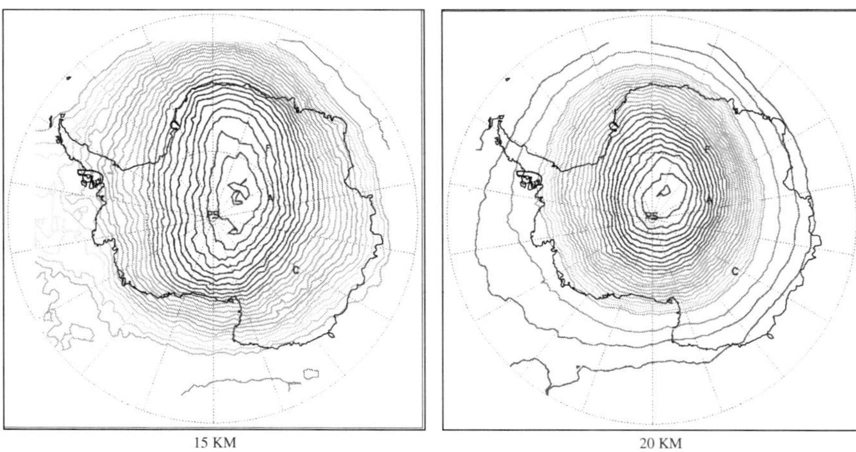

Fig. 2. Median wind speed at 15 km (column 1) and 20 km (column 2) above the Antartic continent (polar stereographic projection), for Winter 2005 (May to Spetember). Units in m.s^{-1}. Isocontours starting from 1 m.s^{-1} and increment of 1 m.s^{-1}.

4. Conclusions

We have studied the Antarctica polar vortex from the point of view of the minimum wind speed with respect to the altitude. For that we post-processed the analyses of the ECMWF between May and September 2005 (winter) at two different levels (15 km and 20 km). We deduced a preferential position of the minimum of wind speed in a zone between South Pole and Dome A, thus far from Dome C. This confirms the conclusion regarding the "position space" of the polar high deduced by Ref. 1, and the link between the position of an Antarctic site with respect to this center and the wind speed in altitude. Dome C appears to have in winter a wind speed in altitude much higher than other sites like South Pole and Dome A, closer to the center of the vortex. Such a quantitative information should be considered by astronomers as a key issue in future plans for astronomical facilities to be set-up in different sites of the Internal Antarctic Plateau.

Acknowledgments

ECMWF analyses are extracted from the Catalog MARS, *http://www.ecmwf.int.* This study has been funded by the Marie Curie Excellence Grant (FOROT) - MEXT-CT-2005-023878.

References

1. S. Hagelin, E. Masciadri, F. Lascaux and J. Stoesz, *MNRAS* **387**, 1499 (2008).
2. P. H. Haynes, *Annual Review of Fluid Mechanics* **37**, 263 (2005).
3. A. Karpetcho, E. Kyro and E. Knudsen, *J. Geophysical R.* **110 - D21109** (2005).
4. D. W. Waugh and W. J. Randel, *J. Atmos. S.* **56**, 1594 (1999).
5. D. W. Waugh, W. J. Randel, S. Pawson, P. A. Newman and E. R. Nash, *J. Geophysical R.* **104**, 27191 (1999).
6. V. L. Harvey, R. B. Pierce, T. D. Fairlie and M. H. Hitchman, *J. Geophysical R.* **107 (D20)**, p. 4442 (2002).

SURFACE LAYER SLODAR

JAMES OSBORN, RICHARD WILSON and TIMOTHY BUTTERLEY

Department of Physics, Centre for Advanced Instrumentation, University of Durham, South Road, Durham DH1 3LE

A prototype of a new SLODAR instrument has been developed at Durham University CfAI and tested at the Paranal observatory. The instrument targets very wide double star targets, with separations of several arc-minutes, to achieve profiling of the surface layer of turbulence with very high resolution in altitude (10m or less). We describe the instrument and the results of preliminary observations made at Paranal.

Keywords: Atmospheric effects – instrumentation; adaptive optics – site testing.

1. Introduction

The low altitude turbulent layers typically contribute a large fraction of the total integrated seeing.[1] Measuring the turbulence profile in the ground layer is important for many reasons. It will enable performance prediction and optimization of adaptive optics (AO) systems such as ground layer AO (GLAO).[2] The profile will also be useful at the site selection and planning stage for future observatories. Profiling will enable an estimation of the thickness of this layer to ensure that future telescopes are built above or protected from the surface layer.

SLODAR (SLOpe Detection and Ranging) is a technique which has been developed to profile the vertical distribution of optical turbulence, $C_n^2(h)$. SLODAR measures the wavefront gradient as a function of position at the telescope pupil and then using triangulation can estimate the strength, altitude and velocity of each turbulent layer up to a maximum altitude, H_{max}.[3] The turbulence profile (strength as a function of altitude) can be recovered from a fit of the cross-correlation in the direction of target separation to the theoretical impulse response functions of the system.[4] The layer velocity can be obtained from calculating the velocity of the peak movement in the centroid cross correlation with an increasing temporal delay between the two channels. Figure 1 shows a diagram of the geometry of SLODAR.

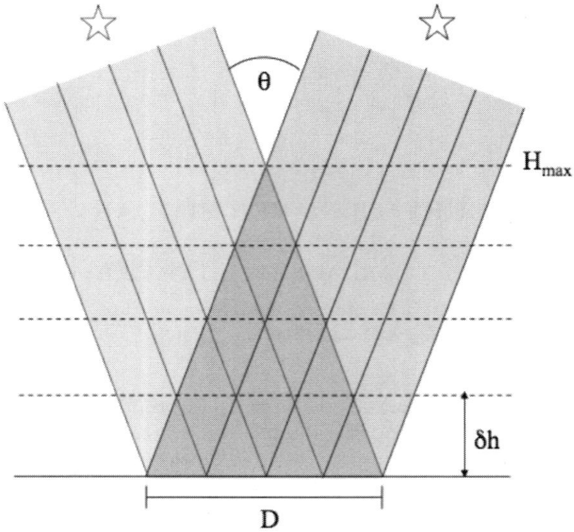

Fig. 1. Slodar geometry. θ is the angular separation of the target stars, D is the Dimeter of the telescope, H_{max} is the maximum altitude that a layer can be resolved and δh is the altitude resolution from the system.

Previous results show that SLODAR often does not resolve the surface layer as the altitude bins are too large.[5] The vertical resolution, δh, of the SLODAR system is given by,

$$\delta \mathrm{h} = \frac{\mathrm{D}}{\mathrm{n}\theta} \tag{1}$$

where D is the telescope pupil diameter, n is the number of subapertures across the pupil and θ is the angular separation of the target. For a given telescope the resolution can be improved by either increasing the number of subapertures across the pupil or by increasing the separation of the target stars. There is a limit to the number of subapertures which can be used, set by photon flux. The target separation is limited by the field of view of the telescope and the imaging camera. To avoid this limitation the system has been modified. A reflective wedge is used to split the two images of the stars onto separate cameras. A simple diagram of the wedge can be seen in figure 2. Using this system high resolution profiles of the surface turbulent layer can be obtained, avoiding the need for a camera with an extremely large field of view and unnecessarily long read out times.

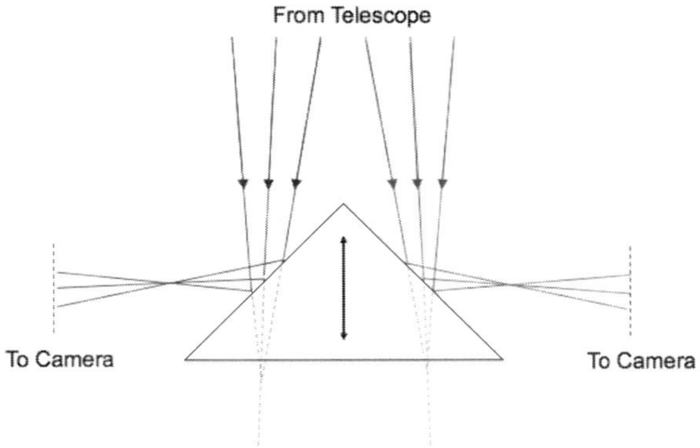

Fig. 2. SL-SLODAR uses a reflective wedge to separate the images of the two stars in order to increase the field of view of the system. The wedge is on a single axis stage so that the correct separation can be selected for the particular target under observation.

2. The Instrument

The instrument has been mounted on a 0.35 m Celestron and a 0.4 m Meade telescope at Paranal. A reflective wedge is used to spit the beams for the two target stars into different wavefront sensor (WFS) channels. Each channel consists of an 8x8 Shack-Hartmann WFS and an Andor Luca EMCCD camera. The cameras have effectively zero read noise whilst running with a frame rate of approximately 60 Hz and an exposure time between 2 and 5 ms. Figure 3 is a photo of the SL-SLODAR instrument on the 0.4 m meade.

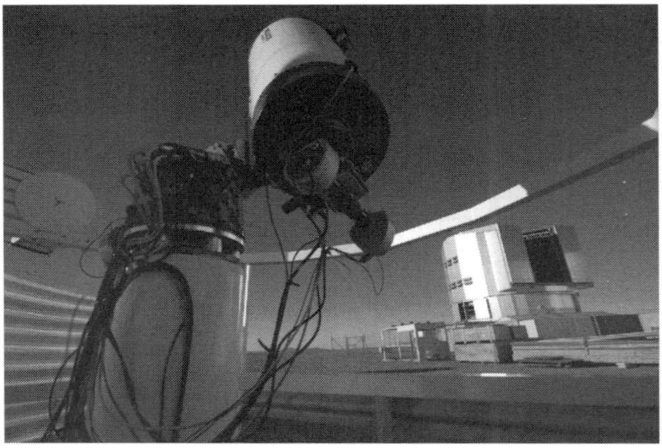

Fig. 3. Photo of the SL-SLODAR on the 16 inch meade at Paranal. (Photo TB)

3. Results

An example profile sequence from SL-SLODAR is shown in figure 4. The plot shows a night dominated by high altitude turbulence. Initially the surface layer appears very weak then at approximately 05:00 UT we see the turbulence up to 30 m dramatically increasing in strength.

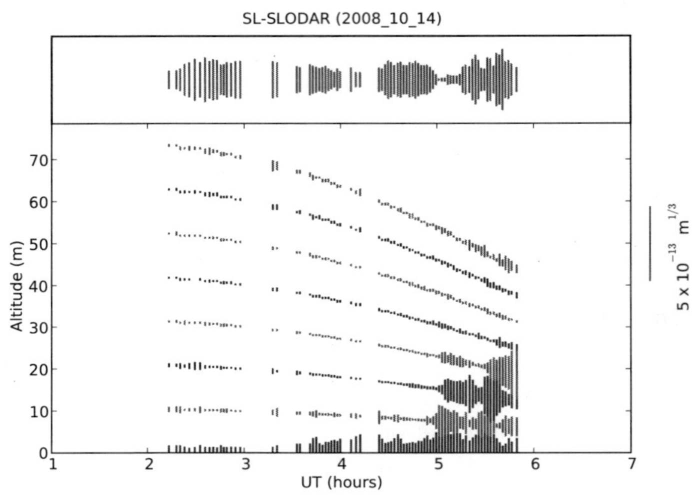

Fig. 4. Example SL-SLODAR profile sequence. The thickness of the line at each point indicates the strength of the turbulence at that altitude and the line in the top box shows the residual 'unsensed' higher altitude turbulence.

Figure 5 shows an example histogram of the mean profile for a night at Paranal. This particular plot was chosen as the turbulence distribution was constant throughout the night (unlike figure 4) and shows a strong surface layer contribution which can be modeled by an exponential decay with a scale height of approximately 11 m. The last bin shows the integrated C_n^2 for the 'unsensed' higher altitude atmosphere and not a layer at 80 m.

Figure 6 shows a plot of the seeing values from the MASS and the DIMM from Paranal's automated site monitor (ASM) on the same night as figure 5. The plot shows a large discrepancy between the DIMM, which calcultes the seeing for the entire volume of atmosphere above it, and the MASS, which only sees the high altitude turbulent layers. The discrepancy suggests a high fraction of low altitude turbulence as seen by SL-SLODAR in figure 5.

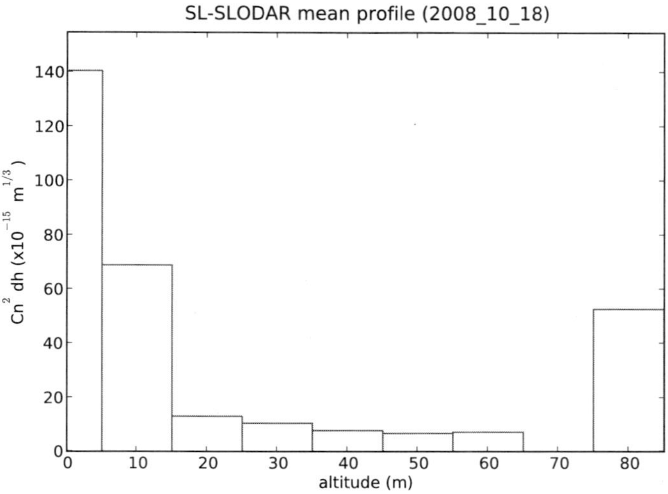

Fig. 5. An example average profile for a whole night at Paranal, Chile. The integrated C_n^2 in the last bin is actually the residual 'unsensed' optical turbulence and not a layer at 80 m.

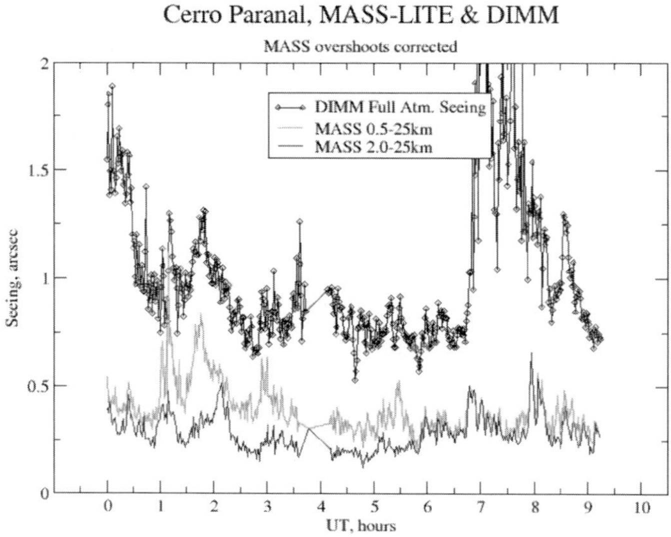

Fig. 6. MASS - DIMM seeing plot from the Paranal ASM for the same night as figure 5. The DIMM seeing is worse than the MASS suggesting that the majority of the optical turbulence is close to the ground. This is supported by SL-SLODAR which resolves a large fraction of the turbulence in the first 20 m (figure 5).

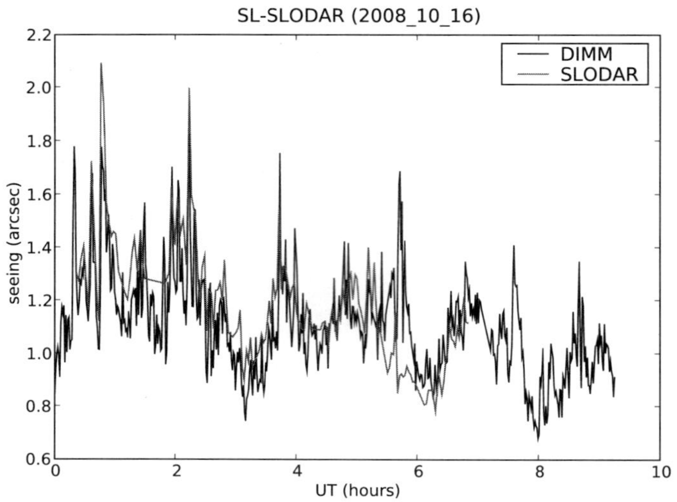

Fig. 7. (color on-line) Seeing comparison between SL-SLODAR and DIMM. The black line shows the seeing from the Paranal DIMM site monitor ad the red line is the seeing as calculated by SLODAR from theoretical fits to the auto-covarinace functions.

Figure 7 compares the total integrated seeing values measured by the DIMM and SL-SLODAR. The SL-SLODAR seeing value (for the whole atmosphere) is calculated by a fit of the theoretical response functions to the measured auto-covariance.

The plot shown was taken on a night with very weak surface layer. The DIMM is located approximately 5 m higher than SL-SLODAR so if there was a strong surface layer we would expect the SL-SLODAR values to be higher than the DIMM. Figure 8 is a scatter plot of SL-SLODAR seeing against the DIMM seeing. The plot shows that usually SL-SLODAR does measure a worse seeing than the DIMM. The seeing is however sometimes better for the SL-SLODAR. This is usually when the the surface layer is weak and so the turbulence in the first 5 m makes little difference. In this case the spatial separation of the two instruments could account for the discrepancy.

4. LuSci and SL-SLODAR Comparison

The Lunar Scintillometer (LuSci)[6] is an instrument designed to profile the ground layer optical turbulence by cross correlating the intensity signals received by a number of detectors. The scintillation pattern caused by a turbulent layer will have a correlation length determined by the distance

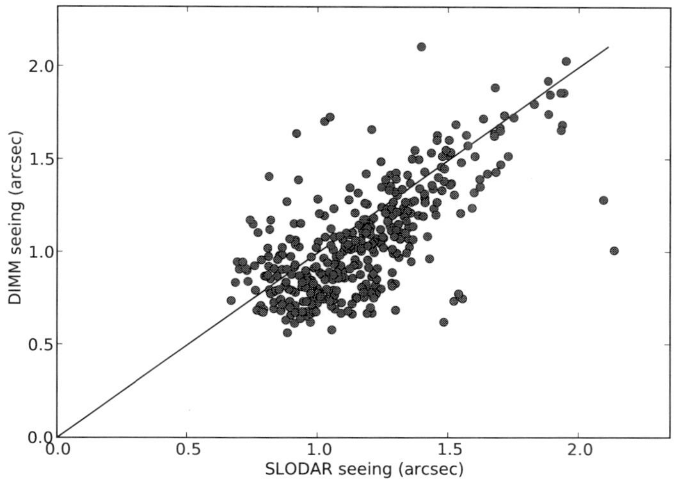

Fig. 8. Seeing comparison between SL-SLODAR and DIMM. The scatter plot shows that generally SL-SLODAR has worse seeing which is expected as it is 5 m lower.

to the layer and the strength of the turbulence in that layer, and so each baseline, as formed by two of the detectors, will have a peak sensitivity to a particular altitude. The ESO LuSci has four resolved layer ranges, 4 m, 16 m, 64 m and 256 m. The first three of which overlap with the altitude range of SL-SLODAR. Contemporaneous data was taken with the two instruments at Paranal in October 2008 and a comparison of the data is currently being made.

5. Conclusion

We have developed a modified SLODAR instrument which can profile the surface turbulent layer with high altitude resolution (less than 10 m). This was accomplished with the addition of a reflective wedge to allow for greater separations and therefore a higher resolution. A few examples from the data collected at Paranal are presented to show the resolution of the system. Early data collected at Paranal are consistent with contemporaneous MASS and DIMM measurements.

Acknowledgements

We would like to thank Marc Sarazin and ESO for the data from the Paranal ASM and support for the SLODAR instrument at ESO. We are grateful to the Science and Technology Facilities Committee (STFC) for financial support (JO).

References

1. A. Tokovinin and T. Travouillon, *MNRAS* **365**, 1235 (2006).
2. A. Tokovinin, *Publications of the Astronomical Society of the Pacific* **116**, 941 (2004).
3. R. W. Wilson, *MNRAS* **337**, 103 (November 2002).
4. T. Butterley, R. W. Wilson and M. Sarazin, *MNRAS* **369**, 835 (May 2006).
5. R. W. Wilson, T. Butterley and J. Osborn, these proceedings.
6. J. Rajagopal, A. Tokovinin, E. Bustos and J. Sebag, Lusci: a lunar scintillometer to study ground layer turbulence *Optical and Infrared Interferometry* **7013** (SPIE, 2008).

GIANT MAGELLAN TELESCOPE SITE TESTING AND CHARACTERIZATION AT LAS CAMPANAS OBSERVATORY

JOANNA E. THOMAS-OSIP* and GABRIEL E. PRIETO

GMT, Las Campanas Observatory,
*La Serena, Chile *E-mail: jet@lco.cl*

Las Campanas Observatory has been designated as the location for the Giant Magellan Telescope (GMT). We report results obtained since the commencement, in 2005, of a systematic site testing campaign at LCO. Measurements of the turbulence profile of the free-atmosphere above LCO have been collected with a MASS/DIMM. We examine the contribution to the seeing arising from turbulence in the ground layer (defined here as below an altitude of 500 m) through the difference between the turbulence integrals in the full atmosphere (as measured by DIMM) and in the free atmosphere (as measured by MASS).

In preparation to characterize the Giant Magellan Telescope site and guide the development of its adaptive optics system, two campaigns to systematically compare the turbulence profiles obtained independently with three different instruments were conducted at Las Campanas Observatory in September, 2007 and January 2008. Slope detection and ranging (SLODAR) was used on the 2.5-m duPont telescope. SLODAR measures the C_n^2 profile as a function of altitude through observations of double stars. The separation of the observed double star sets the maximum altitude and height resolution. Ground layer (altitudes $<$ 1 km) and free atmosphere turbulence profiles are compared with those obtained with a lunar scintillometer (LuSci) and a multi-aperture scintillation sensor (MASS), respectively. In addition, the total atmospheric seeing was measured by both SLODAR and a differential image motion monitor (DIMM).

Keywords: Site characterization; turbulence profile; seeing; precipitable water vapor; meteorological characteristics; ground layer seeing; scintillation; slope detection and ranging; lunar scintillation.

1. Introduction

The Giant Magellan Telescope (GMT) Science Working Group determined that Las Campanas Observatory (LCO) has the potential to meet its science goals.[1] The GMT project is in the fortunate position of having clear access to a developed site with a long history of excellent performance. Light pollution is negligible and likely to remain so for decades to come. The seeing quality is as good or better than other developed sites in Chile. Furthermore, the weather pattern has been stable for more than 30 years. Our site testing effort is, therefore, concentrated on idenepsying the best peak within LCO in terms of seeing, turbulence profile, and wind speeds. Additionally, quantifying the potential impact of precipitable water vapor on GMT mid-infrared science goals requires further characterization in terms of both precision and time variability. In preparation for a report that will form the basis for the site selection, Ref. 2 summarizes our ongoing site testing efforts in the context of known properties based on the history of almost 40 years of operations at the site.

In addition to our site testing efforts, we are preparing to further characterize the selected GMT site and guide the development of its adaptive optics system. Thus, two campaigns to systematically compare the turbulence profiles obtained independently with three different types of instruments (SLODAR, LuSci, and MASS/DIMM) were conducted at Las Campanas Observatory (LCO) in September, 2007 and January 2008. Ref. 3 provides a detailed description of these campaigns.

1.1. *LCO Description*

The Carnegie Institution of Washington established LCO in 1969 to build the Swope 1-m telescope in 1971 and the du Pont 2.5-m telescope in 1977. The purchase of the land was arranged by Horace Babcock, then the Observatory's director. The property (see Figure 1) is approximately 200 km^2 and located just north of the European Southern Observatory's La Silla. LCO became the Observatories primary observing site in 1986 with the transfer of Mt. Wilson, in Southern California, to the Mount Wilson Institute. The twin Magellan 6.5-m telescopes, completed in 2001, are operated by Carnegie for the Magellan consortium including Harvard, MIT, and the Universities of Arizona and Michigan in addition to Cargnegie. Also located at LCO but not operated by Carnegie are the 1.3-m Warsaw telescope, a NASA Telescopes in Education remotely operated 35-cm telescope and one telescope in the Birmingham Solar Oscillations Network (BiSON). Given the long operational baseline, a fair amount was known about the LCO site when the current GMT site testing effort began.[1]

1.2. *Site Evaluation Program*

An extensive site testing program has been underway since mid-2005 to idenepsy the best available location within the LCO property for the GMT.

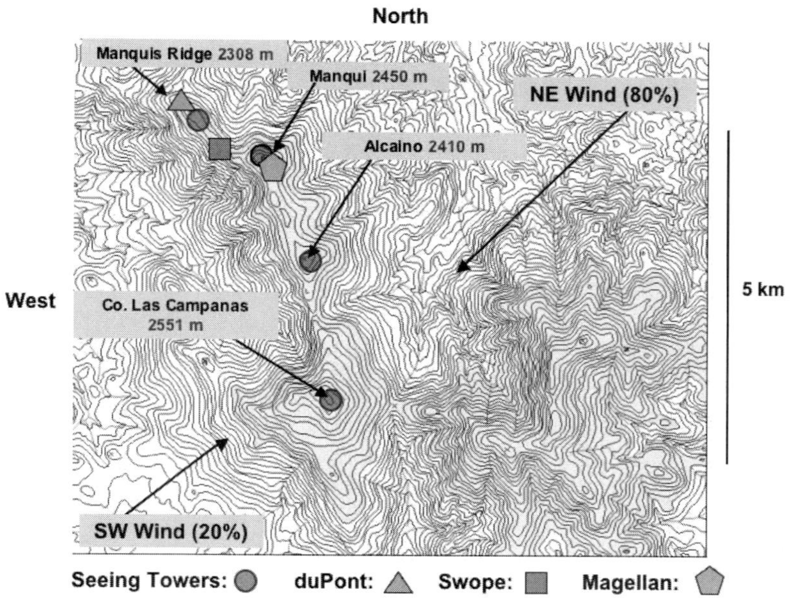

Fig. 1. Topographic map showing the peaks within the LCO property including some current facilities and potential GMT sites. The purple circles show the locations of site testing towers. The sites of the du Pont, Swope, and Magellan telescopes are also indicated. The prevailing wind directions of 80% from the NE and 20% from the SW are also shown.

We are measuring meteorological characteristics (pressure, temperature, humidity, and wind) and seeing at four sites within the LCO property (see Figure 1). The turbulence profile of the free atmosphere (above 500m), precipitable water vapor, cloud cover and light pollution complement the site specific quantities. A description of the sites follows but for brevity the reader is referred to Refs. 2, 3 for details on the instrumentation.

Cerro Manqui is home to the Magellan telescopes. Although there is not sufficient space to add another large telescope, it is the best characterized site at LCO and therefore it serves as a reference. The Manquis Ridge site, between the du Pont and Swope telescopes, is the lowest altitude site. The Magellan site survey found it to have slightly poorer seeing than Manqui but also lower wind speeds. Cerro Las Campanas is the highest of our sites and was found by the Magellan site survey to have seeing similar to Manqui but significantly higher wind speeds. Furthermore, it has sufficient space for a large telescope and support facilities and is the presumed site for GMT. Finally Alcaino, in between Co. Las Campanas and Manqui and nearly the same altitude as Manqui, was the site of the Nagoya 5-m radio telescope until 2004. It was not included in the Magellan site survey and the current studies provide the first detailed examination of its properties.

2. Selected Site Testing Results: Seeing

The DIMM measurements have been filtered to remove poor quality data due three possible criteria: focus, read noise, and the number of measurements during one minute. The MASS data has been reprocessed using a new version of Turbina[4] to correct for the effect of strong scintillation under poor seeing conditions.[5] In order to compare concurrent measurements from each site, the timestamps from the DIMMs and MASS were matched to timestamps of one of the DIMMs to find the closest measurement within one minute. Each instrument takes approximately one measurement per minute so using a shorter timescale significantly reduces the number of matches.

Figure 2 shows the normalized cumulative histograms for the seeing measured by the DIMMs at the four sites during a two year period. We found 318 nights of concurrent data between the four sites. The seeing at Cerros Alcaino, Manqui, and Las Campanas (LCP) have extremely similar distributions while the Manquis Ridge site clearly shows a degradation of about 0.05″ in the median. Uncertainties range between 0.02″ to 0.05″ as determined with side-by-side comparisons of two DIMMs located in the same tower. The free atmosphere median seeing is 0.48″ demonstrating the possibilities in the absence of the contribution from the ground layer.

The seeing, β, contribution by the ground layer (defined here as lower than 500 m) can be inferred from the MASS and DIMM measurements at each matching time in the following manner:

$$\beta_{GL} = |\beta_{DIMM}^{\frac{5}{3}} - \beta_{MASS}^{\frac{5}{3}}|^{\frac{3}{5}} \tag{1}$$

The ground layer can be examined at each site using the MASS located at the Manqui site if we assume that the turbulence above 500 m is similar over the entire property. Furthermore, the thickness of the ground layer probed at each site varies depending on the altitude of each site. The cumulative histograms of the fraction of the total seeing found in the ground layer are shown in the bottom of Figure 2 along with appropriate ground layer thicknesses. Under median conditions, all four sites contain approximately 60% of the seeing within the ground layer. The differences between the sites do not vary strictly with altitude and thus ground layer thickness.

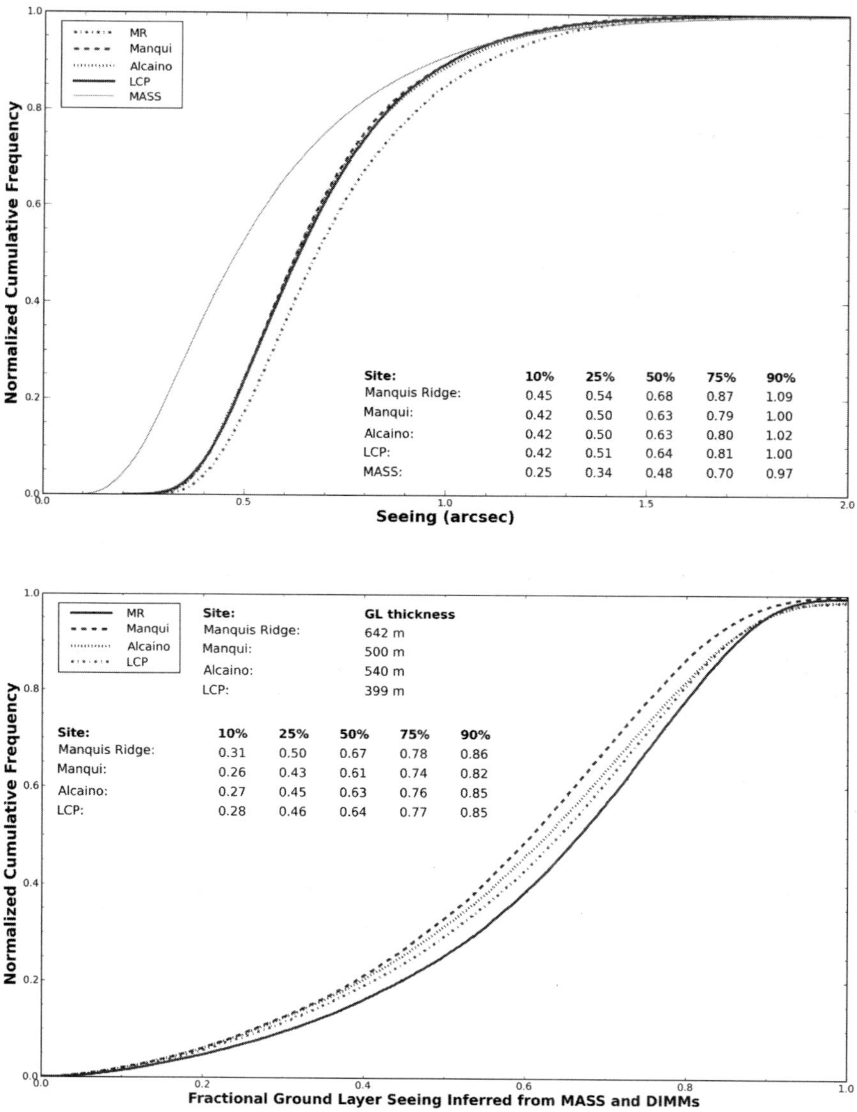

Fig. 2. Top: Normalized cumulative histograms of the MASS free atmosphere seeing and the DIMM seeing at all four sites for a period of two years beginning November 2005. Bottom: Normalized cumulative histogram of the fraction of the total seeing found in the found in the ground layer as inferred by the MASS and the DIMM at each site for a period of two years beginning November 2005.

This could be due to morphological effects but is more likely attributable to noise inherent in the subtraction of two similar numbers.

3. Results Overview: Turbulence Campaigns

A primary motivation for making these instrument comparisons is to understand if a combination of low resolution, portable instruments such as LuSci and MASS/DIMM are capable of giving a similar statistical picture of the turbulence characteristics to those of a higher resolution instrument like SLODAR that demands a larger telescope (on the order of 1 m minimum). Table 1 shows the 10, 25, 50, 75, and 90th percentiles for the distributions of seeing in three altitude ranges (near the ground, free atmosphere, and total) as measured by our suite of instruments.

Table 1. Statistical distributions of the seeing as measured by the instruments in our campaign

Instrument (type)	Number of Points	10%	25%	50%	75%	90%
SLODAR non-Kolmogorov (total)	1666	0.48	0.58	0.70	0.87	1.03
SLODAR DIMM (total)	1666	0.50	0.60	0.74	0.90	1.06
DIMM (total)	1666	0.52	0.60	0.70	0.85	1.02
SLODAR (free atmosphere)	546	0.27	0.34	0.47	0.61	0.77
MASS (free atmosphere)	546	0.22	0.28	0.39	0.59	0.80
SLODAR (ground layer)	405	0.32	0.40	0.49	0.64	0.76
LuSci (ground layer)	405	0.27	0.32	0.42	0.56	0.68
MASS-DIMM (ground layer)	788	0.27	0.40	0.51	0.67	0.86
MASS-DIMM (boundary layer)	805	0.27	0.41	0.52	0.68	0.89
LuSci (ground layer)	788 or 804	0.28	0.36	0.46	0.63	0.82

Each set (SLODAR/DIMM, SLODAR/MASS, SLODAR/LuSci, and LuSci/MASS-DIMM) was binned separately to a 2 minute interval to maximize the number of points in each set as discussed in each of the previous sections. Thus, the distributions for each instrument comparison set (delineated with solid lines in Table 1) are made up of a distinct number of points in time. The effect of this can be seen in the difference between the two different LuSci ground layer distributions. The LuSci data that matched the SLODAR data had a slightly better overall seeing distribution than the LuSci data that matched the MASS-DIMM data. This is not surprising when we consider the fact that there are twice as many data points in the LuSci to MASS-DIMM correlation as in the LuSci to SLODAR correlation.

We can see that the distributions of free atmosphere and total seeing integrals match very closely although they do not have exactly the same shapes. For example, the MASS senses less turbulence in the 10th percentile than SLODAR does but the opposite is true for the 90th percentile. In the comparison of LuSci and SLODAR data, we see a small systematic shift in the distributions with LuSci sensing less turbulence overall. This systematic effect continues in the comparison of LuSci to the ground and boundary layer inferred by the MASS-DIMM. Although the effect is quite small, it is probably real given that it shows up in two different comparisons. Further work is needed to understand the cause or at least to come to a conclusion on the proper combination of the LuSci and MASS profiles.

4. Conclusions

The long history of excellent conditions at LCO, concerns regarding biases in short-term site surveys, and limited resources led the GMT project to focus its site evaluation activities to certain peaks within the LCO property. The final site evaluation at LCO will take into account seeing (both image quality and stability), wind speeds, and ground layer characteristics. However, it is presumed that Cerro Las Campanas will be the site of the GMT given the confirmation of the good seeing and meteorological characteristics found to date.

In our preliminary analysis of the data from the turbulence campaigns, we have found a qualitative agreement of the profiles produced by SLODAR, MASS, and LuSci. There is a general correlation seen in the integral seeing from the ground layer, free atmosphere, and total atmosphere as measured by our variety of instruments. Furthermore, we have seen that the turbulence in the vicinity of Las Campanas Observatory is highly temporally variable and site dependent. It is, therefore, recommended that GMT location be studied in depth. Finally, we find that low resolution profiles would appear sufficient to characterize the statistics of the turbulence characteristics although some extra work is needed to define the process for this.

References

1. *Giant Magellan Telescope Conceptual Design Report* (GMT Cosortium, Pasadena, 2006). http://www.gmto.org/CoDRpublic.
2. J. Thomas-Osip, G. Prieto, M. Johns and M. Phillips, Giant Magellan Telescope site evaluation and characterization at Las Campanas Observatory, in *Astronomical Telescopes and Instrumentation*, Proc. SPIE Vol. 70122008.
3. J. Thomas-Osip, E. Bustos, M. Goodwin, C. Jenkins, G. Prieto and A. Tokovinin, A campaign to compare three turbulence profiling techniques at las campanas observatory, in *Astronomical Telescopes and Instrumentation*, Proc. SPIE Vol. 70142008.

4. V. Kornilov and N. Shatsky, *MASS data reprocessing* (Sternberg Astronomical Institute, Moscow University, 2005). http://dragon.sai.msu.ru/mass/download/doc/remass.pdf.
5. A. Tokovinin, *MASS simulations under strong scintillation* (CTIO internal document, 2006). http://www.ctio.noao.edu/~atokovin/profiler/simul06.pdf.

AUTHOR INDEX

Arcidiacono, C. 128, 299
Ashley, M. C. B. 82, 264
Auger, L. 156
Avila, R. 58
Aviles, J.-L. 58

Baruffolo, A. 128
Beckers, J. M. 1, 23, 307
Berdja, A. 42, 108, 120
Blanco, M. J. 342
Bonatto, L. 342
Bonner, C. S. 264
Borgnino, J. 42, 108, 120
Brangier, M. 283
Brousseau, P. 156
Businger, S. 196
Bustos, E. 108
Butterley, T. 50, 58, 371

Carlberg, R. 291
Castro-Almazan, J. 66
Castro-Almazán, J. A. 350
Chacón, A. 216, 334
Chenegros, G. 283
Cherubini, T. 196
Chun, M. R. 58, 291
Comerón, F. 308, 324
Conan, J.-M. 74
Crotts, A. P. 26
Cuillandre, J.-C. 291
Cuevas, O. 216, 334
Cure, M. 216
Curé, M. 334
Cuxart, J. 144

Dali Ali, W. 108
Delgado, J. M. 108
del Giorgio-Castiglione, C. 358
Dima, M. 128
Diolaiti, E. 128, 299
Donaldson, R. 299

Eff-Darwich, A. 342, 358
Els, S. 90, 98, 216, 248
Els, S. G. 224

Falomo, R. 299
Farinato, J. 128, 299
Fuensalida, J. J. 66, 108, 350, 358

Garcia-Lorenzo, B. 66
García-Lorenzo, B. 342, 350, 358
Gendron, E. 283
Gentile, G. 128
Green, R. F. 316
Gullieuszik, M. 299

Hagelin, S. 34, 184, 208, 240, 366
Hernández-Gutiérrez, L. 342
Hickson, P. 26
Hubert, Z. 283

Lac, C. 144
Lai, O. 291
Lascaux, F. 34, 165, 184, 208, 240, 366
Lawrence, J. S. 82, 264
Le Moigne, P. 165
Lombardi, G. 108, 232
Lombardi, G.-L. 136
Lombini, M. 299
Luong-Van, D. M. 264
Lyman, R. 196

Maire, J. 42, 108, 120
Marchetti, E. 299
Maric, T. 144
Martin, F. 120
Masciadri, E. 34, 165, 184, 208, 240, 366
Masson, V. 144
Melnick, J. 136
Metti, C. 128
Michau, V. 74
Momany, Y. 299
Moretti, A. 299
Muñoz-Tuñón, C. 256, 342

Navarrete, J. 108, 136
Noilhan, J. 165

Ortolani, S. 232
Osborn, J. 50, 371

Pfrommer, T. 26
Pietroni, I. 165
Pinty, J. P. 144
Piotto, G. 299
Pouplard, F. 283
Powers, J. G. 173
Prieto, G. E. 379

Racine, R. 13
Ragazzoni, R. 128, 299
Ramio, H. V. 108
Reyes, M. 108
Richer, H. B. 291
Riddle, R. 98, 216, 224, 248
Riddle, R. L. 90
Robert, C. 74
Rodriguez-Hernandez, M. A. C. 66
Rodriguez-Losada, J. A. 342
Rousset, G. 283

Sarazin, M. 108, 136
Schöck, M. 90, 98, 216, 224, 248
Skidmore, W. 90, 98, 216, 224, 248
Stoesz, J. 34, 184, 208, 240, 366
Storey, J. W. V. 82, 264

Thomas-Osip, J. E. 379
Tokovinin, A. 108
Travouillon, T. 90, 98, 216, 224, 248
Tulet, P. 144
Turolla, R. 299

Varela, A. M. 256
Védrenne, N. 74
Vidal, F. 283
Viñas, R. 342
Viotto, V. 128

Wilson, R. 371
Wilson, R. W. 50, 58
Wizinowich, P. 271

Ziad, A. 42, 108, 120
Zitelli, V. 232